Nonlinear Estimation and Control of Automotive Drivetrains

Hong Chen · Bingzhao Gao

Nonlinear Estimation and Control of Automotive Drivetrains

Science Press
Beijing

Springer

Hong Chen
Jilin University
Changchun, People's Republic of China

Bingzhao Gao
Jilin University
Changchun, People's Republic of China

ISBN 978-3-662-51304-0 ISBN 978-3-642-41572-2 (eBook)
DOI 10.1007/978-3-642-41572-2
Springer Heidelberg New York Dordrecht London

Jointly published with Science Press Beijing
ISBN: 978-7-03-038887-2 Science Press Beijing

Softcover reprint of the hardcover 1st edition 2014

Printed on acid-free paper

Springer is part of Springer Science+Business Media (www.springer.com)

Preface

Motivation

Electronic control has become the core technology in automotive industry to meet the increasingly stringent emission legislation and dynamic performance requirements. Accordingly, automotive electronics account for a larger and larger proportion of the manufacturing cost of the whole vehicle, including not only hardware cost but also the development cost of control software.

Although at present the widely used control algorithms are still based on event-driven (rule-based) feedforward and PID control, the question of how to design a high-performance control program efficiently using advanced control theories has become a hot topic in the fields of both control and automobile engineering. Since 2006, many academic journals, including IEEE T. Control Systems Technology, Control Engineering Practice, Int. J. Control, Vehicle System Dynamics, Int. J. Powertrain, etc. have published their special issues on automotive control. Besides, sessions on automotive control are organized every year at the annual conferences of IFAC, IEEE CDC and ACC, etc.

The application of advanced control theories is attractive because of its potential to reduce the calibration workload and improve the dynamic control performance under numerous driving conditions and large environmental variations.

This text presents an in-depth discussion on the control problems in automotive drivetrains, particularly the types of hydraulic Automatic Transmission (AT), Dual Clutch Transmission (DCT) and Automated Manual Transmission (AMT). The challenging estimation and control problems, such as driveline torque estimation and gear shift control, are addressed by applying the most up-to-date nonlinear control theories, including constructive nonlinear control (Backstepping, Input-to-State Stable) and Model Predictive Control (MPC). The estimation and control performance is improved while the calibration effort is reduced significantly. This book gives a detailed design process of many examples, and thus enables the readers to understand how to successfully combine the "purely theoretical methodologies" with "actual vehicle applications".

Intended Readers

This book should enable graduate and higher-level undergraduate students to understand the control and estimation problems in automotive drivetrains, and how to use control theories to solve these practical problems.

This book is also suited for the professional control engineers in the R&D centers of automobile manufacturers.

The Authors

Dr.-Ing. Hong Chen received the B.S. and M.S. degrees in process control from Zhejiang University, Hangzhou, China, in 1983 and 1986, respectively, and the Ph.D. degree (mit Auszeichnung bestanden—with honors) from the University of Stuttgart, Stuttgart, Germany, in 1997. From 1993 to 1997, she was a "Wissenschaftlicher Mitarbeiter" (research assistant) at the Institut für Systemdynamik und Regelungstechnik, University of Stuttgart. Since 1999, she has been a Professor at Jilin University, where she currently serves as "Tang Aoqing Professor". She is now an IEEE senior member, and serving as a member of international and national technical committees, including IFAC TC Automotive Control, Control Theory of CAA and Process Control of CAA. She was honored and awarded by the National Science Fund of China for Distinguished Young Scholars. Prof. Chen is also the leader of a Program for Changjiang Scholars and Innovative Research Team in University, China. Her main research interests include model predictive control, optimal and robust control, nonlinear control and applications in automobile engineering and mechatronic systems.

Dr. Bingzhao Gao received the B.S. and M.S. degrees in vehicle engineering from Jilin University of Technology, China, in 1998 and Jilin University, China, in 2002, respectively. He received the Ph.D. degree in control engineering under the instructions of Prof. Chen in 2009, and his thesis was honored as an Excellent Doctoral Dissertation of Jilin Province, China. He is also a holder of the Doctor's degree in mechanical engineering of Yokohama National University, Japan. Dr. Gao is currently an associate professor at Jilin University. His research interests include vehicle powertrain control and vehicle stability control.

Acknowledgements

The authors would like to express their great appreciation to many students in Prof. Chen group, chen's group and, in particular, to Dr. Xiaohui Lu and Lu Tian for their hard work and contributions to Chap. 8 and Sect. 4.5, and to Qifang Liu and Fang Xu for their help in manuscript review and proofreading, and also to Dr. Shuyou Yu for his help in the programming of LMI and Nonlinear MPC.

The authors also greatly acknowledge National Nature Science Foundation of China, Ministry of Education of China, Ministry of Science and Technology of China, and Jilin Provincial Science & Technology Department for the financial support.

Changchun, People's Republic of China

Hong Chen
Bingzhao Gao

Contents

Chapter 1
Introduction

1.1 Introduction of Automotive Drivetrain

Generally speaking, the terms of "powertrain" and "drivetrain" (see Fig. 1.1) refer both to the vehicle components which produce and deliver the power and torque. The term "powertrain" sometimes emphasizes the engine and the transmission, while "drivetrain" (or driveline) stresses the clutch (torque converter), transmission, driveshaft, differential gear box, axle shaft and wheels. The drivetrain delivers engine torque to the tires, and makes it possible for the vehicle to accelerate or climb a gradient. Figure 1.2 shows the function of a step-ratio transmission, where the engine torque characteristics are re-distributed, through different gear ratios, to approach a desired pattern of wheel torque.

1.1.1 Engine

Internal combustion engine generates power by converting chemical energy contained in the fuel into heat, and the heat produces then mechanical work. The engine torque T_e is determined by the flow rate of intake air and fuel, and influenced by combustion efficiency and friction losses. In modern vehicular powertrains, high-speed CAN (Controller Area Network) bus connects the control units of the engine and the transmission, and the shared information includes throttle angle, engine torque and engine speed, etc. On the other hand, the transmission sends torque request to the engine through CAN bus. Here, the detailed engine model will not be considered in the context of this book.

The dynamic equation of engine speed is described by

$$I_e\dot{\omega}_e + C_e\omega_e = T_e - T_c, \tag{1.1}$$

where I_e is the inertia moment of the engine crank shaft, C_e is the damping coefficient, ω_e is the engine rotational speed, T_e is the engine torque and T_c denotes the

H. Chen, B. Gao, *Nonlinear Estimation and Control of Automotive Drivetrains*,
DOI 10.1007/978-3-642-41572-2_1,

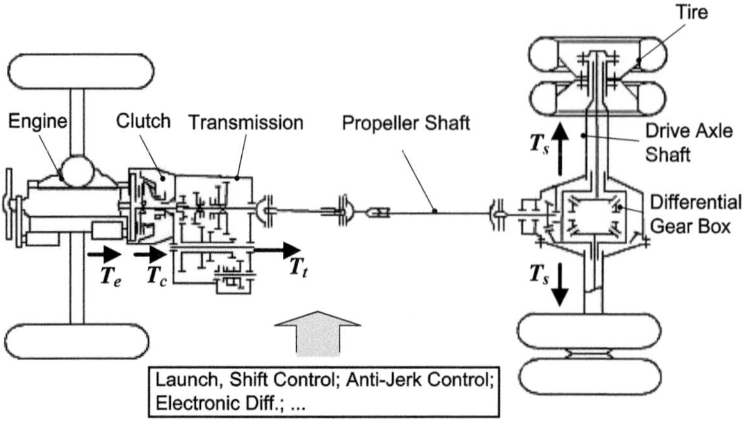

Fig. 1.1 Drivetrain of FR (Front Engine, Rear Wheel Drive) vehicle [30]: T_e, engine torque; T_c, clutch torque; T_t, transmission output torque; T_s, torque of drive axle shaft

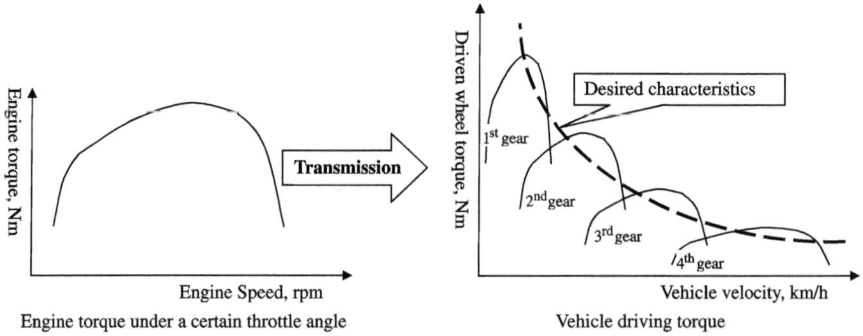

Fig. 1.2 Drive torque characteristics of a vehicle with a step-ratio transmission

clutch torque. The most simple engine model is the static torque map, which is a lookup table with the inputs of the throttle angle and the engine speed, and denoted as

$$T_e = T_e(\omega_e, \theta_{th}), \tag{1.2}$$

where θ_{th} is the engine throttle angle. Various maps in vehicle engineering are obtained from large numbers of experiments in the steady state. As an example, the torque map of a 2000 cc gasoline engine is shown in Fig. 1.3.

1.1.2 Clutch/Torque Converter

A dry clutch, shown in Fig. 1.4 [61], consists of a housing, pressure plates, friction plates, a clutch disc with torsion damper and a release mechanism. In manual

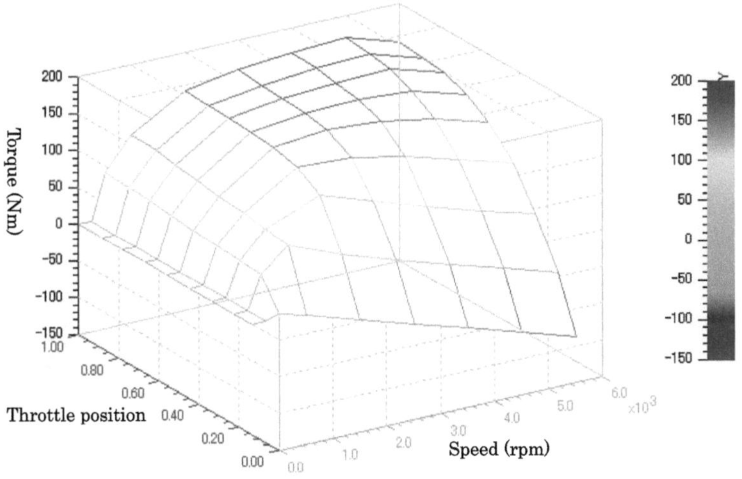

Fig. 1.3 Example of an engine torque map (lookup table)

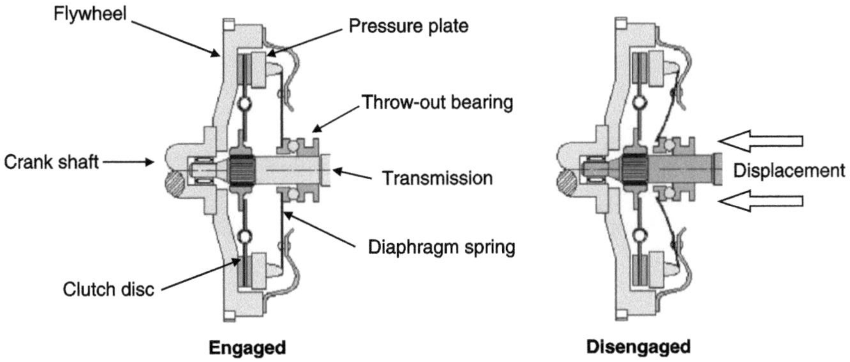

Fig. 1.4 Schematic overview of a dry clutch. Reprinted from [61], copyright 2007, with permission from Taylor & Francis

transmissions, when the vehicle is starting off from standstill, the clutch slips to compensate for the speed difference between the engine and the drivetrain. Moreover, when a gear shift operation takes place, the clutch disengages the engine from the transmission, and then engages them after gear shifting is over.

When the clutch is slipping, the torque delivered through the clutch T_c is determined by the clamping force F_c implemented on the friction disc:

$$T_c = F_c \mu_d R_c \operatorname{sign}(\Delta\omega), \tag{1.3}$$

where F_c is the clamping force, μ_d is the dynamic friction coefficient, R_c is the effective radius, and $\Delta\omega$ is the speed difference between the engine and the drivetrain.

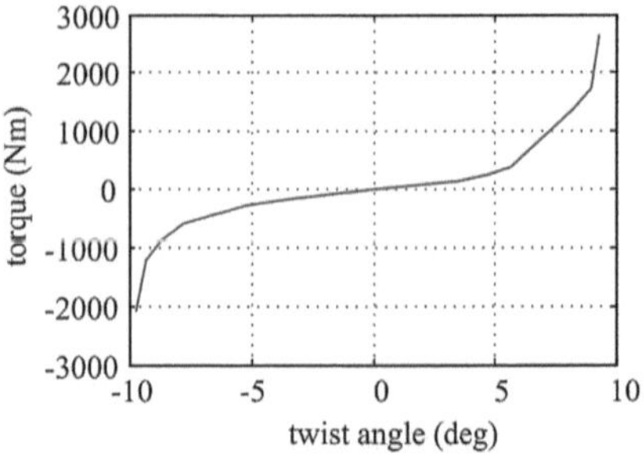

Fig. 1.5 Clutch spring characteristics

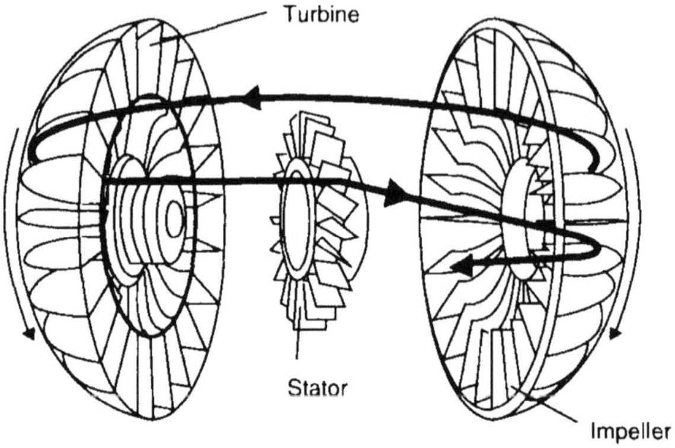

Fig. 1.6 Schematic overview of a torque converter [55]

It is worth noting that μ_d is not time-invariant, but varying with slip speed and the temperature.

Considering the damp spring embedded in the clutch, the torque of the clutch friction plate T_c is also a nonlinear function of the twist angle θ_c as follows:

$$T_c = T_c(\theta_c, \dot{\theta}_c), \tag{1.4}$$

and the equation is applicable for both slipping and locked-up state of the clutch.

The static characteristics of the torsion spring of a 4-ton truck clutch is shown in Fig. 1.5.

On hydraulic automatic transmissions, the torque converter, as shown in Fig. 1.6, assumes the functions of the clutch. When the turbine is driven forward, the dynamics of the torque converter are often characterized as [145]

$$T_p = C(\lambda)\omega_e^2 \tag{1.5}$$

and

$$T_{tb} = t(\lambda)T_p, \tag{1.6}$$

where T_p is the pump torque and T_{tb} is the turbine torque, λ is the speed ratio defined as

$$\lambda = \frac{\omega_{tb}}{\omega_e}, \tag{1.7}$$

with ω_{tb} being the turbine speed.

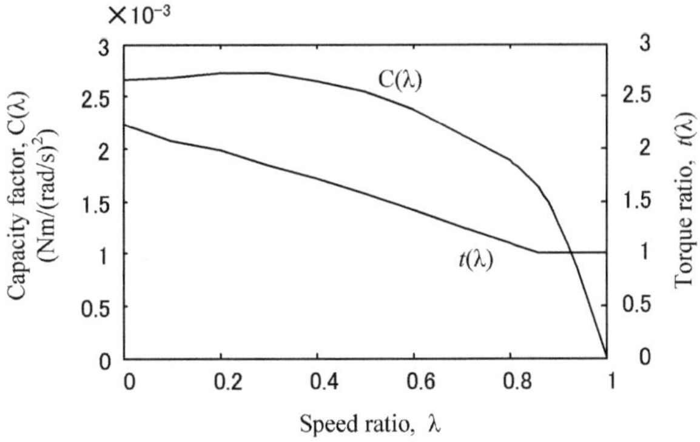

Fig. 1.7 Capacity factor and torque ratio of a torque converter

An example of the capacity factor $C(\lambda)$ and the torque ratio $t(\lambda)$ in a mid-size passenger car are given in Fig. 1.7.

1.1.3 Transmissions

There are many different types of transmissions which will be described in detail in the following sections. The function of the transmission is to modify the engine torque and engine speed with the ith gear ratio R_i, so that the momentary traction requirement could be satisfied. Neglecting the friction and the inertia torques, the transmission could be modeled as

$$T_t = T_c R_i, \tag{1.8a}$$

$$\omega_t = \frac{\omega_c}{R_i}, \tag{1.8b}$$

where T_t is the transmission output torque, T_c is the clutch output torque (transmission input torque), ω_t is the transmission output speed, ω_c is the clutch output speed (transmission input speed).

1.1.4 Propeller Shaft and Differential Gear Box

In FF (Front Engine, Front Wheel Drive) vehicles, the differential box is always combined with the transmission directly, while in FR (Front Engine, Rear Wheel Drive) vehicles, a propeller shaft connects the transmission and the differential box. The stiffness of the propeller shaft is comparatively larger, compared with that of the axle shaft and the clutch torsion spring. However, the clearance in the drivetrain shafts is an important element when modeling the propeller shaft precisely.

The differential unit compensates for the speed difference between the inside and the outside wheels when the vehicle is cornering. Generally speaking, the two output torques of the differential box are equivalent, while the two rotational speeds do not necessarily equal each other. If the twist deflection of the propeller shaft is ignored, we have

$$T_l = T_r, \tag{1.9a}$$

$$\omega_l + \omega_r = \frac{2\omega_t}{R_{df}}, \tag{1.9b}$$

where R_{df} is the gear ratio of the differential box, the subscripts l and r denote the left side and the right side.

At the same time, the rotational dynamic equation from the transmission to the differential is

$$I_p\dot{\omega}_t + C_p\omega_t = T_t - \frac{2T_{l,r}}{R_{df}}, \tag{1.10}$$

where I_p and C_p are the inertia and the damping of the propeller shaft, respectively.

1.1.5 Drive Axle Shaft

The two drive shafts between the differential gear and the driven wheels are represented as a torsion spring with stiffness coefficient K_s and a damping with coefficient C_s as follows:

$$T_s = 2T_{l,r}, \tag{1.11a}$$

$$T_s = K_s\theta_s + C_s\dot{\theta}_s, \tag{1.11b}$$

where T_s is the axle shaft torque and θ_s is the twist angle of the axle shaft satisfying

$$\dot{\theta}_s = \omega_{l,r} - \omega_w, \tag{1.12}$$

with ω_w being the wheel speed.

1.1.6 Tires and Vehicle

The longitudinal tire force F_x, which is usually simplified as a function of the longitudinal slip ratio S_x, rises fast when S_x increases under a threshold and declines slowly after that [55], see Fig. 1.8. The force F_z is vertical load of the tire, and the longitudinal slip is calculated as

$$S_x = \frac{R_w\omega_w - V}{R_w\omega_w} \quad \text{when driving,} \quad \text{and} \tag{1.13a}$$

$$S_x = \frac{V - R_w\omega_w}{V} \quad \text{when braking,} \tag{1.13b}$$

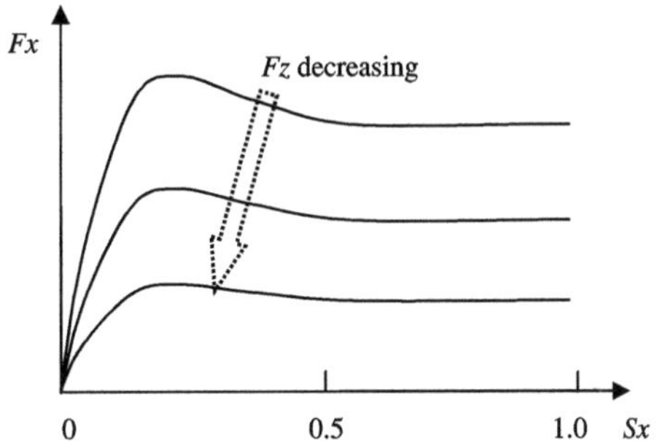

Fig. 1.8 Longitudinal force characteristics of tires

where R_w is the tire radius, ω_w is the wheel rotary velocity and V is the car body velocity.

The road load consists of three parts: the grade force F_G, the rolling resistant moment T_w of tires and the aerodynamics drag F_A. The resistant moment T_w of tires is regarded as constant here. The grade force is calculated as

$$F_G = mg \sin \theta_g, \tag{1.14}$$

where m is the vehicle mass, θ_g is the grade angle of the road. The aerodynamic drag is described as

$$F_A = \frac{1}{2} \rho C_D A_A V^2, \tag{1.15}$$

where C_D is the aerodynamic drag coefficient, A_A is the front area of the vehicle and ρ is the air density.

1.2 Overview of Automotive Transmissions

Automatic transmission, which relieves the driver from shift operation, changes the speed ratio of a drivetrain automatically according to the driver intent, current engine state and road surface condition, so that optimal drivability or fuel economy could be obtained. As mentioned before, many types of transmissions have been developed, and Fig. 1.9 shows the history of automotive transmissions. Different transmission has its own unique features and thereby its own control tasks.

1.2.1 Hydraulic Automatic Transmission (AT)

The predominant form of a hydraulic Automatic Transmission (AT) [1] uses a torque converter, and a set of planetary gearsets to provide a range of gear ratios. The torque converter consists of three rotating elements with curved blades: pump, turbine and stator. The pump and turbine hydraulically connect the engine to the transmission and the stator is used to enhance torque multiplication. The torque converter is followed by a set of planetary gearsets, usually including 2–4 planetary gearsets. Each

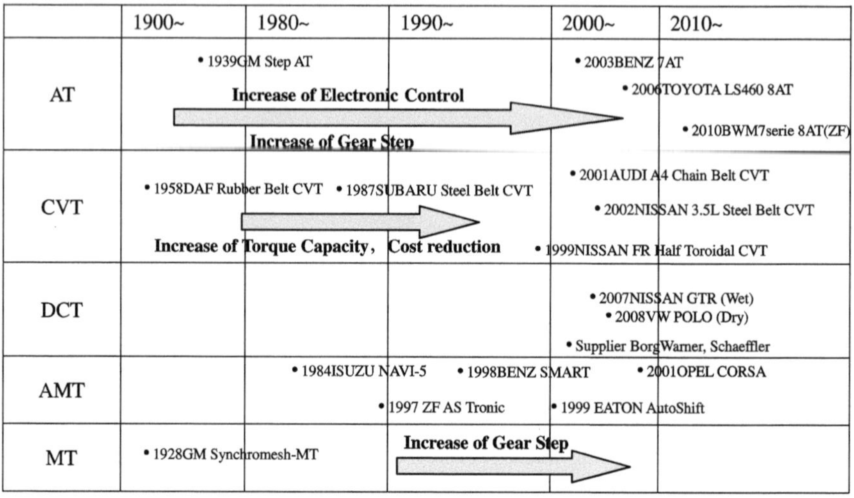

Fig. 1.9 History of automotive transmissions

planetary gearset contains a sun gear, a planetary carrier and a ring gear. These rotational members are connected with hydraulic clutches or brakes, which make particular members of the planetary gearset motionless, while allowing other members to rotate. Thereby different gear ratios could be achieved. The merits of an AT include smooth shifting and comfort driving. However, the dynamic response is relatively slow, compared to manual transmissions, and because of the hydraulic loss of the torque converter, traditional ATs have low transmission efficiency. In addition, the structure of traditional ATs is somehow complex and the maintenance cost is high.

As shown in Fig. 1.9, one of the development trends of ATs is to increase the number of gear ratios, from 4-speed to 5-speed, 6-speed, and at present 8-speed or 9-speed ATs have been developed. The 8-speed AT of ZF, a transmission supplier, is shown in Fig. 1.10. A higher number of transmission ratios is preferred because the more the gear ratios, the better the fuel economy. However, the effectiveness becomes very weak when the number of gears reaches 10.

Another trend of ATs is the introduction of electronic control. For example, the slip control of the lock-up clutch in the torque converter [62] greatly improves the transmission efficiency while preserving the merit of shock absorbing; independent clutch control [104] uses a proportional solenoid valve to control each clutch (or brake), which greatly simplifies the mechanical content of the AT.

1.2.2 Automated Manual Transmission (AMT)

As implied by the name, an Automated Manual Transmission (AMT) [95] can be regarded as a robotized manual transmission where the operations of the clutch and

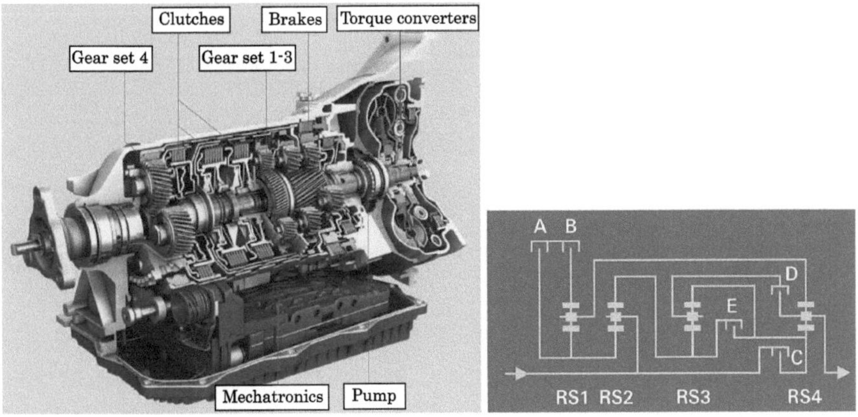

Fig. 1.10 8-speed AT and diagram chart (provided by ZF Friedrichshafen AG [http://www.zf.com])

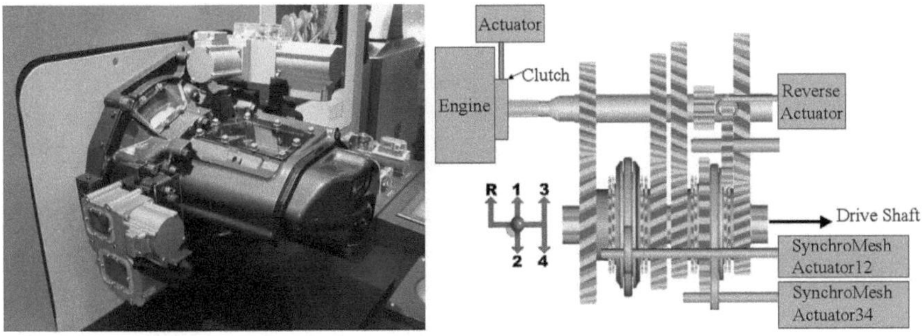

Fig. 1.11 AMT and diagram chart (courtesy of NTN Corporation, at the Tokyo Motor Show in 2008)

shift lever are carried out by electro-mechanical or electro-hydraulic actuators. The actuators are driven by a Transmission Control Unit (TCU). Figure 1.11 shows the AMT and its diagram chart. AMTs offer the advantages of lower weight, lower cost and higher fuel economy. However, one of the limitations of AMTs should be the driving comfort reduction, caused by the lack of traction during gear shift actuation, which is shown in Fig. 1.12.

Because of its inherent characteristics, an AMT is a suitable choice for micro cars, sport cars and heavy duty trucks. In vehicles with AMTs, the integrated control of the engine and clutch becomes an important control issue in order to reduce the affection of the traction interruption.

An AMT is also suitable for a pure electric vehicle which needs a multi-speed transmission to improve the launch performance and to extend the cruising speed. Examples of 2-speed transmissions include a novel seamless transmission proposed in [130] and the I-AMT (Inverse Automated Manual Transmission) proposed by the

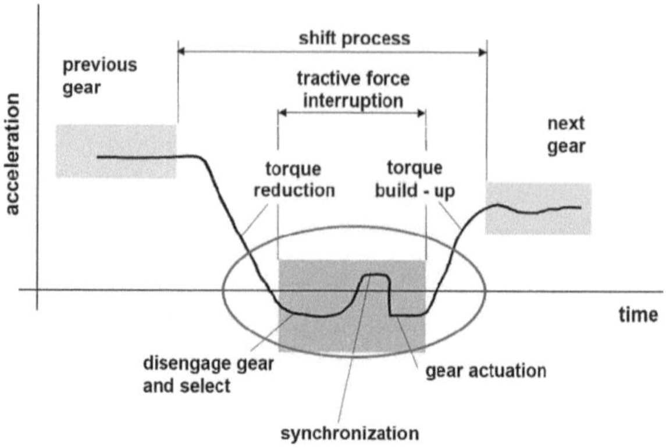

Fig. 1.12 Torque interruption of AMT (courtesy of LUK Corporation [47])

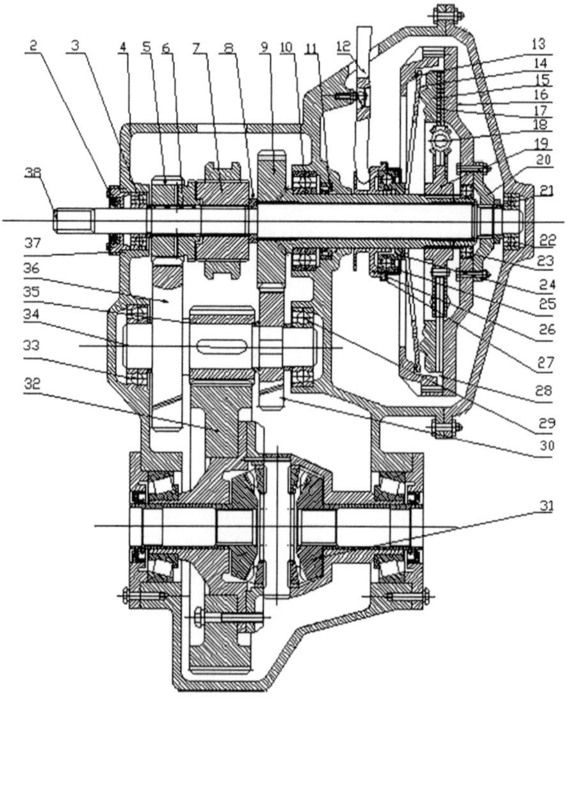

Fig. 1.13 2-speed I-AMT for EV, 38-shaft connected with motor (courtesy of Jilin University, China [91])

authors [91], which have similar architectures and the traction interruption could be eliminated. As shown in Fig. 1.13, the name of "I-AMT" derives from the fact that the dry clutch, which is at the front of a traditional AMT, is put to the end of the transmission.

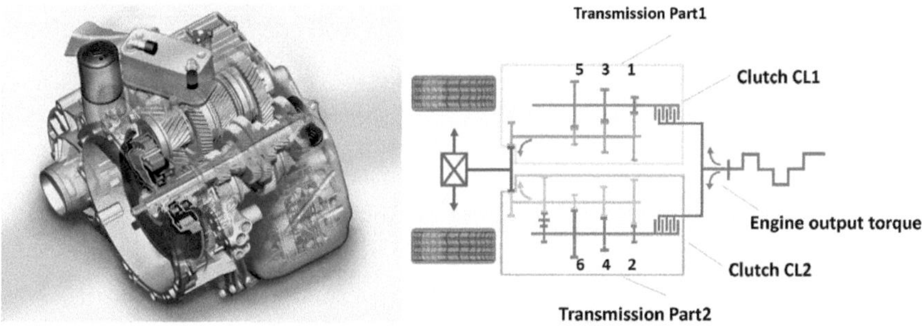

Fig. 1.14 DCT and diagram chart [http://www.volkswagenag.com]

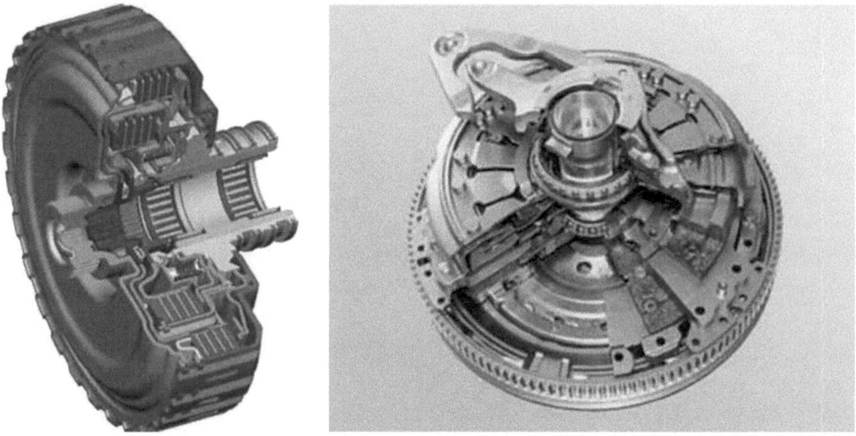

Fig. 1.15 Wet and dry clutch [http://www.borgwarner.com, http://www.schaeffler.cn]

1.2.3 Dual Clutch Transmission (DCT)

A Dual Clutch Transmission (DCT) [56], also referred to as a twin clutch transmission or a double clutch transmission, uses two separate clutches for odd and even gear sets, respectively, as shown in Fig. 1.14. It can be fundamentally described as two separate manual transmissions. When the vehicle is operating with one sub-drive, the control unit is already selecting the next gear in the other sub-drive. Therefore, the shift process takes place through the torque delivery from one clutch to the other, namely we have a clutch-to-clutch shift. Hence, a DCT not only preserves the advantage of an AMT in high efficiency and fast response, it also eliminates the traction interruption of an AMT.

There are two fundamental types of clutches in DCTs: one is the wet multi-plate clutch and the other is the dry single-plate clutch, as shown in Fig. 1.15. Because there is less pumping and friction loss, a DCT with a dry clutch has better efficiency. However, a DCT with a wet clutch can handle higher torque input. The 7-speed DCT with a dry clutch produced by LUK is designed to transmit torque up to 250 Nm,

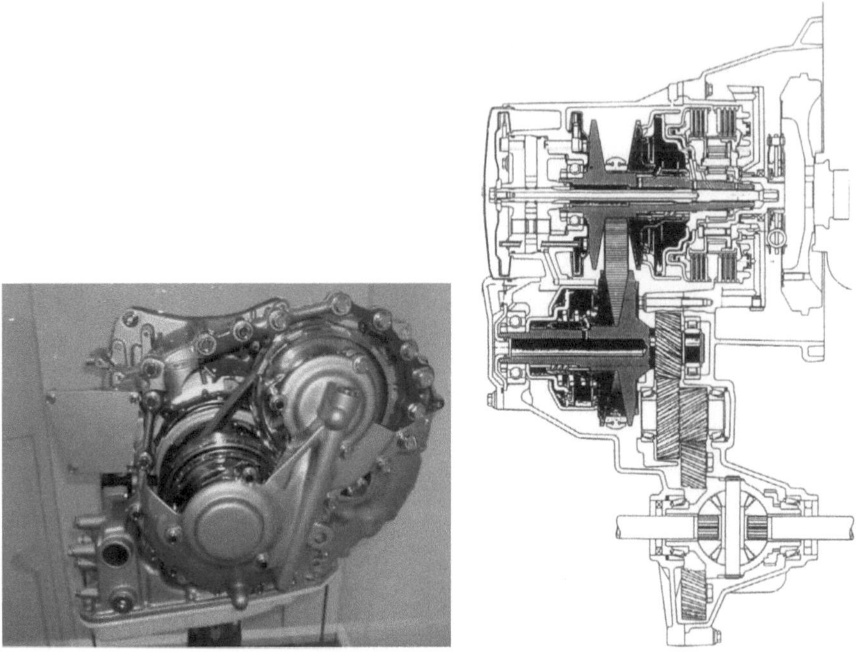

Fig. 1.16 CVT and diagram chart (courtesy of Aisin Seiki Co. Ltd., at the Tokyo Motor Show in 2008)

while the 6-speed DCT with a wet clutch supplied by Borgwarner is used for engines which can generate a torque of 350 Nm.

Because there is no torque converter to absorb the shift shock and the process of the clutch-to-clutch shift is usually finished within hundreds of milliseconds, the two clutches must be controlled precisely enough so that large torque interruption or clutch tie-up can be prevented.

1.2.4 Continuously Variable Transmission (CVT)

Different from the above step-ratio transmissions with a few different distinct gear ratios to be selected, a Continuously Variable Transmission (CVT) [1], see Fig. 1.16, can create a continuously-variable ratio between the higher engine speed and the lower wheel speed. The flexibility of CVTs allows the engine to maintain a constant angular velocity over a range of vehicle speeds, which makes the engine operate in a high-efficiency area. However, a CVT transmits torque through friction or traction drive, its efficiency is worse than that of gear transmissions, and finally, the advantage of fuel economy may not be so significant. A CVT is usually utilized for low-power or mid-power vehicles due to the trade-off of efficiency and power density.

	6-Speed AMT	6-Speed DCT (Wet)	6-Speed Step-AT	CVT
Fuel Efficiency	++	+	0	-
Cost	++	+	0	-
Cost of Ownership	-	++	++	-
OEM Invest US / Asia	0	0	++	0
OEM Invest Europe	++	++	0	- -
Weight / Size	++	++	+	-
Launch Quality	-	++	++	++
Torque Cpacity	++	++	++	-
Shift Quality	- -	+	+	++
Thermal Capability	-	+	+	+
Response	+	++	0	0
Safety	+	+	+	+
Technology Challenge	++	++	++	- -

Fig. 1.17 Characteristics of automotive transmissions [http://www.borgwarner.com]

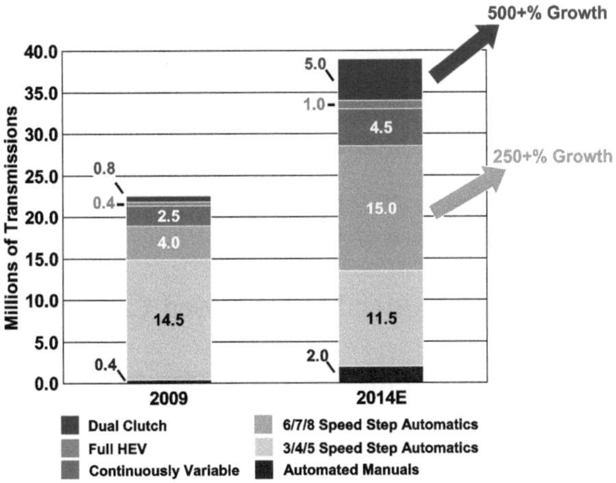

Fig. 1.18 Prediction of the market share of automatic transmissions [http://www.borgwarner.com]

The most widely used two types of CVTs are Belt-CVT and Toroidal-CVT [45]. Normally, a Belt-CVT has less torque transmission capability than a Toroidal-CVT.

1.2.5 Final Remark

Finally, the characteristics of each type of transmission are summarized in Fig. 1.17.

For passenger cars, according to the prediction of the market share which is shown in Fig. 1.18, ATs will still occupy the largest share in the near future and DCTs will have the fastest increase. In both ATs and DCTs, the change of speed ratios can be regarded as a process of one clutch being engaged while another being disengaged, namely, clutch-to-clutch shift. A clutch-to-clutch shift greatly simplifies the transmission mechanical content and increases the control flexibility, however, *it is quite a challenge to obtain robust shift performance because if the shift process is not controlled exactly, clutch tie-up, engine flare-up or traction interruption will be caused.*

In the case of commercial vehicles, the AMT is the most popular automatic transmission used because of its low cost and high efficiency. *In AMT vehicles, control effort is necessary to weaken the traction loss during the shift process.* The reduction of shift time and shift shock should be taken into account, where clutch control plays an important role. Moreover, the driveline of commercial vehicles has a relatively larger torsional vibration compared to passenger cars. When the accelerator pedal is pressed or released, driveline resonances have a larger impact on the driver, and thereby active engine control is always adopted to damp the unintentional driveline jerking. Generally speaking, the precise control of the engine and clutch is crucial to improve the longitudinal dynamical behavior of AMT vehicles.

A CVT has a basically different mechanical topology, and there is no problem of shift shock anymore. The major control challenge is to maintain an optimal clamping force to prevent slipping. At the same time, the speed ration should be controlled fast enough to maximize the fuel economy benefit. If, by combining a CVT with a planetary gear, an infinite gear ratio range including zero ratio can be obtained, which is called an Infinitely Variable Transmission (IVT), at the geared neutral point, the speed ratio control becomes inadequate and the output torque control is necessary [132].

1.3 Why Consider Model-Based Control?

1.3.1 Evolution of Control Systems for Automotive Powertrains

Global warming and other environmental concerns require automotive industry to produce automobiles with more efficiency and less emission. On the other hand, the customer's requirement for driving comfort and fuel economy is more and more demanding. The drivetrain propelled by an engine (together with motors in the case of hybrid powertrains), as shown in Fig. 1.1, plays an important role in these tasks.

In order to meet the above challenges, a lot of new devices and innovative technologies are proposed and applied in the area of automotive powertrain. In the engines, there are several different ways to improve the performance of fuel economy and emission, including using a turbocharged gasoline engine with variable valve timing (VVT) and exhaust gas recirculation (EGR); direct injection stratified charge (DISC) gasoline engine with lean NOx trap (LNT) aftertreatment system; and a diesel engine with variable geometry turbocharger (VGT), higher common-rail pressure (up to 200 MPa) and EGR [37, 79]. In addition, homogeneous charge compression ignition (HCCI) has the characteristics of both the gasoline and diesel engine, and provides higher fuel efficiency and almost negligible NOx emission [131]. The implementation of all these technologies depends on the development of electronic control systems. Along with the increasing complexity of engines, interactions between different devices become more and more substantiated. For example, lean burn of spark ignition gasoline engine brings about worse emission; in diesel engines the actuator of EGR affects the characteristics of VGT. In other words, a newly introduced control input has an influence on different system outputs [60, p. 11].

On the other hand, the transmission box delivers and adapts engine torque to the following jointed drive shafts and tires. To improve efficiency and drivability, different kinds of automatic transmission have been introduced, such as automatic transmission (AT) with torque converter, automated manual transmission (AMT), dual clutch transmission (DCT), continuously variable transmission (CVT), electrical variable transmission (EVT) [132], and some variants of traditional AMTs, including "Power-Shift AMT" [106], seamless transmission proposed in [130] and I-AMT [91]. These transmissions greatly improve the drivetrain performance, and yet require higher-performance actuator hardware and control software. For instance, as aforementioned, DCTs improve the drivability of AMTs by eliminating the torque interruption during the shift process. The two clutches, however, have to be controlled precisely during the clutch-to-clutch shift to avoid tie-up or traction interruption [56].

In addition, the hybridization of the vehicle propulsion system shows significant potential in reducing fuel consumption and exhaust emissions. Hybrid electric vehicles (HEV), firstly released en masse into market by Japanese makers Toyota and Honda, can provide fuel efficiency improvement and CO_2 reduction of about one third [25]. Plug-in hybrid electric vehicle (PHEV) [22], a hybrid vehicle with rechargeable batteries that can be connected to an external electric power source, further reduces the well-to-wheel CO_2 emission significantly if the car is driven in an urban area. Introducing electric motors into the propulsion system brings about many new control issues. Besides the optimization of the energy management, there are also some highly transient dynamic control problems, such as mode-switching control and active damping of drivetrain oscillations [8, 63, 87].

Some other drivetrain control systems include a 4-wheel-drive (4WD) with electronic torque control [43], and brake systems, such as anti-lock brake system (ABS) and electronic stability program (ESP) [1], which are more likely to belong to the area of vehicle safety control [2].

In other words, new functions and legislations are forcing automotive control systems to become more and more complex. *Although automotive control systems had been developed separately in the past, at present, these systems have to be designed by considering the interactions and communications between them.* One of the most important interactions in ground vehicles happens between steering and brake systems [21] because the characteristics of the tire force (or moment) in one direction is not independent from the others. In the drivetrain system, on the other hand, perfect drivability will never be achieved without a close cooperation of the engine and transmission.

Along with the increment of functions, the proportion of electronic components used in ground vehicles has been increasing steeply in recent years. In a 2007-model car, electronic components amounted to 20–30 % of the total production cost, and this figure is expected to reach 40 % or so by 2015 [70]. Moreover, the software development occupies a high percentage of the total expenses of automotive electronics. Figure 1.19 shows the increase of the size of automotive software, and it is

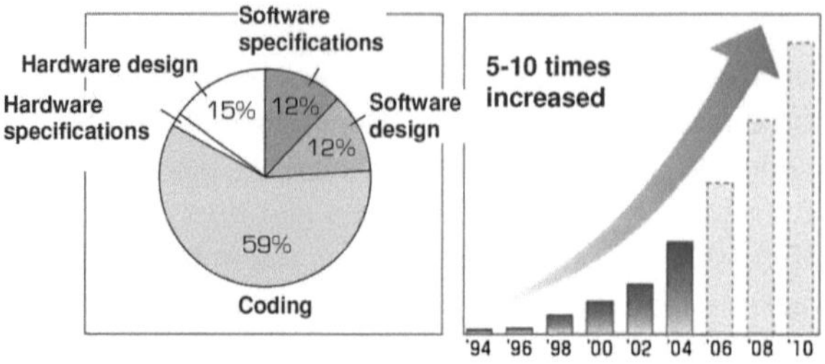

Fig. 1.19 Increasing of software development cost [70]

pointed out that, at present, the software accounts for 80 % of the total development cost of an electronic control unit (ECU) [70].

Model-based control is introduced and developed rapidly under such a technical background. With the growing intricacy of drivetrain systems, the conventional methodology of developing control systems, normally based on event-driven (rule-based) control or feedforward control, needs a large amount of cost- and time-consuming calibrations in order to obtain reasonable over-all control performance. It is unfavorable for automotive makers because the competition of this industry is becoming increasingly intense.

1.3.2 Introduction of Model-Based Design

"Model-based design" [11, 16, 89, 110, 123] is a methodology applied in designing embedded software which provides an efficient approach for establishing a common framework for communication throughout the design process while supporting the development cycle (such as "V" diagram) [71, 122]. Model based design answers the requirement of developing automotive control systems which should satisfy the following attributes:

- Safety;
- Low cost;
- Fuel economy & Emission;
- Drivability;
- Time to market.

Graphical modeling tools, such as MATLAB/Simulink, AMESim, Dymola and Easy5, etc., provide generic and unified graphical modeling environment, and there are many different levels of test tools, including MIL (Model In the Loop), RCP (Rapid Controller Prototyping), SIL (Software In the Loop), PIL (Processor In the Loop) and HIL (Hardware In the Loop). Although there are numerous test tools, the models can be approximately classified into 3 categories: simulation model, control-oriented model, and controller model.

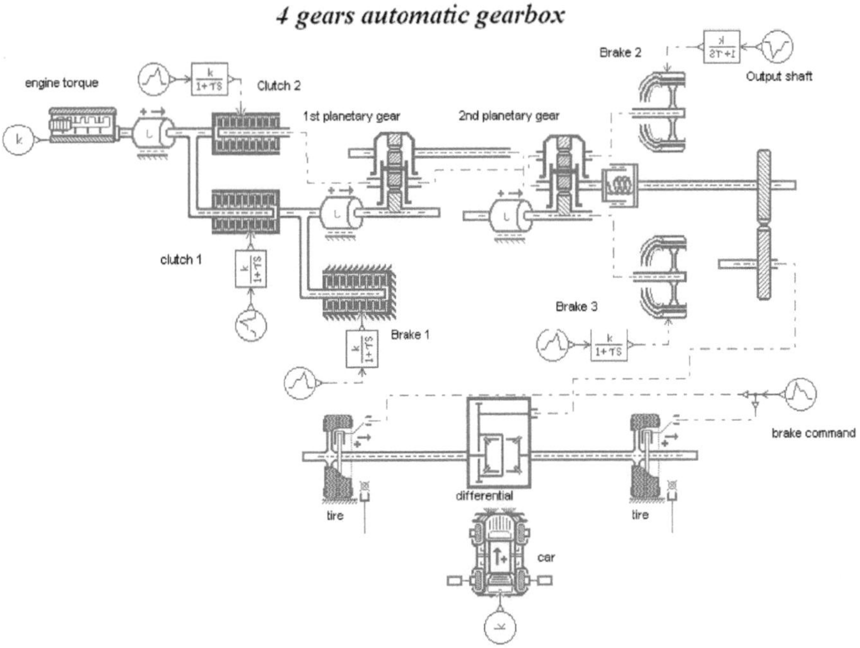

Fig. 1.20 Example of a drivetrain simulation model

Simulation Model

The term simulation model refers to the model required to be as accurate as possible to represent the real system dynamics, which makes it possible to test the control software before the hardware prototype is available. Of course, real time simulation models, which are used in HILS, also belong to this category. In some large-size real time simulation models, however, some relatively "high-frequency" dynamics, such as the oil compressibility in hydraulic actuators, may be ignored to assure the successful implementation of real time computation. Figure 1.20 gives an example of a drivetrain simulation model established in AMESim.

In this example, the rotational motion of the driveline, including the parts from the clutch to wheels, is modeled by dynamic and kinematic disciplines. For example, the dynamics of the drive shaft is described as

$$T_s = K_s\theta_s + D_s\dot{\theta}_s, \tag{1.16}$$

where T_s denotes the shaft torque, θ_s is the twist angle, K_s is the stiffness of the drive shaft, and D_s the damping coefficient. The user needs to give the value of K_s and D_s in a graphical interface, and AMESim will generate the dynamic equations automatically, which are based on the power flow analysis of the Bond Graph technique [77].

The engine torque characteristics, on the other hand, are modeled based on experimental identification, which is shown in Fig. 1.3. The torque is identified as a static map with the inputs of the engine throttle and the engine speed. This is the simplest engine model [4, 148]; some other engine models with different time-scales are the mean-value model (MVM) [60], 1D engine model and CFD (Computational Fluid Dynamics) model [3, 80]. Another identified part is the tire model, where the Magic Formula model [108] and UniTire model [59] are always adopted.

HILS is always introduced when a system or a part of it is too difficult to be modeled accurately, where the real physical hardware is integrated with mathematical models through a real time interface. One of the most popular HILSs is the driving simulator, where the "real part" of the system is the driver, which is also called a human-in-the-loop simulation. Some other HILSs are always relevant to electro-hydraulic actuators, the characteristics of which are highly nonlinear and time-varying.

Control-Oriented Model

The term control-oriented model refers to the model based on which controllers are synthesised. A control-oriented model must be simple enough to guarantee easy implementation of the derived controller. On the other hand, it should also be able to capture the dominant system dynamics. For example, in an automotive driveline with a dry clutch, see Fig. 1.1, there are distributed compliances including the damping spring of the clutch, the stiffness of the propeller shaft and the stiffness of the drive shaft/axle half shaft, etc. However, in a control-oriented model for the problem of active damping control [109], these rotational freedoms are not all included, but only the compliance of the drive half-shaft is considered because it dominates the fundamental vibration of the driveline. In other words, the drive half-shaft has the "softest" twist characteristics compared to those of the others. A control-oriented model can be given in the form of *state-space equations*. For example, the following set of differential equations

$$\dot{\omega}_c = \frac{1}{I_i}\left(T_e - \frac{1}{R_i R_{df}} T_s\right), \tag{1.17a}$$

$$\dot{\omega}_w = \frac{1}{I_v}(T_s - T_v), \tag{1.17b}$$

$$\dot{T}_s = K_s\left(\frac{1}{R_i R_{df}}\omega_c - \omega_w\right) \tag{1.17c}$$

is used to describe the driveline dynamics for anti-jerk control, where the rotational damping is ignored. In model (1.17a), (1.17c), ω_c is the output speed of the clutch; ω_w is the wheel speed; T_s is the axle shaft torque; I_i denotes the equivalent inertia moment from the engine to the axle shaft, at the ith gear position, $i = 1, 2, \ldots, 6$; I_v is the equivalent inertia of the vehicle; T_e is the engine torque, and T_v is the driving resistance torque; R_i denotes the gear ratio of the ith gear position, and R_{df} is the ratio of the differential gear box; K_s is the stiffness of the axle shaft.

Examples of a *transfer function model* could be found in [120], where the dynamics of the proportional pressure control valve is identified as a first-order lag, and the transfer function from the valve current u to the brake speed ω is given as

$$G_0(s) = \frac{\Omega(s)}{U(s)} = -\frac{\mu RNA}{Is + C_d} \frac{K_{rv}a}{s + a}, \tag{1.18}$$

where μ is the friction coefficient, R is the effective radius of the brake plate, N is the number of friction plates, A is the piston area, C_d is the coefficient of viscous friction, K_{rv} is the gain of the proportional reducing valve and a is the identified parameter of the first-order lag.

Controller Model

A controller model represents the control algorithm which is designed to achieve the desired performance of the control system. It is singled out especially because in recent years a controller model established by MATLAB/Simulink can be converted to embedded code very conveniently using tools of automatic code generation.

Traditionally, when developing embedded control systems, text-based programming was used and it was time-consuming and prone to error. This problem was overcome by introducing graphical modeling tools, such as MATLAB/Simulink. If the controller algorithm is represented by a graphical model, the designers of different departments can grasp the entire controller construction quickly, and it is easy to transport the model from one stage to another in the whole design process. Some examples of code generation software are dSPACE/TargetLink, Real-Time Workshop, Embedded Coder.

1.3.3 Application Examples of Model-Based Control

Until now, simulation and automatic code generation have been widely applied in control system development of automotive drivetrains. There are, however, relatively fewer published reports about successful applications of model-based controller design in production drivetrain systems (in this book, *the model-based controller design refers to synthesizing a controller based on a control-oriented model*). Actually, at present the widely used control algorithms are still based on event-driven (rule-based) or feedforward control. One of the reasons may be the relevance of various uncertainties in automotive drivetrains, such as large variations of the operating condition, part-to-part variability and long-term aging.

Although there exist many difficulties, the automotive and supplier companies are making strenuous attempts to a establish model-based controller synthesis framework because of its merit of reducing parameter calibrations while achieving higher dynamic performances. In engines, the model-based air/fuel (A/F) ratio control [60] and model-based cylinder torque estimation [26] have been tested practically. In the area of the driveline, one successful example is reported in [62], where a high performance and highly reliable slip control system for a torque converter clutch is realized through the application of the H_∞ control theory. It was claimed that this is the first case of practically applying the H_∞ theory to mass-produced automotive components. Another example is INVECS-II, a 5-speed AT (Automatic Transmission) produced by Mitsubishi in Japan, where a model-based feedback control is adopted to control the clutch speed during the shift process [119]. Shift shock was reduced by carefully designing the reference trajectory of the clutch speed.

1.4 Why Consider Nonlinear Control?

1.4.1 Necessity of Nonlinear Control

When applying model-based control to automotive drivetrains, the following control methodologies have been used:

- Open-loop control, event-driven control [42];
- Linear control, PID control [148];
- Robust control, H_∞, μ-synthesis [62, 120];
- Nonlinear control, sliding mode control, model predictive control, etc. [13, 98].

Among these methodologies, open-loop control and feedback control based on linear models have been applied in the mass-production of drivetrains, while the majority of nonlinear control designs are still in the stage of academic research.

Nonlinear control approaches are attracting more and more attention because of the inherent characteristics of automotive drivetrains, namely,

- Large operation range;
- Large modeling uncertainty;
- Strong actuator nonlinearity.

First, automotive drivetrains work in a wide range of speed and torque. For a typical mid-size passenger car, the engine speed may vary from 500 to 8000 rpm, while its torque changes from negative (engine braking) to positive about 300 Nm. When the operation range of a control system is large, linear controllers behave poorly and gain scheduling is always necessary, which introduces a large amount of calibration works. Nonlinear control can handle nonlinearities directly, and it makes the control system work well at every operating point without laborious tuning of parameters.

Second, automotive drivetrains work in an uncertain environment, such as altitude, temperature, road slope/grade, road surfacing condition and vehicle mass. These uncertain variables yield modeling errors when a control-oriented model is established. For example, when a fully loaded vehicle is climbing a slope, the drivetrain dynamics is much different from an empty vehicle driving on a flat road. In addition, long-term aging and part-to-part variability also bring about inevitable uncertainties.

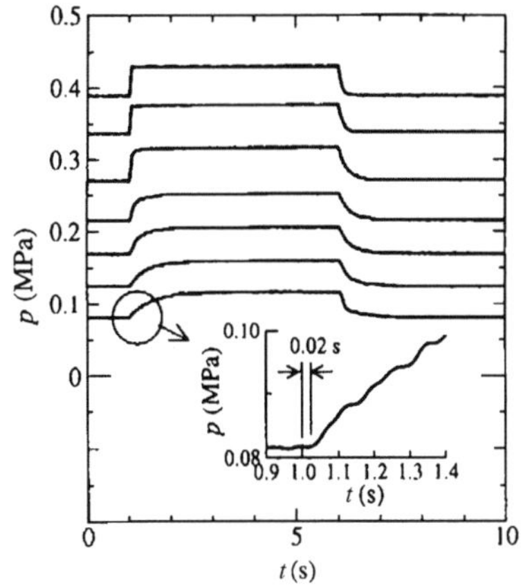

Fig. 1.21 Pressure response of wet clutch valve. Reprinted from [120], copyright 1998, with permission from Elsevier

Finally, automotive drivetrains are inherently nonlinear. Engine torque characteristics, torque converter characteristics and air drag characteristics are all nonlinear. Furthermore, the characteristics of drivetrain actuators are highly nonlinear. For example, the clutch plays a crucial role in drivetrain dynamics. However, the friction characteristics of the clutch plates and the dynamics characteristics of a clutch actuator are very complex.

Figure 1.21 shows the pressure response of a clutch actuator, i.e., a hydraulic cylinder controlled by a proportional pressure control valve. The valve current is stepped at different levels of pressure, and it is shown that the higher the operating pressure, the quicker the step response.

Not only the dynamic response from the valve current to the clutch pressure is nonlinear, the response from the clutch pressure to the delivered torque is also highly nonlinear. Figure 1.22 gives the experimental results of the torque response of a wet clutch. The thick solid line is the delivered torque under the input of the clutch pressure, the thin-color line. It is shown that the responses are quite different along with the variations of the temperature and the clutch speed.

In the case of dry clutch, the delivered torque is a nonlinear function of the throwout bearing position and the slip speed. Figure 1.23 gives an example of the torque mapping of a dry clutch.

From above we can see that the complex friction and actuator characteristics make drivetrain control a highly nonlinear control problem.

1.4.2 State-of-the-Art of Applied Nonlinear Control

Nonlinear control consists of a variety of powerful methods, including

- Sliding mode control;
- Feedback linearization;
- Differential flatness;
- Backstepping;

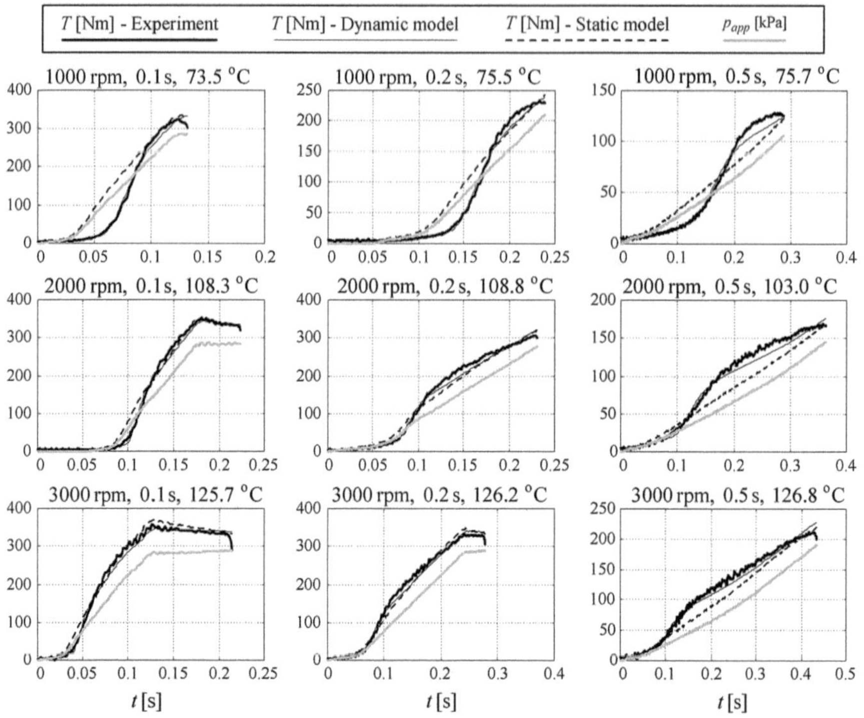

Fig. 1.22 Torque response of a wet clutch for different slip speeds and temperatures. Reprinted from [39], copyright 2006, with permission from IEEE

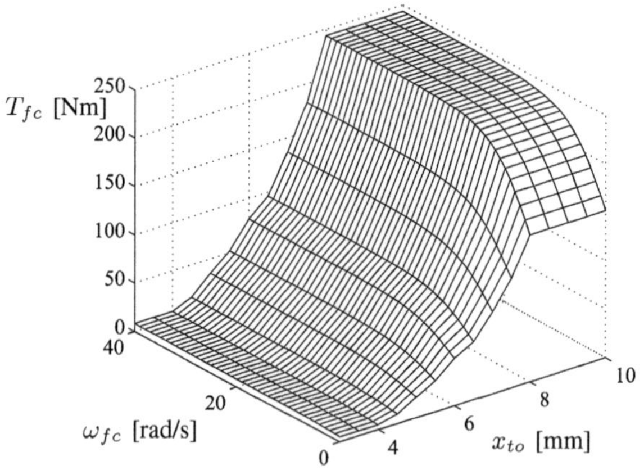

Fig. 1.23 Torque map of a dry clutch for different slip speeds and throwout bearing positions. Reprinted from [138], copyright 2011, with permission from IEEE

- Input-to-state stability (ISS);
- Model predictive control;
- Data-driven control, etc.

Each of them has its own characteristics and successful industrial applications.

Sliding Mode Control

Sliding mode control, initiated and prompted by the works of Itkis [73], Utkin [136], Slotine [124], Hedrick [126] and others has been extensively developed in both theory and applications in the last few decades. Sliding mode control is essentially a form of variable structure control, it accounts for nonlinearity and provides good robustness against modeling imprecision. The control law is not a continuous function, but switches from one structure to another according to the current system state. The multiple control structures are designed so that state trajectories always move toward a switching condition, and finally the trajectories are forced to slide along the preferred surface boundaries. The motion of sliding along the boundaries is called sliding mode, and the boundaries are called sliding hypersurface.

The main advantage of sliding mode control is its robustness. If the "matching condition" is satisfied, the sliding mode can be achieved in spite of disturbances and parametric variations. This makes the variable structure control have the following merits: fast response, insensitive to disturbances and uncertainties, no need of online identification, and easy implementation. The shortcoming of sliding mode control is that, once the state trajectory arrives at the sliding surface, it is difficult to make it strictly slide along the surface. Under hard sliding-mode-control action, the state inevitably moves across the surface repeatedly, which is called "chatter".

Along with the recent development of sliding mode control, the control discontinuities can be eliminated while the concept of "attractive" surface is retained [133]. Sliding mode control has been successfully applied to robot manipulators, underwater vehicles, automotive transmissions and engines, high-performance electric motors, and power systems [125]. In automotive engines, sliding mode control has been used in EGR (Exhaust Gas Recirculation) and VGT (Variable Geometry Turbocharger) [136], and in the driveline area, sliding mode observer has been designed to estimate the torque of the drive axle shaft and the pressure of the clutch [98, 141]. There is also research on the clutch speed control using the sliding mode method [146].

Feedback Linearization

Feedback linearization was proposed and developed in the works of Brockett [19], Hunt et al. [69], Isidori [72], Nijmeijer and Van der Schaft [107], and others. The approach stems from the theory of differential geometry, and the central idea is to algebraically transform nonlinear dynamics into a linear one, which is achieved exactly by a state transformation and a state feedback, rather than by a linear approximation [125].

Feedback linearization has been successfully used in some practical control areas, including helicopters, high performance aircrafts, industrial robots, and biomedical devices [125]. In other industries, the application of feedback linearization is also extended. For example, in [94], feedback linearization is used to transform the nonlinear dynamics of an active magnetic bearing into a linear one and then a high-performance controller for the feedback linearized plant is designed with μ-synthesis to guarantee the compliance performance specification of the beam. In [20], feedback linearization is applied in an automotive suspension system where a somehow intelligent feedback linearization controller is suggested through online estimation of nonlinear parameters. Reference [18] investigates the stop-and-go cruise control of a heavy duty truck, and an extended feedback linearization based on the nonlinear Smith predictive method is proposed. This method not only converts the nonlinear system to a linear canonical form, but also can compensate for the variable time delay effectively. Then, a tracking control algorithm provides precise acceleration/deceleration tracking under low-speed driving.

Differential Flatness

It has been proved that the existence of dynamic exact feedback linearization is equivalent to the property of flatness. The concept of Differential Flatness was first proposed by M. Fliess et al. [49], and then further researched by P. Martin and R.M. Murray [48, 137]. A dynamic system is called differentially flat if the system input and state variables can be expressed as functions of the system output and a finite number of its time derivatives [49]. The system output is then called a flat output. For a differentially flat system, if the trajectory of the flat output is given, the desired control input can then be derived directly as a function of the flat output and its derivatives, which can be served as a feedforward control for tracking problems. In other words, the use of differential flatness can improve the performance of an existing linear feedback control system by introducing a nonlinear feedforward compensator. The key issue of applying the differential flatness technique is to check if the considered system is flat and to find a reasonable flat output.

The concept of differential flatness is widely used for trajectory planning and tracking control. In [34], for the solenoid valve actuator used for gas exchange in internal combustion engines, flatness is used for the motion control of the armature. In [64], an electro-hydraulic clutch position control system of an AMT is considered, and based on the flatness approach, a nonlinear feedforward control is designed to combine with a linear feedback control.

Backstepping

In recent years, robust nonlinear control has attracted a great deal of research interest. Many synthesis approaches were proposed where the controlled variable is chosen to make the time derivative of a control Lyapunov function (CLF) [10, 127] negative definite [133]. One important methodology is "integrated backstepping", and a systematic design procedure has been developed for the backstepping approach in the book by M. Krstic et al. [83].

The term backstepping refers to a technique developed by Petar V. Kokotovic and others [81, 82] to design stabilizing controllers for a class of nonlinear dynamical systems in the strict-feedback form. One can start the design process from designing a stabilizing controller for the inner subsystem by viewing the state of the upstream subsystem as virtual control and "back out" new controllers that progressively stabilize each outer subsystem. The process terminates when the real external control is finally reached. In this sense, the process is called backstepping. In each recursive step, one constructs an augmented CLF and renders its derivative negative to obtain the virtual control, therefore, the control law obtained by backstepping is asymptotically stabilizing.

The backstepping technique has wide-area application, such as in the control of an electric motor [67] and an electro-hydraulic system [6, 76]. It has been demonstrated that backstepping is a suitable method to deal with the nonlinearity introduced by hydraulic actuators, such as the nonlinear orifice flow characteristics [135].

Input-to-State Stability (ISS)

For a linear time-invariant system, the zero-input response decays to the origin exponentially, while the zero-state response is bounded for every bounded input, namely is a bounded-input bounded-state property. For a general nonlinear system, however, it should not be surprising that these properties may not hold [78]. The concept of input-to-state stability (ISS) extends the notion of the global asymptotical stability (GAS) to nonlinear systems, and provides a natural framework in which stability is formulated with respect to the input.

The notion of input-to-state stability (ISS) was originally introduced by E.D. Sontag [128, 129], and it has been proved that ISS can be stated in several equivalent manners using, e.g., dissipation, robustness margins, and classical Lyapunov-like definitions. This indicates that ISS is a mathematically natural concept. The concept of ISS has been employed by several authors in deriving results on control of nonlinear systems, including discrete-time nonlinear systems [75], switched systems [139], model predictive control [97], nonlinear observer design [5, 52] and neural networks [121]. The industries wherein ISS has been applied include at least automotive [53, 66] and robot [9].

Model Predictive Control

Model Predictive Control (MPC), also referred to as moving horizon control or receding horizon control, has become an attractive feedback strategy for controlling constrained systems. The main idea of MPC is to use a mathematical model to predict the future dynamic behavior of the to-be-controlled system over a prediction horizon, and to determine then the control input over a finite control horizon such that a predetermined open-loop performance objective function is optimized. Hence, the MPC problem is formulated as solving online a finite horizon optimal control problem subject to (linear or nonlinear) system dynamics and time-domain constraints involving states and inputs [24, 116]. At each sampling instant, the optimization problem is solved to obtain the optimal control sequence, and only the first element is applied to the system. The procedure is repeated at the next sampling instant, updated with the new measurement.

MPC was introduced and developed at the end of the 1970s under the different names such as Model Algorithm Control (MAC) [117, 118], Dynamic Matrix Control (DMC) [38], Generalized Predictive Control (GPC) [35, 36] and Moving Horizon Control [84, 85], for different application areas and from different viewpoints. Over the past few years, academic research on MPC has produced significant progresses on issues of both stability and robustness (see, for example, [14, 28, 29, 31, 32, 57, 58, 86, 92, 100, 101]). For a complete survey, we refer to, for example, [7, 12, 27, 46, 54, 93, 96, 99, 102, 103, 114, 115].

The main advantages of MPC are the functionality of performing multi-variable optimal control, the ability to account for nonlinear dynamics and to handle time-domain system constraints in an explicit and optimal way. With the rapid development of computing, MPC is successfully applied not only in refining and petrochemicals where slow dynamics are dominant [111, 112], but also in aerospace and defense (see, for example, [17, 44, 74]). In the area of automotive powertrains, application examples include Homogeneous Charge Compression Ignition (HCCI) engine [15], idle speed control [23] and engagement control of a dry clutch [13]. The to-be-coordinated requirements are, for example, drivability, fuel economy and ride comfort, while time-domain constraints could include

- Safety constraints;
- Emission regulation;
- Actuator saturation;
- Frequency response limitation of actuators;
- Trajectory constraints.

Data-Driven Control

It should be noted that automotive systems are inherently complex, highly nonlinear, switching and strongly coupling. Moreover, in recent years, for catering to customer's requirements and stringent emission regulations, new technologies and new actuators were introduced, which increased the degree of freedom and coupling of the automotive control system and made automotive systems more and more complex. At the same time, the system characteristics change along with the variation of driving conditions and long-term aging. For example, the damping coefficients of rotational shafts change greatly with the environmental temperature. Long-term aging and variation of driving conditions still bring about significant modeling errors. Therefore, one of the direct consequences is the difficulty of decision making or realizing the system's control and optimization with model-based approaches, which need mathematically building the dynamic models of the system [65, 140, 147]. Fortunately, computer technologies, digital sensor technologies, and networking techniques are widely used, which generate a great quantity of historical and real-time data related to the modern industries. In this case, technologies for data management such as data mining, data collection, and data fusion have emerged [41, 143]. All of these have led to the development of data-driven methods which have great interest from the system and control communities, especially in the research field of control techniques [134, 140, 144].

The data-driven modeling method has received considerable attention, stemming from artificial intelligence and machine learning [68, 105]. Data-driven methods present not only a new avenue but also new challenges both in theories and applications [88, 90, 113, 144]. The data-driven model-free control method implies that the controller design is merely based on the input/output measurement data of the controlled plant, without explicitly or implicitly using the plant structure or dynamics information of the controlled plant, and whether the plant is linear or nonlinear. Although the data driven approach focuses on input–output relations and avoids derivation of differential equations, it is a consensus that a physically-based model and a data-driven model are not necessarily mutually exclusive, and sometimes they play complementary roles, which will be seen in the example of vehicle launch control in Chap. 8.

1.5 Structure of the Text

As stated above, the function of an automotive drivetrain is to provide the thrust force to induce the longitudinal motion of a vehicle, and it determines the fuel economy and the longitudinal dynamic performance. Even though technical performance of automotive vehicles has been improved, customers demand global comfort and fuel economy more and more. To remain competitive, vibratory comfort and drivability have become a main preoccupation for car manufacturers.

The definition of drivability is somehow complex because it implies subjective perception of the driver. Vibratory discomfort means all low frequency sensations which cause trouble to passengers. One important aspect of drivability is the first resonant mode of the driveline in the frequency range of 0–10 Hz, which is called "shuffle". The book will focus on this frequency range of the drivetrain vibration, which involves the shift shock and other unintentional driveline jerks.

Because CVTs offer continuously variable gear ratios between established minimum and maximum limits, the mechanism of the dynamic control is quite different from the step-ratio transmissions. In this book, we will concentrate on the dynamic control of step-ratio transmissions, including AT, DCT and AMT, and the outline of the content is shown in Fig. 1.24.

First, the clutch-to-clutch shift control [33], which is the shift technique of ATs and DCTs, will be investigated in detail. The clutch-to-clutch shift process can be approximately divided into the torque and inertia phases. In the torque phase, the traction torque of the engine or the turbine of the torque converter is transferred from the off-going clutch to the on-coming clutch, where the precise timing of releasing and applying of clutches is crucial for the prevention of the clutch tie-up and traction

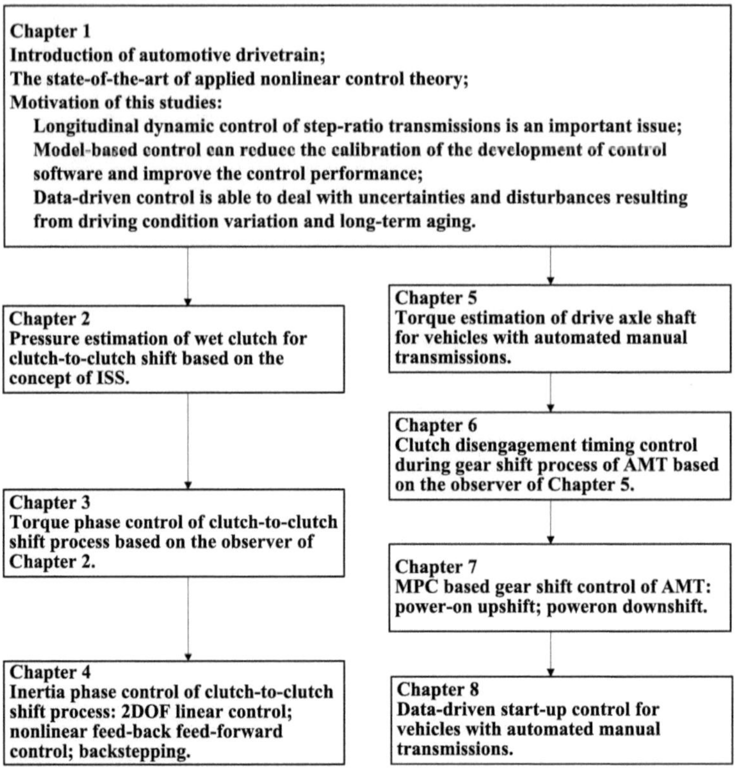

Fig. 1.24 Outline of the book

interruption. In the inertia phase, the on-coming clutch is synchronized through the engagement slip, where the clutch speed control [40] or engine speed control [56] can be adopted to guarantee good shift quality. The context of this part (the left column of Fig. 1.24) is separated as Chaps. 2, 3 and 4.

In traditional ATs, a logic hydraulic circuit and a one-way clutch are used to guarantee the smooth torque transfer between two shifting clutches, and the off-going clutch can be disengaged automatically when the torque delivered in it reaches zero. However, in newly developed ATs with independent clutch control, one-way clutches are always eliminated in order to simplify the mechanical content and to improve the control flexibility. Therefore, for the vehicles with a hydraulic cylinder as a clutch actuator, which is ubiquitous in the present transmissions, the cylinder pressure control becomes important for good shift quality. Sensors measuring the clutch cylinder pressure, however, are seldom used. Hence, it is required to estimate the shaft torque or the cylinder pressure, in order to enhance control performance [142]. In Chap. 2, a reduced-order nonlinear observer is proposed for estimating the clutch pressure in the framework of input-to-state stability [129]. Based on the pressure estimation, a feedback control strategy is designed for the shift torque phase of the clutch-to-clutch shift process in Chap. 3.

In Chap. 4, special attention will be put on the clutch speed control during the inertia phase of the clutch-to-clutch shift. The clutch slip control during the inertia phase greatly influences the shift shock and shift time. Because the clutch engagement is expected to satisfy different and sometimes conflicting objectives, namely minimizing the clutch lock-up time, minimizing the friction losses during the slipping phase and ensuring a smooth acceleration of the vehicle, the integrated control of the engine and the clutch is necessary. In this text, the control scheme is designed to make the clutch speed track a reference trajectory. Although in the speed tracking control scheme one does not consider the above multiple control objectives directly, the required shift time and shift comfort can be achieved by selecting a proper reference trajectory and the friction loss can also be reduced by a suitable engine torque coordination during the shift process [51].

Then, in the second part of the book (the right column of Fig. 1.24), driveline torque estimation, shift control and launch control of AMTs will be addressed in Chaps. 5, 6, 7 and 8.

As aforementioned, for commercial vehicles with an AMT, active damping of the driveline oscillation is an important control issue. Knowing the torque information of the driveline helps to attenuate the driveline vibration. It has been pointed out that the drive axle shaft is the most dominant compliance of the driveline, and the torque of the drive axle shaft can be used to evaluate the overall torque delivered in the driveline [109]. Although the information of axle shaft torque can help restrain the driveline oscillation, torque sensors are seldom used in production vehicles because of the cost and durability. Hence, it is required to estimate the torque of the drive axle shaft. The estimation of the axle shaft torque will be discussed in Chap. 5.

In Chaps. 6 and 7, the clutch disengagement and engagement control will be respectively addressed. Based on the observer designed in Chap. 5, a clutch disengagement strategy is proposed in Chap. 6 to achieve a fast and smooth clutch disengagement process. In Chap. 7, the clutch engagement control in both power-on upshift and downshift of an AMT will be studied. The processes of upshift and downshift are quite different because the engine speed has to be reduced to reach the synchronization speed for the gear upshift while it has to be increased for the gear downshift. However, the engine speed cannot be controlled equally fast in both directions [50]. In other words, in the case of the downshift, the synchronization speed can be reached in a sufficiently short time, while for the upshift it is difficult for the engine to decelerate in a short time. Hence, in order to obtain a short shift time with a small shift shock optimally, the processes of gear upshift and gear downshift will be addressed in different control schemes. When considering the conflicting requirements of small friction wear and comfort for the shift process, Model Predictive Control (MPC) is adopted because of its ability to deal with multiple objectives in an optimal sense and to handle time-domain constraints in an explicit fashion.

Besides gear shifting, starting-up is also an important control issue for vehicles with an AMT. The characteristics of the AMT clutch during the start-up process are fast and complex. Moreover, the system characteristics change along with the variation of the driving conditions and long-term aging. Therefore, in Chap. 8, a data-driven predictive controller will be designed directly from the input–output data and will not require an explicit model of the AMT clutch system.

References

1. BOSCH (2007) Bosch automotive handbook, 7th edn. Society of Automotive Engineers, Warrendale
2. Abe M (1999) Vehicle dynamics and control for improving handling and active safety: from four-wheel steering to direct yaw moment control. Proc Inst Mech Eng, Proc, Part K, J Multi-Body Dyn 213(2):1464–4193
3. Albrecht A, Corde G, Knop V, Boie H, Castagne M (2005) 1D simulation of turbocharged gasoline direct injection engine for transient strategy optimization. SAE technical paper 2005-01-0693
4. Albrecht A, Knop V, Corde G, Simonet L, Castagne M (2006) Observer design for downsized gasoline engine control using 1D engine simulation. Oil Gas Sci Technol Rev IFP 61(1):165–179
5. Alessandri A (2004) Observer design for nonlinear systems by using input-to-state stability. In: Proceedings of the 43th IEEE conference on decision and control, Paradise Island, Bahamas, vol 4, pp 3892–3897
6. Alleyne A, Liu R (2000) A simplified approach to force control for electro-hydraulic systems. Control Eng Pract 8(12):1347–1356
7. Allgöwer F, Badgwell TA, Qin JS, Rawlings JB, Wright SJ (1999) Nonlinear predictive control and moving horizon estimation—an introductory overview. In: Frank PM (ed) Advances in control, highlights of ECC'99. Springer, Berlin, pp 391–449
8. Amann N, Bocker J, Prenner F (2004) Active damping of drive train oscillations for an electrically driven vehicle. IEEE/ASME Trans Mechatron 9(4):697–700
9. Angeli D (1999) Input-to-state stability of pd-controlled robotic systems. Automatica 35(7):1285–1290
10. Artstein Z (1983) Stabilisation with relaxed controls. Nonlinear Anal 7:1163–1173
11. Balluchi A, Benvenuti L, di Benedetto MD, Pinello C, Sangiovanni-Vincentelli AL (2000) Automotive engine control and hybrid systems: challenges and opportunities. Proc IEEE 88(7):888–912
12. Bemporad A, Morari M (1999) Robust model predictive control: a survey. In: Vicino A, Garulli A, Tesi A (eds) Robustness in identification and control. Lecture notes in control and information sciences, vol 245. Springer, Berlin, pp 207–226
13. Bemporad A, Borrelli F, Glielmo L, Vasca F (2001) Hybrid control of dry clutch engagement. In: Proceedings of the European control conference, Porto, Portugal
14. Bemporad A, Morari M, Dua V, Pistikopoulos EN (2002) The explicit linear quadratic regulator for constrained systems. Automatica 38(1):3–20
15. Bengtsson J, Strandh P, Johansson R (2006) Multi-output control of a heavy duty HCCI engine using variable valve actuation and model predictive control. SAE technical paper 2006-01-0873
16. Bertram T, Bekes F, Greul R, Hanke O, Haß C, Hilgert J, Hiller M, Öttgen O, Opgen-Rhein P, Torlo M, Ward D (2003) Modelling and simulation for mechatronic design in automotive systems. Control Eng Pract 11(2):179–190
17. Bhattacharya R, Balas GJ, Kaya MA, Packard A (2002) Nonlinear receding horizon control of an F-16 aircraft. J Guid Control Dyn 25(5):924–931
18. Bin Y, Li KQ, Ukawa H, Handa M (2006) Modelling and control of a non-linear dynamic system for heavy-duty trucks. Proc Inst Mech Eng, Part D, J Automob EngMech 220(10):1423–1435
19. Brockett RW (1978) Feedback invariants for nonlinear systems. In: Proc 7th IFAC World Congress, Helsinki, pp 1115–1120
20. Buckner GD, Schuetze KT, Beno JH (2001) Intelligent feedback linearization for active vehicle suspension control. ASME J Dyn Syst Meas Control 123(4):727–733

21. Burgio G, Zegelaar P (2006) Integrated vehicle control using steering and brakes. Int J Control 79(5):534–541
22. Burke AF (2007) Batteries and ultra-capacitors for electric, hybrid, and fuel cell vehicles. Proc IEEE 95(4):806–820
23. Cairano SD, Yanakiev D, Bemporad A, Kolmanovsky IV, Hrovat D (2008) An MPC design flow for automotive control and applications to idle speed regulation. In: Proceedings of the 47th IEEE conference on decision and control, pp 5692–5697
24. Camacho EF, Bordons C (2004) Model predictive control. Springer, London
25. Chan CC (2007) The state of the art of electric, hybrid, and fuel cell vehicles. Proc IEEE 95(4):704–718
26. Chauvin J, Corde G, Moulin P, Petit N, Rouchon P (2006) High frequency individual cylinder estimation for control of diesel engines. Oil Gas Sci Technol Rev IFP 61(1):57–72
27. Chen H, Allgöwer F (1998) Nonlinear model predictive control schemes with guaranteed stability. In: Berber R, Kravaris C (eds) Nonlinear model based process control. Kluwer Academic, Dordrecht, pp 465–494
28. Chen H, Allgöwer F (1998) A quasi-infinite horizon nonlinear model predictive control scheme with guaranteed stability. Automatica 34(10):1205–1217
29. Chen H, Scherer CW (2006) Moving horizon H_∞ control with performance adaptation for constrained linear systems. Automatica 42(6):1033–1040
30. Chen JR, Zhang JW (2005) Automobile structure. China Machine Press, Beijing. In Chinese
31. Chen H, Gao X-Q, Wang H (2006) An improved moving horizon H_∞ control scheme through Lagrange duality. Int J Control 79(3):239–248
32. Chisci L, Rossiter JA, Zappa G (2001) Systems with persistent disturbances: predictive control with restricted constraints. Automatica 37(7):1019–1028
33. Cho D (1987) Nonlinear control methods for automotive powertrain systems. PhD Thesis, MIT
34. Chung SK, Koch CR, Lynch AF (2007) Flatness-based feedback control of an automotive solenoid valve. IEEE Trans Control Syst Technol 15(2):394–401
35. Clarke DW, Mohtadi C, Tuffs PS (1987) Generalized predictive control—part I. The basic algorithm. Automatica 23(2):137–148
36. Clarke DW, Mohtadi C, Tuffs PS (1987) Generalized predictive control—part II. Extensions and interpretations. Automatica 23(2):149–160
37. Cook JA, Sun J, Buckland JH, Kolmanovsky IV, Peng H, Grizzle JW (2006) Automotive powertrain control—a survey. Asian J Control 8(3):237–260
38. Cutler CR, Ramaker BL (1980) Dynamic matrix control—a computer control algorithm. In: Proceedings of joint automatic control conference, San Francisco, CA
39. Deur J, Petric J, Asgari J, Hrovat D (2006) Recent advances in control-oriented modeling of automotive powertrain dynamics. IEEE/ASME Trans Mechatron 11(5):513–523
40. Dolcini P, Wit CC, Béchart H (2008) Lurch avoidance strategy and its implementation in amt vehicles. Mechatronics 18(5–6):289–300
41. Dong XK, Fang YC, Zhang YD (2011) An improved AFM dynamic imaging method based on data fusion of neighboring point set. Acta Autom Sin 37(2):214–221
42. Dourra H, Mourtada A (2008) Adaptive nth order lookup table used in transmission double swap shift control. SAE technical paper 2008-01-0538
43. Dudzinski PA (1986) The problems of multi-axle vehicle drives. J Terramech 23(2):85–93
44. Dunbar WB, Milam MB, Franz R, Murray RM (2002) Model predictive control of a thrust-vectored flight control experiment. In: Proc IFAC World Congress
45. Fellows TG, Greenwood CJ (1991) The design and development of an experimental traction drive CVT for a 2.0 litre FWD passenger car. SAE technical paper 910408
46. Findeisen R, Imsland L, Allgöwer F (2003) State and output feedback nonlinear model predictive control: An overview. Eur J Control 9(2–3):179–195

47. Fischer R, Berger R (1998) Automation of manual transmissions. In: The 6th LuK symposium. http://www.casacuomo.com.ar/download/folien_asg.pdf
48. Fliess M, Lévine J, Martin P, Rouchon P (1995) Design of trajectory stabilizing feedback for driftless flat systems. In: Proceedings of the 3rd European control conference ECC'95, Rome, Italy, pp 1882–1887
49. Fliess M, Lévine J, Martin P, Rouchon P (1995) Flatness and defect of nonlinear systems: introductory theory and examples. Int J Control 61:1327–1361
50. Fredriksson J, Egardt B (2003) Active engine control for gearshifting in automated manual transmissions. Int J Veh Des 32(3/4):216–230
51. Gao B-Z, Chen H, Sanada K (2008) Two-degree-of-freedom controller design for clutch slip control of automatic transmission. SAE technical paper 2008-01-0537
52. Gao B-Z, Chen H, Zhao H-Y, Sanada K (2010) A reduced-order nonlinear clutch pressure observer for automatic transmission. IEEE Trans Control Syst Technol 18(2):446–453
53. Gao B-Z, Chen H, Sanada K, Hu Y-F (2011) Design of clutch slip controller for automatic transmission using backstepping. IEEE/ASME Trans Mechatron 16(3):498–508
54. García CE, Prett DM, Morari M (1989) Model Predictive Control: theory and practice—a survey. Automatica 25(3):335–347
55. Gillespie TD (1992) Fundamentals of vehicle dynamics. Society of Automotive Engineers, New York
56. Goetz M, Levesley MC, Crolla DA (2005) Dynamics and control of gearshifts on twin-clutch transmissions. Proc Inst Mech Eng, Part D, J Automob EngMech 219(8):951–963
57. Grimm G, Messina MJ, Tuna SE, Teel AR (2004) Examples when nonlinear model predictive control is nonrobust. Automatica 40:1729–1738
58. Grimm G, Messina MJ, Tuna SE, Teel AR (2007) Nominally robust model predictive control with state constraints. IEEE Trans Autom Control 52(5):1856–1870
59. Guo D, Lu KH, Chen SK, Lin WC, Lu XP (2005) The UniTire model: a nonlinear and non-steady-state tyre model for vehicle dynamics simulation. Veh Syst Dyn 43(1):341–358
60. Guzzella L, Onder CH (2004) Introduction to modeling and control of internal combustion engine systems. Springer, Berlin
61. Heijden ACVD, Serrarens AFA, Camlibel MK, Nijmeijer H (2007) Hybrid optimal control of dry clutch engagement. Int J Control 80(11):1717–1728
62. Hibino R, Osawa M, Yamada M, Kono K, Tanaka M (1996) H-infinite control design for torque-converter-clutch slip system. In: Proceedings of the 35th IEEE conference on decision and control, pp 1797–1802
63. Hohn BR, Pflaum H, Lechner C, Draxl T (2010) Efficient CVT hybrid driveline with improved drivability. Int J Veh Des 53(1/2):70–88
64. Horn J, Bamberger J, Michau P, Pindl S (2003) Flatness-based clutch control for automated manual transmissions. Control Eng Pract 11(12):1353–1359
65. Hou ZS, Xu JX (2009) On data-driven control theory: the state of the art and perspective. Acta Autom Sin 35(6):650–667
66. Hu YF, Chen H, Guo HY (2010) Output feedback control of electronic throttle based on observer. In: Proceedings of the 29th Chinese control conference (CCC), Beijing, China, pp 6016–6019
67. Huang C-I, Fu L-C (2007) Adaptive approach to motion controller of linear induction motor with friction compensation. IEEE/ASME Trans Mechatron 12(4):480–490
68. Huang B, Kadali R (2008) Dynamic modeling, predictive control and performance monitoring: a data-driven subspace approach. Lecture notes in control and information sciences. Springer, Berlin
69. Hunt LR, Su R, Meyer G (1983) Global transformations of nonlinear systems. IEEE Trans Autom Control AC-28(1):24–31
70. Information Quarterly, Japanese version 5(2):10–17 (2007). http://www.jp.arm.com/document/magazine/pdf/IQ_2007autumnP10-17.pdf
71. Isermann R (2008) Mechatronic systems—innovative products with embedded control. Control Eng Pract 16(1):14–29

72. Isidori A (1985) Nonlinear control systems: an introduction. Springer, Berlin
73. Itkis V (1976) Control systems of variable structure. Wiley, New York
74. Jadbabaie A, Hauser J (2002) Control of a thrust-vectored flying wing: a receding horizon—LPV approach. Int J Robust Nonlinear Control 12:869–896
75. Jiang ZP, Wang Y (2001) Input-to-state stability for discrete-time nonlinear systems. Automatica 37(6):857–869
76. Kaddissi C, Kenne J-P, Saad M (2007) Identification and real-time control of an electro-hydraulic servo system based on nonlinear backstepping. IEEE/ASME Trans Mechatron 12(1):12–22
77. Karnopp DC, Rosenberg RC, Margolis DL (1990) System dynamics: a unified approach. Wiley, New York
78. Khalil HK (2002) Nonlinear Systems. Prentice Hall, New York
79. Kiencke U, Nielsen L (2005) Automotive control systems: for engine, driveline, and vehicle, 2nd edn. Springer, Berlin
80. Knop V, Jay S (2006) Latest developments in gasoline auto-ignition modelling applied to an optical CAITM engine. Oil Gas Sci Technol Rev IFP 61(1):121–137
81. Kokotovic PV (1992) The joy of feedback: nonlinear and adaptive. In: IEEE control systems
82. Kokotovic PV, Krstic M, Kanellakopoulos I (1992) Backstepping to passivity: recursive design of adaptive systems. In: Proc 31st IEEE conf decision contr. IEEE Press, New Orleans, pp 3276–3280
83. Krstić M, Kanellakopoulos I, Kokotović P (1995) Nonlinear and adaptive control design. Wiley, New York
84. Kwon WH, Byun DG (1989) Receding horizon tracking control as a predictive control and its stability properties. Int J Control 50(5):1807–1824
85. Kwon WH, Bruckstein AM, Kailath T (1983) Stabilizing state-feedback design via the moving horizon method. Int J Control 37(3):631–643
86. Lazar M, Muñoz de la Peña D, Heemels W, Alamo T (2008) On input-to-state stabilizing of min–max nonlinear model predictive control. Syst Control Lett 57(1):39–48
87. Lee HC (2006) Driveline vibration control of electric vehicle. MSc Thesis, Cranfield University, England
88. Lee JM, Lee JH (2005) Approximate dynamic programming-based approaches for input-output data-driven control of nonlinear processes. Automatica 41:1281–1288
89. Lee W, Park S, Sunwoo M (2004) Towards a seamless development process for automotive engine-control system. Control Eng Pract 12(8):977–986
90. Lespinats S, Verleysen A, Giron M, Fertil B (2007) DD-HDS: a method for visualization and exploration of high-dimensional data. IEEE Trans Neural Netw 18(5):1265–1279
91. Liang Q, Gao BZ, Chen H (2012) Gear shifting control for pure electric vehicle with Inverse-AMT. Appl Mech Mater 190(191):1286–1289
92. Limón D, Álamo T, Salas F, Camacho EF (2006) Input to state stability of min–max MPC controllers for nonlinear systems with bounded uncertainties. Automatica 42(5):797–803
93. Limon D, Alamo T, Raimondo DM, Peña D, Bravo JM, Camacho EF (2009) Input-to-state stability: a unifying framework for robust model predictive control. In: Magni L, Raimondo D, Allgöwer F (eds) Nonlinear model predictive control—towards new challenging applications. Lecture notes in control and information sciences. Springer, Berlin, pp 1–26
94. Lindlau JD, Knospe CR (2002) Feedback linearization of an active magnetic bearing with voltage control. IEEE Trans Control Syst Technol 10(1):21–31
95. Lucente G (2007) Modelling of an automated manual transmission system. Mechatronics 17(2–3):73–91
96. Magni L, Scattolini R (2007) Robustness and robust design of MPC for nonlinear discrete-time systems. In: Allgower F, Findeisen R, Biegler LT (eds) Assessment and future directions of nonlinear model predictive control. Lecture notes in control and information sciences. Springer, Heidelberg, pp 239–254

97. Marruedo DL, Alamo T, Camacho EF (2002) Input-to-state stable MPC for constrained discrete-time nonlinear systems with bounded additive uncertainties. In: Proceedings of the 41th IEEE conference on decision and control, Las Vegas, NV, vol 4, pp 4619–4624
98. Masmoudi RA, Hedrick K (1992) Estimation of vehicle shaft torque using nonlinear observers. ASME J Dyn Syst Meas Control 114:394–400
99. Mayne DQ, Rawlings JB, Rao CV, Scokaert POM (2000) Constrained model predictive control: stability and optimality. Automatica 36(6):789–814
100. Mayne DQ, Seron MM, Rakovic SV (2005) Robust model predictive control of constrained linear systems with bounded disturbances. Automatica 41(2):219–224
101. Mayne DQ, Kerrigan EC, van Wyk EJ, Falugi P (2011) Tube-based robust nonlinear model predictive control. Int J Robust Nonlinear Control 21(11):1341–1353
102. Morari M, Lee JH (1991) Model predictive control: the good, the bad, and the ugly. In: Arkun Y, Ray W (eds) Proc 4th international conference on chemical process control—CPC IV, pp 419–444. AIChE, CACHE
103. Morari M, Lee JH (1999) Model predictive control: past, present and future. Comput Chem Eng 23(4–5):667–682
104. Morimoto Y (2006) Mechanism and control of AT. Grand Prix Publishing, Tokyo. In Japanese
105. Namburu SM, Chigusa S, Prokhorov D, Qiao L, Choi K, Pattipati K (2007) Application of an effective data driven approach to real-time fault diagnosis in automotive engines. In: Aerospace conference, 2007 IEEE, pp 1–9
106. Ngo DV, Hofman T, Steinbuch M, Serrarens A, Merkx L (2010) Improvement of fuel economy in power-shift automated manual transmission through shift strategy optimization—an experimental study. In: Proceedings of 2010 IEEE vehicle power and propulsion conference (VPPC), Lille, France, pp 1–5
107. Nijmeijer H, van der Schaft AJ (1990) Nonlinear dynamical control systems. Springer, New York
108. Pacejka HB (2005) Tire and vehicle dynamics, 2nd edn. SAE International/Elsevier, Warrendale/Amsterdam
109. Pettersson M (1997) Driveline modeling and control. PhD Thesis, Linköping University, Sweden
110. Powers WF, Nicastri PR (2000) Automotive vehicle control challenges in the 21st century. Control Eng Pract 8(6):605–618
111. Qin SJ, Badgwell TA (2000) An overview of nonlinear model predictive control applications. In: Allgöwer F, Zheng A (eds) Nonlinear model predictive control. Birkhäuser, Basel, pp 369–392
112. Qin SJ, Badgwell TA (2003) A survey of industrial model predictive control technology. Control Eng Pract 11(7):733–764
113. Qin SJ, Lin W, Ljung L (2005) A novel subspace identification approach with enforced causal models. Automatica 41(12):2043–2053
114. Raimondo DM, Limon D, Lazar M, Magni L, Camacho EF (2009) Min-max model predictive control of nonlinear systems: a unifying overview on stability. Eur J Control 15(1):5–21
115. Rawlings JB (2000) Tutorial overview of model predictive control. IEEE Control Syst Mag 20(3):38–52
116. Rawlings JB, Mayne DQ (2009) Model predictive control: theory and design. Nob Hill Publishing, Madison
117. Richalet J, Rault A, Testud JL, Papon J (1976) Algorithmic control of industrial processes. In: Proc 4th IFAC symposium on identification and system parameter estimation, Tbilisi, pp 1119–1167
118. Richalet J, Rault A, Testud JL, Papon J (1978) Model predictive heuristic control: application to industrial processes. Automatica 14:413–428
119. Sakamoto K (2005) Basic of automatic transmission. Grand Prix Publishing, Tokyo. In Japanese

120. Sanada K, Kitagawa A (1998) A study of two-degree-of-freedom control of rotating speed in an automatic transmission, considering modeling errors of a hydraulic system. Control Eng Pract 6:1125–1132
121. Sanchez EN, Perez JP (1999) Input-to-state stability (ISS) analysis for dynamic neural networks. IEEE Trans Circuits Syst I, Fundam Theory Appl 46(11):1395–1398
122. Schäuffele J, Zurawka T (2005) Automotive software engineering: principles, processes, methods, and tools. SAE International, Warrendale
123. Schöner H-P (2004) Automotive mechatronics. Control Eng Pract 12(11):1343–1351
124. Slotine J-JE (1984) Sliding controller design for nonlinear systems. Int J Control 40(2):421–434
125. Slotine J-JE, Li W (1991) Applied nonlinear control. Prentice Hall, Englewood Cliffs
126. Slotine J-JE, Hedrick JK, Misawa EA (1987) On sliding observers. ASME J Dyn Syst Meas Control 109:245–252
127. Sontag ED (1983) A Lyapunov-like characterisation of asymptotic controllability. SIAM J Control Optim 21:462–471
128. Sontag ED (1989) Smooth stabilization implies coprime factorization. IEEE Trans Autom Control 34:435–443
129. Sontag ED (2008) Input to state stability: basic concepts and results. In: Cachan JM, Groningen FT, Paris BT (eds) Nonlinear and optimal control theory. Lecture notes in mathematics. Springer, Berlin, pp 163–220
130. Sorniotti A, Loro Pilone G, Viotto F, Bertolotto S (2011) A novel seamless2-speed transmission system for electric vehicles: principles and simulation results. SAE Int J Engines 4(2):2671–2685
131. Stanglmaier RH, Roberts CE (1999) Homogeneous charge compression ignition (HCCI): benefits, compromises, and future engine applications. SAE Transact 108(3):2138–2145
132. Sun Z, Hebbale K (2005) Challenges and opportunities in automotive transmission control. In: Proceedings of American control conference, vol 5, pp 3284–3289
133. Swaroop D, Hedrick JK, Yip PP, Gerdes JC (2000) Dynamic surface control for a class of nonlinear systems. IEEE Trans Autom Control 45(10):1893–1899
134. Tan S, Zhang JF (2007) Adaptive measured-data based linear quadratic optimal control of stochastic systems. Int J Control 80(10):1676–1689
135. Ursu I, Ursu F, Popescu F (2006) Backstepping design for controlling electrohydraulic servos. J Franklin Inst 343(1):94–110
136. Utkin VI, Chang HC (2002) Sliding mode control on electro-mechanical systems. Math Probl Eng 8(4):451–473
137. van Nieuwstadt MJ, Murray RM (1998) Real-time trajectory generation for differentially flat systems. Int J Robust Nonlinear Control 8(11):995–1020
138. Vasca F, Iannelli L, Senatore A, Reale G (2011) Torque transmissibility assessment for automotive dry-clutch engagement. IEEE/ASME Trans Mechatron 16(3):564–573
139. Vu L, Chatterjee D, Liberzon D (2007) Input-to-state stability of switched systems and switching adaptive control. Automatica 43(4):639–646
140. Wang Z, Liu DR (2011) Data-based controllability and observability analysis of linear discrete-time systems. IEEE Trans Neural Netw 22(12):2388–2392
141. Watechagit S, Srinivasan K (2003) On-line estimation of operating variables for stepped automatic transmissions. In: IEEE conference on control applications (CCA 2003), Istanbul, Turkey, vol 1, pp 279–284
142. Watechagit S, Srinivasan K (2005) Implementation of on-line clutch pressure estimation for stepped automatic transmissions. In: Proc American control conference, vol 3, pp 1607–1612
143. Xiang M, Shi WR (2010) A cluster data management algorithm based on data correlation of wireless sensor networks. Acta Autom Sin 36(9):1343–1350
144. Xu JX, Hou ZS (2009) Notes on data-driven system approaches. Acta Autom Sin 35(6):668–675

145. Yi K, Shin BK, Lee KL (2000) Estimation of turbine torque of automatic transmissions using nonlinear observers. ASME J Dyn Syst Meas Control 122:276–283
146. Yokoyama M (2008) Sliding mode control for automatic transmission systems. J Jpn Fluid Power Syst Soc 39(1):34–38. In Japanese
147. Young PC (2006) The data-based mechanistic approach to the modelling, forecasting and control of environmental systems. Annu Rev Control 30(2):169–182
148. Zheng Q, Srinivasan K, Rizzoni G (1999) Transmission shift controller design based on a dynamic model of transmission response. Control Eng Pract 7(8):1007–1014

Chapter 2
Pressure Estimation of a Wet Clutch

2.1 Introduction

In both DCTs and new ATs [19], the change of the speed ratio is regarded as the process of one clutch being engaged with the other being disengaged, namely, the clutch-to-clutch shift. Furthermore, smart proportional valves with a large flow rate are developed for direct clutch pressure control, without using the pilot duty solenoid valve [3]. These valves can be used in new ATs to improve the ability of adapting to different driving conditions, as well as to reduce cost and to improve packaging. For vehicles with a hydraulic cylinder as clutch actuator, the cylinder pressure control becomes important for good shift quality. Sensors measuring the clutch cylinder pressure, however, are seldom used because of the cost and durability. Hence, it is required to estimate the shaft torque or the cylinder pressure, in order to enhance control performance [22].

There have been some studies on the estimation of the transmission shaft torque and the clutch pressure. A sliding mode observer is designed to estimate the torque of an automotive drive shaft in [13, 14]. An adaptive sliding mode algorithm is proposed to estimate the turbine torque of a torque converter in [23]. Furthermore, [22] uses the sliding mode method to estimate the clutch pressure in a hydraulically powered stepped AT. The extended algorithm in [21] is used to estimate the clutch pressure and the transmission output shaft torque simultaneously. In [9, 17], a neural network is suggested to estimate the turbine torque, in which the engine speed, the turbine speed and the oil temperature are inputs. Reference [9] also designs a driving load observer by assuming that the driving load is slowly-varying. In [20], a recursive least squares method with multiple forgetting factors is used to estimate the road grade and the vehicle mass. In [10], a full-order observer is proposed for the pressure monitoring of a torque converter's lock up clutch, where a state-dependent term is appended in the conventional Luenberger state observer to eliminate the effect of possible parameter variations in some sense. The question of how to design this term is crucial for the performance and the implementation of the observer.

A new AT with clutch-to-clutch shift technology is considered in this chapter, in which electro-hydraulic actuators are adopted to control the clutches independently.

H. Chen, B. Gao, *Nonlinear Estimation and Control of Automotive Drivetrains*,
DOI 10.1007/978-3-642-41572-2_2,

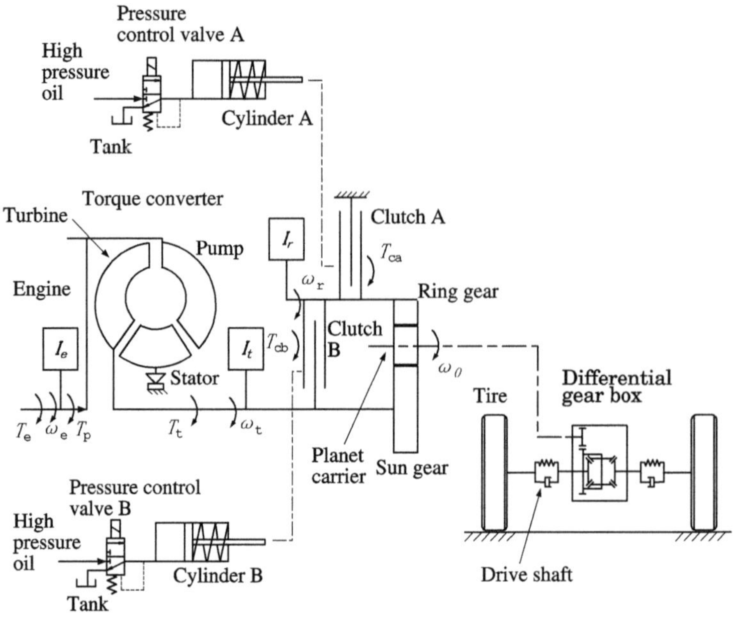

Fig. 2.1 Schematic graph of an automatic transmission

Because of the complex nonlinearities in an automotive powertrain, such as the speed–torque relationship of engines and the characteristics of torque converters, it is very hard to model the whole dynamics with physical principles. Lookup tables, which are obtained from many experiments in the steady state, are widely used to describe the nonlinear characteristics. There inherently exist model uncertainties, such as steady-state error and unmodeled dynamics. Moreover, the variation of the vehicle mass and the road grade also bring uncertainties to the powertrain dynamics. Therefore, the clutch pressure/torque estimator must be robust against the variation of powertrain parameters and the uncertainties.[1]

2.2 Description and Modeling of a Powertrain System

We consider the powertrain in passenger vehicles with a two-speed AT, as schematically shown in Fig. 2.1. A planetary gear set is adopted as the shift gear. Two clutches are used as the actuators, and two proportional pressure valves are used to control the two clutches. When clutch A is engaged and clutch B disengaged, the

[1]This chapter uses the content of [4], with permission from IEEE.

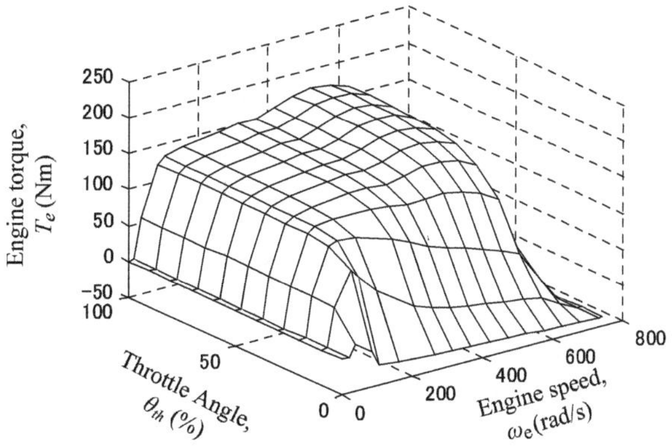

Fig. 2.2 Engine torque map with speed and throttle opening

powertrain operates in the first gear and the speed ratio is given by

$$i_1 = 1 + \frac{1}{\gamma}, \tag{2.1}$$

where γ is the ratio of the teeth number of the sun gear to that of the ring gear. While clutch A is disengaged and clutch B engaged, the vehicle is driven in the second gear with a speed ratio of

$$i_2 = 1. \tag{2.2}$$

The powertrain simulation model is established by the commercial simulation software AMESim. Except for the simplified 2-speed transmission, the simulation model represents a typical front-wheel-drive mid-size passenger car equipped with a 2000 cc injection gasoline engine. The constructed model captures the important transient dynamics during the vehicle shift process, such as the drive shaft oscillation and the tire slip. Moreover, the time-delays in control and time-varying parameters are also considered in the simulation model of the proportional valves [16], which are neglected in the observer design.

Engine The work reported here is primarily concerned with shift transients, and therefore a simple engine model is used. The dynamic equation of the engine speed is represented by

$$I_e\dot{\omega}_e + C_e\omega_e = T_e - T_p, \tag{2.3}$$

where T_e is the engine output torque and T_p is the output torque of the converter pump. The engine output torque is simplified as a nonlinear function of the engine rotational speed ω_e and the engine throttle angle θ_{th}, i.e., $T_e = T_e(\omega_e, \theta_{th})$. This map is shown in Fig. 2.2.

Torque Converter The capacity factor $C(\lambda)$ and the torque ratio $t(\lambda)$ of the considered torque converter are given in Fig. 2.3.

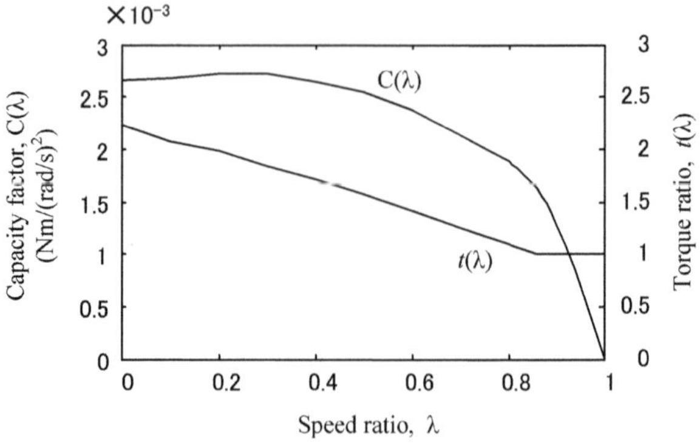

Fig. 2.3 Capacity factor $C(\lambda)$ and torque ratio $t(\lambda)$ of the torque converter

Planetary Gear Set Using the submodel provided by commercial software, such as AMESim, the planetary gear set can be modeled conveniently. The following parameters are required for the modeling setting: the inertia moment of the torque converter turbine I_t; the inertia moment of the ring gear I_r; the teeth number of the sun gear Z_s and the teeth number of the ring gear Z_r.

Differential Gear Box and Drive Shaft The gear ratio of the differential gear box is denoted as R_{df}. The two drive shafts between the differential gear and the front wheels are represented as a torsion spring with stiffness coefficient K_s and a torsion damping with damping coefficient C_s.

Tires The longitudinal tire force F_x, which is usually simplified as a function of the longitudinal slip ratio S_x, rises fast when S_x increases under a threshold dS_x and declines slowly after that [7]. Here it is represented approximately as a tanh function of $F_x = F_{x\max} \tanh(\frac{2S_x}{dS_x})$. The longitudinal slip S_x has been defined in Sect. 1.1.

Road Loads The road load consists of three parts: the grade force F_G, the rolling resistant moment of tires T_w and the aerodynamic drag F_A, which has been introduced in Sect. 1.1.

Clutches and Valves The friction coefficient μ is a nonlinear function of $\Delta\omega$ shown in Fig. 2.5. In the design of the pressure observer in Sect. 2.3.2, we assume the parameters (τ_{cv}, K_{cv}) of the proportional pressure control valve as constant, and we also ignore the time-delay of the valve. Actually, the valve has a time-delay and the parameters vary according to different operating points [16]. Hence, the dynamics of the proportional valve in the powertrain simulation is given by

$$\tilde{\tau}_{cv}\dot{p}_{cb}(t) = -p_{cb}(t) + \tilde{K}_{cv} i_b(t - \tilde{L}_{cv}). \tag{2.4}$$

Finally, the values of the parameters used in the powertrain simulation are listed in Table 2.1. Nonlinear functions $T_e(\omega_e, \theta_{th})$, $C(\lambda)$, $t(\lambda)$, $\mu(\Delta\omega)$, $\tilde{\tau}_{cv}$, $\tilde{L}_{cv}$ and $\tilde{K}_{cv}$ are given in the lookup tables.

Table 2.1 Values of simulation model parameters

A_A	Front area of vehicle	$2\ \text{m}^2$
C_D	Aerodynamic drag coefficient	0.3
C_e	Damping coefficient of engine crane	0.047 Nm/rad
C_l	Damping coefficient of drive shaft	10 Nm/rad
dS_x	Longitudinal slip threshold of tire	0.1
F_s	Return spring force of clutch B	600 N
$F_{x\max}$	Maximum longitudinal force of tire	3200 N
i_{df}	Gear ratio of the differential gear box	3.0
I_e	Inertia of crane and pump	$0.17\ \text{kg}\,\text{m}^2$
I_r	Inertia moment of ring gear	$0.01\ \text{kg}\,\text{m}^2$
I_t	Inertia moment of turbine	$0.06\ \text{kg}\,\text{m}^2$
$\tilde{K}_{cv}$	Gain of valve B	0–1.7 MPa/A
K_l	Stiffness of drive shaft	6500 Nm/rad
$\tilde{L}_{cv}$	Time-delay of valve B	0–0.03 s
m	Vehicle mass	1500 kg
R_w	Tire radius	0.3 m
T_w	Moment of resistance of tires	110 Nm
Z_r	Teeth number of ring gear	60
Z_s	Teeth number of sun gear	40
θ_g	Road grade	0 deg
ρ	Air density	$1.2\ \text{kg/m}^3$
$\tilde{\tau}_{cv}$	Time constant of valve B	0.02–0.20 s

2.3 Clutch Pressure Estimation Without Consideration of Drive Shaft Stiffness

A reduced-order clutch pressure observer based on the concept of input-to-state stability (ISS) [12, 18] is proposed, where the rotational speeds are the measured outputs and the special structure of the clutch pressure system is exploited. The lookup tables of the nonlinear characteristics of powertrain systems appear in their original map form, and the model uncertainties are considered as additional disturbance inputs. A systematic procedure is given to design the nonlinear clutch pressure observer such that

- The error dynamics is input-to-state stable, where modeling errors are the inputs. This means that the initial estimation error decays exponentially and the estimation error is guaranteed to be bounded for the bounded modeling errors;
- The requirements on estimation performance, such as decay rate and error offset, are easily and explicitly considered during the design process;
- The implementation of the designed observer benefits from the reduced order and the time-invariant gains of the observer;

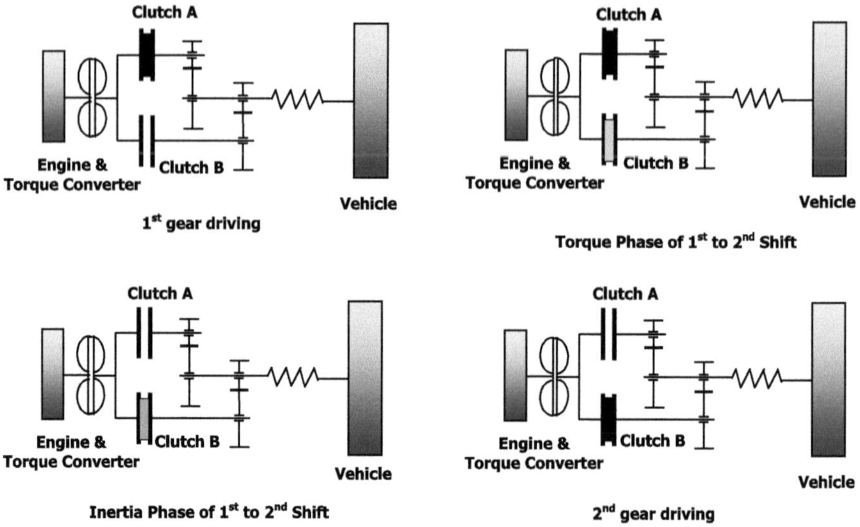

Fig. 2.4 1st-to-2nd gear shift (1st gear driving—torque phase—inertia phase—2nd gear driving)

- Lower observer gains are obtained through convex optimization, which increases the robustness against noises and reduces the estimated upper bound of the error offset.

2.3.1 Clutch System Modeling and Problem Statement

During the shift process, the on-coming and off-going clutches are controlled by the two valves through separate controllers, which are assumed to be well-designed. The controllers discussed in Chap. 3 and Chap. 4 can be viewed as such controllers. The power-on 1st-to-2nd upshift is considered here as an example of the shift process. The gear shift process is generally divided into the torque phase, as shown in Fig. 2.4, where the turbine torque is transferred from clutch A to clutch B and the inertia phase where clutch B is synchronized [8].

Note that 4–8-speed ATs are extensively used in production cars, a 2-speed AT is just adopted here as an example to exploit the design process, and it can be applied to other ATs if the parameters of the clutch-to-clutch shift process are replaced. We start to describe the modeling of the inertia phase and then obtain a model for the torque phase by taking into account the general fact that there is no gear ratio change in the torque phase.

Inertia Phase

In the inertia phase, where the two clutches are both slipping, because the planetary gear set is a two-degree-of-freedom system, two states variables, such as the turbine speed ω_t and the speed ω_r of the ring gear, can be used to describe its movement.

The basic kinematic equation of the planetary gear set is

$$\frac{\omega_r - \omega_0}{\omega_t - \omega_0} = \gamma \tag{2.5}$$

where γ is the ratio of the teeth number of the sun gear to that of the ring gear, ω_r is the ring gear speed, ω_0 is the planetary gear carrier speed, i.e., the output speed of the transmission, ω_t is the sun gear speed, i.e., the turbine speed of the torque converter.

The torque of the planetary gear set should also satisfy the torque balance equations:

$$T_{SG} = \frac{\gamma}{1+\gamma} T_0, \tag{2.6a}$$

$$T_{RG} = \frac{1}{1+\gamma} T_0, \tag{2.6b}$$

where T_{SG} is the sun gear torque, T_{RG} is the ring gear torque, T_0 is the planetary gear carrier torque.

With the driving traction of torque T_0, the vehicle body will move according to the equation

$$I_{ve}\dot{\omega}_0 = T_0 - T_{ve}. \tag{2.7}$$

where I_{ve} is the equivalent inertia of the vehicle body, T_{ve} is the equivalent driving resistant torque.

From Fig. 2.1, the equations of the sun gear and the ring gear are given as

$$I_t\dot{\omega}_t = T_t - T_{cb} - T_{SG}, \tag{2.8a}$$

$$I_r\dot{\omega}_r = T_{cb} + T_{ca} - T_{RG}, \tag{2.8b}$$

where I_t is the inertia moment of the torque converter turbine, I_r is the inertia moment of the ring gear and parts connected, T_{ca} is the torque delivered by clutch A, T_{cb} is the torque delivered by clutch B.

Substituting (2.6a), (2.6b) and (2.7) into (2.8a), (2.8b), we have

$$I_t\dot{\omega}_t + \frac{1}{R_g} I_{ve}\dot{\omega}_0 = T_t - T_{cb} - \frac{1}{R_g} T_{ve}, \tag{2.9a}$$

$$I_r\dot{\omega}_r + \frac{R_g - 1}{R_g} I_{ve}\dot{\omega}_0 = T_{cb} + T_{ca} - \frac{R_g - 1}{R_g} T_{ve} \tag{2.9b}$$

with $R_g = 1 + \frac{1}{\gamma}$.

Using (2.5) and rearranging the above equations, the dynamic equations of the transmission can be obtained as follows:

$$\dot{\omega}_t = C_{11}T_t + C_{12}T_{ca} + C_{13}T_{cb} + C_{14}T_{ve}, \tag{2.10a}$$

$$\dot{\omega}_r = C_{21}T_t + C_{22}T_{ca} + C_{23}T_{cb} + C_{24}T_{ve}, \tag{2.10b}$$

with the coefficients

$$C_{11} = \frac{R_g^2 I_r + (R_g - 1)^2 I_{ve}}{R_g^2 I_t I_r + (R_g - 1)^2 I_t I_{ve} + I_r I_{ve}},$$

$$C_{12} = -C_{11} \frac{(R_g - 1) I_{ve}}{R_g^2 I_r + (R_g - 1)^2 I_{ve}},$$

$$C_{13} = -C_{11} \left[\frac{(R_g - 1) I_{ve}}{R_g^2 I_r + (R_g - 1)^2 I_{ve}} + 1 \right],$$

$$C_{14} = C_{11} \frac{-R_g I_r}{R_g^2 I_r + (R_g - 1)^2 I_{ve}},$$

$$C_{21} = -\frac{(R_g - 1) I_{ve}}{R_g^2 I_t I_r + (R_g - 1)^2 I_t I_{ve} + I_r I_{ve}},$$

$$C_{22} = -C_{21} \frac{R_g^2 I_t + I_{ve}}{(R_g - 1) I_{ve}},$$

$$C_{23} = -C_{21} \left[\frac{R_g^2 I_t + I_{ve}}{(R_g - 1) I_{ve}} + 1 \right],$$

$$C_{24} = C_{21} \frac{R_g I_t}{I_{ve}}.$$

The turbine torque T_t is calculated from the steady-state characteristics of the torque converter as

$$T_t = t(\lambda) C(\lambda) \omega_e^2, \tag{2.11}$$

where $C(\lambda)$ denotes the capacity factor of the torque converter, $t(\lambda)$ is the torque ratio, ω_e is the engine speed and λ is the speed ratio defined as

$$\lambda = \frac{\omega_t}{\omega_e}. \tag{2.12}$$

On the other hand, the transferred torque T_{cb} during clutch slipping is determined by the cylinder pressure. If the force of the return spring is treated as constant, the relationship between the clutch torque and the cylinder pressure is described as

$$T_{cb} = \mu(\Delta\omega) RN \cdot (A p_{cb} - F_s), \tag{2.13}$$

where R is the effective radius of the push force acted on the plates of clutch B, N and A are the plate number and the piston area of clutch B, p_{cb} is the pressure of cylinder B, F_s is the return spring force of clutch B and μ is the friction coefficient of clutch plates depending on the speed difference. The speed difference $\Delta\omega$ is defined as

$$\Delta\omega = \omega_t - \omega_r. \tag{2.14}$$

The cylinder pressure is determined by the input current of the proportional pressure control valve. The dynamics of the proportional valve can be simplified as a first-order system [16],

$$\tau_{cv}\dot{p}_{cb} = -p_{cb} + K_{cv}i_b, \tag{2.15}$$

where τ_{cv} is the time constant of valve B, K_{cv} is the gain of valve B, and i_b is the electric current of valve B.

Moreover, if the torsion dynamics of the drive shaft, the tire slip and the road grade are ignored, the resistant torque T_{ve}, delivered from the tire to the drive shaft, can be calculated as

$$T_{ve} = \frac{T_w}{R_{df}} + \frac{C_A R_w^3}{R_{df}^3}\omega_0^2, \tag{2.16}$$

where T_w denotes the rolling resistance moment of the tire, R_w is the tire radius, R_{df} is the gear ratio of the differential gear box, ω_0 is the output speed of the transmission and C_A is a constant coefficient depending on air density, aerodynamic drag coefficient and the front area of the vehicle.

By selecting the turbine speed ω_t, the speed difference $\Delta\omega$ of clutch B, and the pressure p_{cb} of cylinder B as state variables, denoted as x_1, x_2, x_3, respectively, and ignoring the pressure of clutch A because it is small enough, the inertia phase of the 1st-to-2nd gear upshift process is described in the following state space form:

$$\dot{x}_1 = C_{13}\mu(x_2)RNAx_3 + f_1(\omega_e, x_1, x_2), \tag{2.17a}$$

$$\dot{x}_2 = (C_{13} - C_{23})\mu(x_2)RNAx_3 + f_2(\omega_e, x_1, x_2), \tag{2.17b}$$

$$\dot{x}_3 = -\frac{1}{\tau_{cv}}x_3 + \frac{K_{cv}}{\tau_{cv}}u, \tag{2.17c}$$

with

$$f_1(\omega_e, x_1, x_2) = C_{11}T_t(\omega_e, x_1) + C_{14}T_{ve}(x_1, x_2) - C_{13}\mu(x_2)RNF_s, \tag{2.18a}$$

$$f_2(\omega_e, x_1, x_2) = (C_{11} - C_{21})T_t(\omega_e, x_1) + (C_{14} - C_{24})T_{ve}(x_1, x_2) - (C_{13} - C_{23})\mu(x_2)RNF_s, \tag{2.18b}$$

where $u = i_b$ is the current of valve B and viewed as control input. In order to estimate the pressure of clutch B, the rotational speeds of the transmission are used as the measured outputs, i.e.,

$$y = [x_1 \quad x_2]^T. \tag{2.19}$$

The parameter values can be found in Table 2.2.

Table 2.2 Parameters for observer design

C_{11}	Coefficient in (2.17a)–(2.17c)	$15.52\ \frac{1}{\mathrm{kg\,m^2}}$
C_{13}	Coefficient in (2.17a)–(2.17c)	$-25.85\ \frac{1}{\mathrm{kg\,m^2}}$
C_{14}	Coefficient in (2.17a)–(2.17c)	$-0.011\ \frac{1}{\mathrm{kg\,m^2}}$
C_{21}	Coefficient in (2.17a)–(2.17c)	$-10.33\ \frac{1}{\mathrm{kg\,m^2}}$
C_{23}	Coefficient in (2.17a)–(2.17c)	$17.38\ \frac{1}{\mathrm{kg\,m^2}}$
C_{24}	Coefficient in (2.17a)–(2.17c)	$-0.10\ \frac{1}{\mathrm{kg\,m^2}}$
R	Effective radius of plates of clutch B	0.13 m
N	Plate number of clutch B	3
A	Piston area of clutch B	0.01 $\mathrm{m^2}$
τ_{cv}	Time constant of valve B	0.04 s
K_{cv}	Gain of valve B	1.0 MPa/A
$\mu_{\min}$	Minimum friction coefficient	0.10
$\mu_{\max}$	Maximum friction coefficient	0.16
c_{11}	Coefficient in (2.21a), (2.21b)	$0.40\ \frac{1}{\mathrm{kg\,m^2}}$
c_{13}	Coefficient in (2.21a), (2.21b)	$-0.40\ \frac{1}{\mathrm{kg\,m^2}}$
c_{14}	Coefficient in (2.21a), (2.21b)	$-0.16\ \frac{1}{\mathrm{kg\,m^2}}$

Torque Phase

In the 1st-to-2nd upshift torque phase, there is no slip in clutch A, hence, there is no gear ratio change of the transmission. The motion of the drive line during this phase can be described by the following equation:

$$\dot{\omega}_t = c_{11}T_t + c_{13}T_{cb} + c_{14}T_{ve}, \tag{2.20}$$

with the constant coefficients

$$c_{11} = \frac{R_g^2}{R_g^2 I_t + I_{ve}},$$

$$c_{13} = -c_{11},$$

$$c_{14} = -\frac{c_{11}}{R_g},$$

where ω_t is the turbine speed, T_{cb} is the torque delivered by clutch B, T_{ve} is the resistant torque delivered from the tire to the drive shaft, I_t is the inertia moment of the torque converter turbine, I_{ve} is the equivalent inertia moment of the vehicle body and $R_g = 1 + \frac{1}{\gamma}$ again.

The torque phase of the 1st-to-2nd gear upshift process is then described in the following state space form:

$$\dot{x}_1 = c_{13}\mu(\Delta\omega)RNAx_3 + f_{t1}(\omega_e, x_1, \Delta\omega), \tag{2.21a}$$

$$\dot{x}_3 = -\frac{1}{\tau_{cv}}x_3 + \frac{K_{cv}}{\tau_{cv}}u, \tag{2.21b}$$

with the measured output

$$y = x_1 \tag{2.22}$$

and

$$f_{t1}(\omega_e, x_1, \Delta\omega) = c_{11}T_t(\omega_e, x_1) + c_{14}T_{ve}(x_1, \Delta\omega) - c_{13}\mu(\Delta\omega)RNF_s. \tag{2.23}$$

Note that c_{ij} are different coefficients from C_{ij} in (2.18a), (2.18b). Moreover, although there is no obvious change of the clutch speed during the torque phase, the speed difference $\Delta\omega$ of the clutch is different for various driving maneuvers. Hence $\Delta\omega$ is also considered as an input for the torque phase model.

Estimation Problem

Due to the extreme complexity of the torque converter and the aerodynamic drag, the nonlinear functions in (2.18a), (2.18b) and (2.23) are generally given as lookup tables (i.e., maps), which are obtained by a series of steady-state experiments and inherently contain errors. Other modeling uncertainties include variations of parameters, such as the vehicle mass, the road grade and the damping coefficient of shafts.

Hence, the problem considered here is to estimate the pressure of clutch B in the presence of model errors, given the valve electric current i_b and the measured rotational speeds of the transmission ω_e, ω_t and $\Delta\omega$.

2.3.2 Reduced-Order Nonlinear State Observer

Reduced-Order Nonlinear Observer with ISS Property

In this section, the special structure of the clutch pressure system is considered to derive a reduced-order pressure observer. The robustness of the designed observer with respect to model errors is achieved in the sense of ISS property. To do this, we denote the variable to be estimated as z, and rewrite the system dynamics for estimating the clutch pressure as follows:

$$\dot{\boldsymbol{y}} = \boldsymbol{F}(\boldsymbol{y}, u) + \boldsymbol{G}(\boldsymbol{y}, u)z + \boldsymbol{w}(\boldsymbol{y}, u, z), \tag{2.24a}$$

$$\dot{z} = A_{22}(u, p)z + B_2(u, p), \tag{2.24b}$$

where $\boldsymbol{y}$ is the measured output, u is the control input and p is the vector of parameters which may include t and others, $\boldsymbol{w}(\boldsymbol{y}, u, z)$ summarizes model uncertainties, and in particular

$$F(y,u) = \begin{pmatrix} f_1(\omega_e, x_1, x_2) \\ f_2(\omega_e, x_1, x_2) \end{pmatrix}, \tag{2.25a}$$

$$G(y,u) = \begin{pmatrix} C_{13}\mu(x_2)RNA \\ (C_{13} - C_{23})\mu(x_2)RNA \end{pmatrix}, \tag{2.25b}$$

$$A_{22}(u,p) = -\frac{1}{\tau_{cv}}, \tag{2.25c}$$

$$B_2(u,p) = \frac{K_{cv}}{\tau_{cv}}u \tag{2.25d}$$

for the inertia phase. The expressions for the torque phase are

$$F(y,u) = f_{t1}(\omega_e, x_1, \Delta\omega), \tag{2.26a}$$

$$G(y,u) = c_{13}\mu(\Delta\omega)RNA, \tag{2.26b}$$

$$A_{22}(u,p) = -\frac{1}{\tau_{cv}}, \tag{2.26c}$$

$$B_2(u,p) = \frac{K_{cv}}{\tau_{cv}}u. \tag{2.26d}$$

Remark 2.1 We exploit the more general form of (2.24a), (2.24b) to derive the pressure observer such that the suggested design method might be useful if the time-varying property of the proportional pressure valves is taken into account (where $p = t$), or if other kinds of valves are used to control clutch pressures. For example, if a PWM valve is used, the pressure dynamics can be described by [22]

$$\dot{z} = C_z\sqrt{0.01P_l u - z}, \tag{2.27}$$

where C_z is a positive constant, u is the pulse duty cycle, and P_l is the main line pressure. Then, for all given u, we may linearize (2.27) at a fixed operating point of $z = P_s$ to approximate the pressure dynamics in the form of (2.24b) with

$$A_{22}(p) = -\frac{C_z}{2\sqrt{0.01P_l u - P_s}},$$

$$B_2(p) = C_z\sqrt{0.01P_l u - P_s} + \frac{C_z P_s}{2\sqrt{0.01P_l u - P_s}}.$$

In this case, $p = [P_s\ P_l]^T$.

Because the shaft torque affects the related shaft accelerations directly, the difference between the true accelerations $\dot{y}$ and the estimated values $F(y,u) + G(y,u)\hat{z}$ is used to constitute the correction term. Hence, let the observer be designed in the form of

$$\dot{\hat{z}} = A_{22}(u, p)\hat{z} + B_2(u, p) + \boldsymbol{L}\big(\dot{\boldsymbol{y}} - \boldsymbol{F}(\boldsymbol{y}, u) - \boldsymbol{G}(\boldsymbol{y}, u)\hat{z}\big), \tag{2.28}$$

where $\boldsymbol{L} \in \mathbb{R}^{1\times 2}$ in the inertia phase (or $\boldsymbol{L} \in \mathbb{R}^{1\times 1}$ in the torque phase) is the observer gain to be determined.

By defining the observer error as

$$e = z - \hat{z}, \tag{2.29}$$

the error dynamics can then be described by

$$\dot{e} = A_{22}(u, p)z + B_2(u, p) \tag{2.30}$$

$$- \big(A_{22}(u, p)\hat{z} + B_2(u, p) + \boldsymbol{L}\big(\dot{\boldsymbol{y}} - \boldsymbol{F}(\boldsymbol{y}, u) - \boldsymbol{G}(\boldsymbol{y}, u)\hat{z}\big)\big)$$

$$= \big(A_{22}(u, p) - \boldsymbol{L}\boldsymbol{G}(\boldsymbol{y}, u)\big)e - \boldsymbol{L}\boldsymbol{w}. \tag{2.31}$$

We define $V = \frac{1}{2}e^T e$ and differentiate it along the solution of (2.30) to infer

$$\dot{V} = e^T \dot{e}$$

$$= e^T \big(A_{22}(u, p) - \boldsymbol{L}\boldsymbol{G}(\boldsymbol{y}, u)\big)e - e^T \boldsymbol{L}\boldsymbol{w}. \tag{2.32}$$

Applying Young's Inequality [12] (see Lemma B.1 in Appendix B) to the last term of the above equality leads to

$$-e^T \boldsymbol{L}\boldsymbol{w} \le \kappa_1 e^T e + \frac{1}{4\kappa_1}\boldsymbol{w}^T \boldsymbol{L}^T \boldsymbol{L}\boldsymbol{w} \tag{2.33}$$

with $\kappa_1 > 0$. Then, (2.32) becomes

$$\dot{V} \le e^T \big(A_{22}(u, p) - \boldsymbol{L}\boldsymbol{G}(\boldsymbol{y}, u) + \kappa_1\big)e + \frac{1}{4\kappa_1}\boldsymbol{w}^T \boldsymbol{L}^T \boldsymbol{L}\boldsymbol{w}. \tag{2.34}$$

We now choose $\boldsymbol{L}$ to satisfy the following inequality

$$A_{22}(u, p) - \boldsymbol{L}\boldsymbol{G}(\boldsymbol{y}, u) + \kappa_1 \le -\kappa_2 \tag{2.35}$$

with $\kappa_2 > 0$, then we arrive at

$$\dot{V} \le -\kappa_2 e^T e + \frac{1}{4\kappa_1}\boldsymbol{w}^T \boldsymbol{L}^T \boldsymbol{L}\boldsymbol{w}, \tag{2.36}$$

which implies that the error dynamics admits the input-to-state stability property if the model error $\boldsymbol{w}$ is supposed to be bounded in amplitude (see Lemma B.2 in Appendix B).

Moreover, it follows from (2.36) that

$$\dot{V} \le -2\kappa_2 V + \frac{1}{4\kappa_1}\boldsymbol{w}^T \boldsymbol{L}^T \boldsymbol{L}\boldsymbol{w}. \tag{2.37}$$

Upon multiplication of (2.37) by $e^{2\kappa_2 t}$, it becomes

$$\frac{d}{dt}\left(Ve^{2\kappa_2 t}\right) \leq \frac{1}{4\kappa_1}\boldsymbol{w}^T\boldsymbol{L}^T\boldsymbol{L}\boldsymbol{w}e^{2\kappa_2 t}. \tag{2.38}$$

Integrating it over $[0, t]$ leads to

$$V(t) \leq V(0)e^{-2\kappa_2 t} + \frac{1}{4\kappa_1}\int_0^t e^{-2\kappa_2(t-\tau)}\boldsymbol{w}(\tau)^T\boldsymbol{L}^T\boldsymbol{L}\boldsymbol{w}(\tau)\,d\tau. \tag{2.39}$$

Hence, the properties of the error dynamics of the designed observer (2.28) are described as follows:

Theorem 2.1 *Suppose that*

- $\kappa_1 > 0, \kappa_2 > 0$;
- *The observer gain* $\boldsymbol{L}$ *is chosen to satisfy* (2.35).

Then, the error dynamics of the observer (2.28) *is*

(a) *Input-to-state stable, if* $\boldsymbol{w}$ *is bounded in amplitude, i.e.,* $\boldsymbol{w} \in \mathcal{L}_\infty$;
(b) *Exponentially stable with* κ_2 *for* $\boldsymbol{w} = 0$.

Proof It follows from (2.36) that $\dot{V} \leq -\kappa_2\|e\|^2 + \frac{1}{4\kappa_1}\lambda_{\max}(\boldsymbol{L}^T\boldsymbol{L})\|\boldsymbol{w}\|^2$, which shows that the error dynamics admits the input-to-state stability property [12, p. 503] (see Lemma B.2 in Appendix B) if the model error $\boldsymbol{w}$ is supposed to be bounded in amplitude, as property (a) required.

By taking $\boldsymbol{w} = 0$, we obtain from (2.39) that $|e(t)| \leq |e(0)|e^{-\kappa_2 t}, \forall t \geq 0$ which proves property (b). □

Remark 2.2 By the equivalences of the ISS property listed in Appendix B, property (a) implies that the error dynamics of the designed observer (2.28) is robustly stable if $\boldsymbol{w}$ is viewed as the effect of model uncertainties. This is the case in Sect. 2.3.3.

Remark 2.3 Now we give a discussion on the parameters κ_1 and κ_2. From property (b), κ_2 is chosen according to the required decay rate of the error. If $\boldsymbol{w}$ is bounded in amplitude, i.e., $\boldsymbol{w} \in \mathcal{L}_\infty$, then (2.39) becomes

$$\frac{1}{2}\left\|e(t)\right\|^2 \leq \frac{1}{2}\left\|e(0)\right\|^2 e^{-2\kappa_2 t} + \frac{\|\boldsymbol{w}\|_\infty^2 \sup\limits_{[0,t]}\lambda_{\max}(\boldsymbol{L}^T\boldsymbol{L})}{4\kappa_1}\int_0^t e^{-2\kappa_2(t-\tau)}\,d\tau, \tag{2.40}$$

which implies that

$$\left\|e(\infty)\right\|^2 \leq \frac{\|\boldsymbol{w}\|_\infty^2 \sup(\lambda_{\max}(\boldsymbol{L}^T\boldsymbol{L}))}{2\kappa_1}\lim_{t\to\infty}\int_0^t e^{-2\kappa_2(t-\tau)}\,d\tau, \tag{2.41}$$

and furthermore,

$$\left\| e(\infty) \right\|^2 \leq \frac{\| \boldsymbol{w} \|_\infty^2 \sup(\lambda_{\max}(\boldsymbol{L}^T \boldsymbol{L}))}{4\kappa_1 \kappa_2}. \tag{2.42}$$

Hence, one may choose a larger κ_1 to reduce the offset. From (2.35), however, one should also notice that the larger the κ_1, the higher the observer gain.

Remark 2.4 Inequality (2.42) gives just an upper bound on the estimation error offset if a bound on the model error is given. The real offset could be much smaller, due to the multiple use of inequalities in the above derivation.

Remark 2.5 Besides satisfying (2.35), we do not impose other assumptions on the observer gain $\boldsymbol{L}$. This implies that $\boldsymbol{L}$ can be designed theoretically to depend on $(\boldsymbol{y}, u)$, while one may choose it as time-invariant (constant) in practice. A solution with time-invariant $\boldsymbol{L}$ will be discussed in the next section.

Implementation Issues

In order to avoid taking derivatives of the measured variables, let

$$\eta = \hat{z} - \boldsymbol{L}\boldsymbol{y}, \tag{2.43}$$

then, we can infer for a time-invariant $\boldsymbol{L}$ that

$$\begin{aligned} \dot{\eta} &= \dot{\hat{z}} - \boldsymbol{L}\dot{\boldsymbol{y}} \\ &= A_{22}(u, p)\hat{z} + B_2(u, p) + \boldsymbol{L}\big(\dot{\boldsymbol{y}} - \boldsymbol{F}(\boldsymbol{y}, u) - \boldsymbol{G}(\boldsymbol{y}, u)\hat{z}\big) - \boldsymbol{L}\dot{\boldsymbol{y}}, \end{aligned} \tag{2.44}$$

to arrive at

$$\dot{\eta} = \big(A_{22}(u, p) - \boldsymbol{L}\boldsymbol{G}(\boldsymbol{y}, u)\big)(\eta + \boldsymbol{L}\boldsymbol{y}) + B_2(u, p) - \boldsymbol{L}\boldsymbol{F}(\boldsymbol{y}, u). \tag{2.45}$$

Equations (2.43) and (2.45) constitute the reduced-order observer for the nonlinear clutch slip control system. We notice that the nonlinearities of the powertrain system appear in their original form in the observer. Therefore, the merits arise—the characteristics of powertrain mechanical systems, such as the characteristics of the engine and torque converter, are represented in the form of lookup tables which are easy to process on a computer.

According to Theorem 2.1 and Remark 2.3, a systematic procedure is given to design the reduced-order nonlinear clutch pressure observer in the form of (2.43) and (2.45) as follows:

Step 1 Choose parameter κ_2 according to the required decay rate of the estimation error;

Step 2 Choose parameter κ_1, where it is suggested to start from some smaller values (according to Remark 2.3);

Step 3 Determine the observer gain $\boldsymbol{L}$ such that (2.35) is satisfied;
Step 4 For a given model error bound, use (2.42) to compute the estimated upper bound of the offset and check if the offset bound is acceptable.
Step 5 If the offset bound is acceptable, end the design procedure. If not, go to Step 2.

It is well known that getting model error bounds is in general very difficult, if not impossible. As mentioned in Remark 2.4, for a given model error bound, (2.42) gives just an upper bound of the estimation error offset, which might be much larger than the real offset. Hence, the stopping rule of iterating Step 1–Step 5 is somehow a "rule of thumb".

We now give a solution of (2.35) for choosing $\boldsymbol{L}$ to be time-invariant, where the requirement for low observer gains can be considered through optimization. If $A_{22}(u, p)$ and $\boldsymbol{G}(\boldsymbol{y}, u)$ in (2.35) vary in a polytope with r vertices, i.e.,

$$\begin{aligned}&\left(A_{22}(u,p) \quad \boldsymbol{G}(\boldsymbol{y},u)\right)\\&\quad\in \mathrm{Co}\left\{\left(A_{22,1} \quad \boldsymbol{G}_1\right),\left(A_{22,2} \quad \boldsymbol{G}_2\right),\ldots,\left(A_{22,r} \quad \boldsymbol{G}_r\right)\right\},\end{aligned} \tag{2.46}$$

where $\mathrm{Co}\{\cdot\}$ denotes the convex hull of the polytope, then, there exist

$$\beta_1 \geq 0, \beta_2 \geq 0, \ldots, \beta_r \geq 0 \tag{2.47}$$

satisfying

$$\sum_{i=1}^{r} \beta_i = 1 \tag{2.48}$$

such that

$$\left(A_{22}(u,p) \quad \boldsymbol{G}(\boldsymbol{y},u)\right) = \sum_{i=1}^{r} \beta_i \left(A_{22,i} \quad \boldsymbol{G}_i\right). \tag{2.49}$$

Hence, the result is given as follows.

Theorem 2.2 *Suppose that $A_{22}(u, p)$ and $\boldsymbol{G}(\boldsymbol{y}, u)$ vary in a polytope as* (2.46). *Then, any time-invariant $\boldsymbol{L}$ satisfying the following Linear Matrix Inequalities (LMIs)*

$$A_{22,i} - \boldsymbol{L}\boldsymbol{G}_i + \kappa_1 + \kappa_2 \leq 0, \quad i = 1, 2, \ldots, r \tag{2.50}$$

meets the observer gain condition (2.35).

Proof Since $A_{22}(u, p)$ and $\boldsymbol{G}(\boldsymbol{y}, u)$ vary in a polytope as (2.46), then, we have (2.49) with (2.47) and (2.48). By the convexity of

$$A_{22}(u, p) - \boldsymbol{L}\boldsymbol{G}(\boldsymbol{y}, u) + \kappa_1 + \kappa_2$$

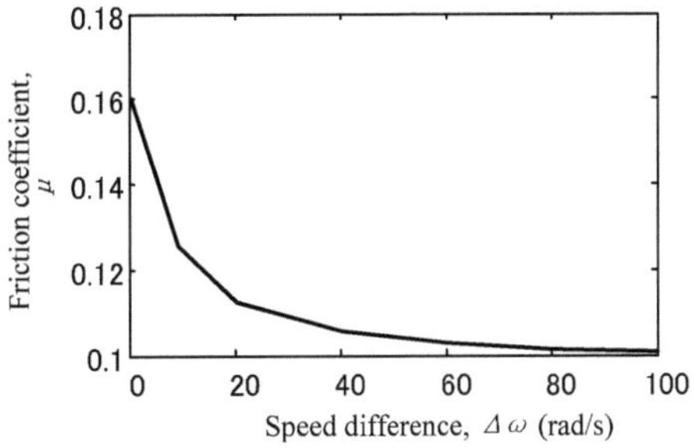

Fig. 2.5 Friction characteristics of clutch plates

in A_{22} and G, and by the use of Jensen's inequality (see Appendix E), we infer

$$A_{22}(u,p) - \boldsymbol{L}\boldsymbol{G}(\boldsymbol{y},u) + \kappa_1 + \kappa_2 \leq \sum_{i=1}^{r} \beta_i (A_{22,i} - \boldsymbol{L}\boldsymbol{G}_i + \kappa_1 + \kappa_2). \tag{2.51}$$

Hence, $\beta_i \geq 0$ and the satisfaction of (2.50) guarantees

$$A_{22}(u,p) - \boldsymbol{L}\boldsymbol{G}(\boldsymbol{y},u) + \kappa_1 + \kappa_2 \leq 0, \tag{2.52}$$

as required. □

In (2.50), $\boldsymbol{G}_i$ and $A_{22,i}$ are known and bounded, κ_1 and κ_2 are selected to be bounded and $r = 2^m$ (where m is the number of time-varying parameters in $\boldsymbol{G}(\boldsymbol{y},u)$ and A_{22}) is bounded, too. Hence, some constant $\boldsymbol{L}$ is always found to render make it hold. Moreover, we prefer low observer gains, due to robustness against noises and the reduction of the upper bound of the error offset, which is estimated by (2.42). Hence, by the use of Schur complement (see Appendix E), $\boldsymbol{L}$ is obtained through the following LMI optimization

$$\min_{\alpha,\boldsymbol{L}} \alpha \quad \text{subject to LMIs (2.50) and} \begin{pmatrix} \alpha & \boldsymbol{L} \\ \boldsymbol{L}^T & \boldsymbol{I} \end{pmatrix} \geq 0. \tag{2.53}$$

Given κ_1 and κ_2, the solution of (2.53) gives then a constant observer gain with the lowest possible gains satisfying condition (2.35).

Observer Design for Clutch Pressure

The inertia phase is taken as an example to show the detailed design procedure. The parameters (τ_{cv}, K_{cv}) are regarded as constants for simplicity, which are listed in Table 2.2 together with the other parameters. Nonlinear functions f_1, f_2 and μ are given as lookup tables for the observer. The map of μ is shown in Fig. 2.5, while f_1, f_2 are given by third-order maps and examples when $\omega_e = 500$ rad/s are shown in Fig. 2.6. These parameters are derived from the nominal setting of an AMESim simulation model of the AT shown in Fig. 2.1.

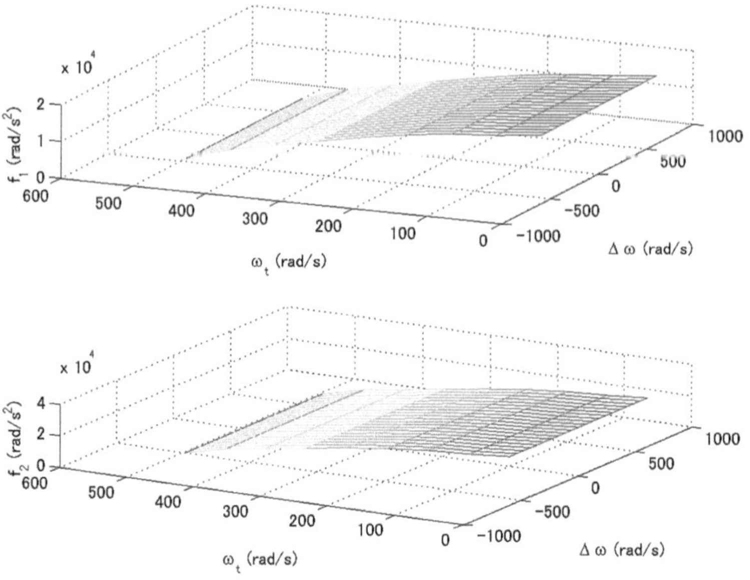

Fig. 2.6 MAPs of f_1, f_2 when $\omega_e = 500$ rad/s

Following the procedure given in Sect. 2.3.2, κ_2 is chosen to meet the requirement for the desired decay rate of the estimation error. It is desired that the error converges in 0.1 s. Then, taking the settling time as 4 times the time constant [15, p. 221] leads to $\frac{4}{\kappa_2} = 0.1$ and results in $\kappa_2 = 40$.

Then, κ_1 is chosen with the purpose of achieving a smaller offset of the estimation error. Start with $\kappa_1 = 1$ and obtain $\boldsymbol{L} = (-718 \; -1201)$ and $e(\infty) \leq 0.175$ MPa (see the following for the detailed calculation). The offset bound is too large for real applications. According to Step 2 of the procedure given in the above subsection, we enlarge the value of κ_1, and finally, the value being used is $\kappa_1 = 15$.

We now solve the optimization problem (2.53) to obtain the lowest possible observer gain. Since $A_{22} = -\frac{1}{\tau_{cv}}$ is considered as constant, the polytope in (2.46) is given by $\boldsymbol{G}(\boldsymbol{y}, u) = \text{Co}\{\boldsymbol{G}_1, \boldsymbol{G}_2\}$, where the two vertices are computed by (2.25b) with $\mu_{\min} \leq \mu(x_2) \leq \mu_{\max}$. The solution reads $\boldsymbol{L} = (-783 \; -1310)$.

In order to check if the estimation offset is acceptable, we now roughly compute the bound of modeling errors. Since powertrain systems admit highly nonlinear, complex dynamics and various uncertainties, it is indeed difficult, if not impossible, to obtain a comprehensive estimate of the modeling error bound. Hence, some major uncertainties are taken as examples to estimate the value of $\boldsymbol{w}$. The major uncertainties here are calculation errors of $\boldsymbol{F}(\boldsymbol{y}, u)$ in (2.24a), (2.24b), which contains the turbine torque T_t and the vehicle driving load T_{ve}. The change of the vehicle mass affects also the coefficients C_{ij} in $\boldsymbol{F}(\boldsymbol{y}, u)$. From numerous simulations of different powertrain settings, a bound on $\boldsymbol{w}$ is determined as $\|\boldsymbol{w}\|_\infty = 1600$ rad/s^2. According to (2.42), an upper bound of the offset is obtained for the designed observer.

The result is $e(\infty) \leq 0.05$ MPa, which is less that 10 % of the variation range of the working pressure of the valve and is acceptable.

Similarly, following the procedure given in the above subsection, the observer gain for the torque phase is calculated, and the result is $\boldsymbol{L} = -5.02 \times 10^4$.

2.3.3 Simulation Results

The proposed clutch pressure observer is programmed using MATLAB/Simulink and combined with the above complete powertrain simulation model through co-simulations. The two clutch valves are controlled by a pre-designed clutch slip controller to ensure a rapid and smooth shift process.

In this study, the major concern is put on the power-on 1st-to-2nd gear upshift process. Figure 2.7 gives the simulation results of the shift process with the driving condition of Table 2.1, i.e., the condition for the observer design. During the shift process, the engine throttle angle is adjusted to cooperate with the transmission shift.

In both torque and inertia phases, the pressure of cylinder B is estimated by the designed observers. After the inertia phase (after 8.34 s), because the clutch B has been locked up, the pressure is computed from the simplified control valve dynamics (2.17c).

During the torque phase (between 7.7 and 7.94 s), the rotational speeds do not change much, whereas during the inertia phase (between 7.94 and 8.34 s), the rotational speeds change intensively because of the clutch slip. Hence, the estimation performance in the inertia phase is much better, although it is also acceptable in the torque phase. The estimation error is plotted in the bottom of Fig. 2.7 as the solid line, where the result for $\boldsymbol{L} = 0$ is also given for comparison. The error peak is reduced by about 35 % and the average error is reduced by about 31 %. Note that the shift process operates in the nominal driving condition, but the stiffness of the drive shaft and the tire slip are considered in the simulation model, while these are ignored in the model for designing the observer. Moreover, the time-delay in control and time-varying parameters are also considered in the simulation model of the proportional valve.

The proposed observer is now tested under the driving conditions which deviate from the nominal setting, where the vehicle mass, road grade, torque characteristics of the engine and the torque converter are varied. We increase or decrease each of the items, and carry out simulations under different combination of these changes. The results with relatively large errors are shown in Fig. 2.8, where the driving condition setting is as follows: the torque characteristic of the engine is enlarged by 15 %, and subsequently the capacity of the torque converter is also enlarged; the vehicle mass is increased from 1500 to 1725 kg, and the road grade angle is varied from 0 to 5 degrees.

Due to the large model errors, the pressure estimation error becomes larger in the torque phase. The reason is that there is no slip in clutch A during the torque

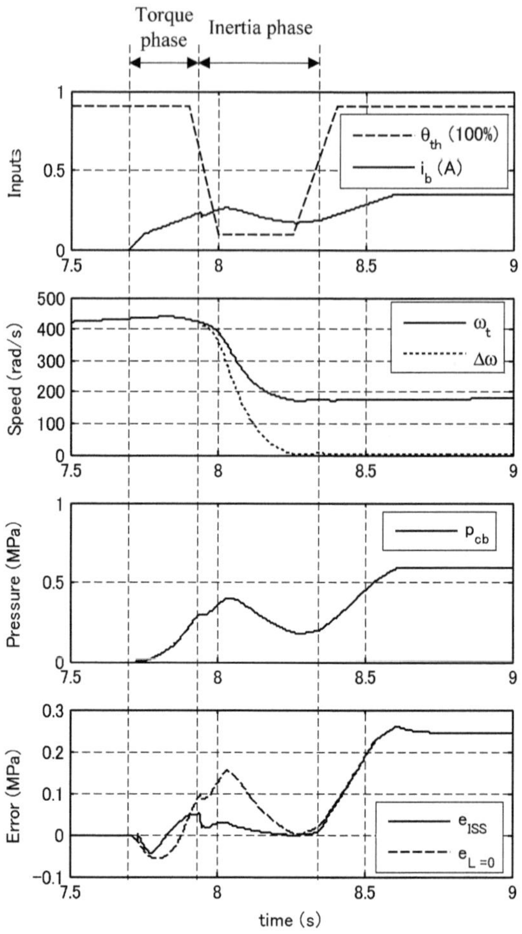

Fig. 2.7 Results of the nominal driving condition

phase, and no large change of the transmission speeds for the large vehicle inertia. Therefore, the torsion of the drive shaft and the tire slip play important roles in the drive line. The omission of these terms in the observer design deteriorates the estimation performance. In the inertia phase, because of the clutch slip, the designed observer still works well and the pressure estimation error is acceptable.

2.3.4 Design of Full-Order Sliding Mode Observer and Comparison

As a comparison, a full-order sliding mode observer is designed according to [13, 22]. Taking the inertia phase as an example, we rewrite system equa-

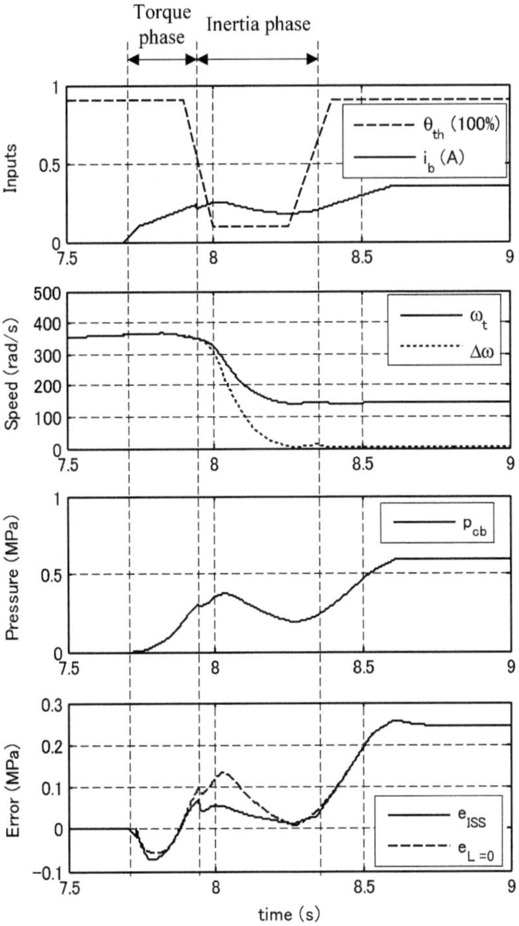

Fig. 2.8 Results of different driving condition

tions (2.24a), (2.24b) as

$$\dot{y}_1 = C_{13}\mu(y_2)RNAz + f_1, \tag{2.54a}$$

$$\dot{y}_2 = (C_{13} - C_{23})\mu(y_2)RNAz + f_2, \tag{2.54b}$$

$$\dot{z} = -\frac{1}{\tau_{cv}}z + \frac{K_{cv}}{\tau_{cv}}u. \tag{2.54c}$$

Following [13], the sliding mode observer can be designed in the following form:

$$\dot{\hat{y}}_1 = C_{13}\mu(\hat{y}_2)RNA\hat{z} + \hat{f}_1 + \kappa_{s1}\,\mathrm{sign}(\tilde{y}_1), \tag{2.55a}$$

$$\dot{\hat{y}}_2 = (C_{13} - C_{23})\mu(\hat{y}_2)RNAz + \hat{f}_2 + \kappa_{s2}\,\mathrm{sign}(\tilde{y}_2), \tag{2.55b}$$

$$\dot{\hat{z}} = -\frac{1}{\tau_{cv}}\hat{z} + \frac{K_{cv}}{\tau_{cv}}u + \kappa_{s3}\,\mathrm{sign}(\tilde{y}_1) + \kappa_{s4}\,\mathrm{sign}(\tilde{y}_2), \tag{2.55c}$$

where $\kappa_{s1}, \kappa_{s2}, \kappa_{s3}$ and κ_{s4} are observer gains, and

$$\tilde{y}_1 = y_1 - \hat{y}_1,$$
$$\tilde{y}_2 = y_2 - \hat{y}_2.$$

Gains κ_{s1} and κ_{s2} should satisfy the following sliding condition:

$$\kappa_{s1} > \|\tilde{f}_1\| + \|C_{13}\mu RNA\tilde{z}\| \approx 1920, \tag{2.56a}$$
$$\kappa_{s2} > \|\tilde{f}_2\| + \left\|(C_{13} - C_{23})\mu RNA\tilde{z}\right\| \approx 3180 \tag{2.56b}$$

with

$$\tilde{f}_1 = f_1 - \hat{f}_1,$$
$$\tilde{f}_2 = f_2 - \hat{f}_2,$$
$$\tilde{z} = z - \hat{z}.$$

Thus, κ_{s1} and κ_{s2} are selected as

$$\kappa_{s1} = 2000, \tag{2.57a}$$
$$\kappa_{s2} = 3200. \tag{2.57b}$$

According to the desired estimation offset and error decay rate, gains κ_{s3} and κ_{s4} can be calculated as

$$\kappa_{s3} = -4.2 \times 10^6, \tag{2.58a}$$
$$\kappa_{s4} = -7 \times 10^6. \tag{2.58b}$$

Similarly, the sliding mode observer for the torque phase can be designed in the following form

$$\dot{\hat{y}}_1 = c_{13}\mu(\Delta\omega)RNA\hat{z}_1 + \hat{f}_{t1} + \kappa_{st1}\,\mathrm{sign}(\tilde{y}_1), \tag{2.59a}$$
$$\dot{\hat{z}}_1 = -\frac{1}{\tau_{cv}}\hat{z}_1 + \frac{K_{cv}}{\tau_{cv}}u + \kappa_{st2}\mathrm{sign}(\tilde{y}_1), \tag{2.59b}$$

and the gains are calculated as

$$\kappa_{st1} = 100, \tag{2.60a}$$
$$\kappa_{st2} = -7.5 \times 10^6. \tag{2.60b}$$

The sampling frequency of the sliding mode observer is chosen to be 100 Hz, in order to test the feasibility of the resulting observer for real applications [9]. In the discrete implementation, the observer gains have to be reduced in order to restrain oscillations resulting from sampling and two sets of the tuned values are given in

Table 2.3 Gains of discrete observers

ISS	Torque phase $L = -1.8 \times 10^4$	Inertia phase $L = (-320 \; - 540)$
Sliding 1 (large gains)	$\kappa_{st1} = 80$ $\kappa_{st2} = -1.5 \times 10^6$	$\kappa_{s1} = 1000$ $\kappa_{s2} = 1600$ $\kappa_{s3} = -8 \times 10^5$ $\kappa_{s4} = -1.2 \times 10^6$
Sliding 2 (small gains)	$\kappa_{st1} = 80$ $\kappa_{st2} = -3.5 \times 10^5$	$\kappa_{s1} = 1000$ $\kappa_{s2} = 1600$ $\kappa_{s3} = -4 \times 10^5$ $\kappa_{s4} = -7 \times 10^5$

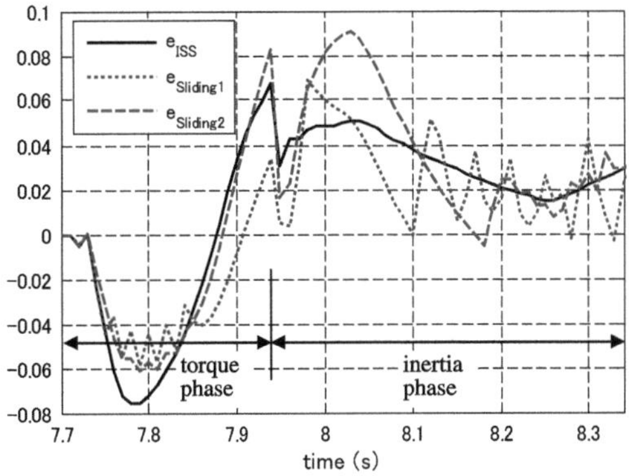

Fig. 2.9 Comparison between ISS observer and sliding mode observer (torque converter capacity is enlarged by 15 %; $m = 1725$ kg; $\theta_g = 5°$)

Table 2.3. Hence, the proposed ISS observer is also discretized by the same sampling frequency and the tuned gains are also listed in Table 2.3.

The comparison results of these three observers are shown in Fig. 2.9, where the driving condition is the same as that of Fig. 2.8. In Fig. 2.9, the solid line represents the error of the reduced-order observer, while the dotted and dashed lines represent the error of the full-order sliding mode observers with the large and small gains, respectively. It is seen that the proposed reduced-order observer works well in the inertia phase. The sliding mode observer with large gains (Sliding 1) tracks true values without large errors but with chatters, while the other sliding mode observer (Sliding 2) achieves few chatters at the cost of the large estimation errors. As for robustness, the proposed observer achieves robustness in the sense of input-to-state stability, where the model errors are represented as external inputs.

2.4 Clutch Pressure Estimation when Considering Drive Shaft Stiffness

In the above, a reduced-order clutch pressure observer was proposed when considering the concept of input-to-state stability (ISS). However, it is pointed out that during the torque phase, the estimation error becomes somehow unacceptable.

During the torque phase, the engine torque is transferred from the off-going clutch to the on-coming clutch. If the clutch pressure can be estimated accurately, the precise timing of releasing and applying clutches can be guaranteed to prevent the clutches from tying-up and the traction interruption. Hence, in order to improve the estimation precision of the clutch pressure during the shift torque phase, the methodology proposed in the above section is extended to design an observer when considering the driveline stiffness. Because the drive axle shafts are the main components of the whole driveline, the rotational freedom of the drive shaft is introduced into the model-based design. The newly designed observer can simultaneously estimate the drive shaft torque as well as improve the accuracy of the clutch pressure estimation [6].

2.4.1 Clutch System Modeling when Considering the Drive Shaft

The power-on 1st-to-2nd upshift is still considered as the example, and the pressure observer is designed to estimate the clutch pressure during the shift process. When considering the drive shaft compliance, the system models can be constructed as follows.

Torque Phase

In the 1st-to-2nd upshift torque phase, it is assumed that there is no slip in clutch A, and the motion of the drive line during this phase is represented by the following equations:

$$\dot{\omega}_t = c_{11}T_t + c_{13}\mu(\Delta\omega)RN(Ap_{cb} - F_s) + c_{14}\frac{T_s}{i_{df}}, \tag{2.61a}$$

$$\dot{\omega}_w = c_{34}T_s + c_{35}T_l, \tag{2.61b}$$

$$\dot{T}_s = \frac{K_s}{i_{df}i_1}\omega_t - K_s\omega_w, \tag{2.61c}$$

$$\dot{p}_{cb} = -\frac{1}{\tau_{cv}}p_{cb} + \frac{K_{cv}}{\tau_{cv}}u, \tag{2.61d}$$

where ω_t is the turbine speed, ω_w is the speed of the driving wheel (front wheel), T_s is the drive shaft torque, p_{cb} is pressure of cylinder B, T_t is the turbine torque,

Table 2.4 Parameters for observer design

c_{13}	Coefficient of clutch torque in (2.69c)	$-11.90\ \frac{1}{\text{kg m}^2}$
c_{14}	Coefficient of clutch torque in (2.69c)	$-4.76\ \frac{1}{\text{kg m}^2}$
c_{34}	Coefficient of clutch torque in (2.69c)	$0.0074\ \frac{1}{\text{kg m}^2}$
C_{13}	Coefficient of clutch torque in (2.70b)	$-24.51\ \frac{1}{\text{kg m}^2}$
C_{14}	Coefficient of clutch torque in (2.70b)	$-0.98\ \frac{1}{\text{kg m}^2}$
C_{23}	Coefficient of clutch torque in (2.70b)	$29.41\ \frac{1}{\text{kg m}^2}$
C_{24}	Coefficient of clutch torque in (2.70b)	$-8.82\ \frac{1}{\text{kg m}^2}$
C_{34}	Coefficient of clutch torque in (2.70b)	$0.0074\ \frac{1}{\text{kg m}^2}$
γ	Gear ratio of sun gear to ring gear	0.667
R	Effective radius of plates of clutch B	0.13 m
N	Plate number of clutch B	3
A	Piston area of clutch B	$0.01\ \text{m}^2$
τ_{cv}	Time constant of valve B	0.04 s
K_{cv}	Gain of valve B	1.0 MPa/A
$\mu_{\min}$	Minimum friction coefficient	0.10
$\mu_{\max}$	Maximum friction coefficient	0.16
i_{df}	Gear ratio of the differential box	3
K_s	Stiffness of drive shaft	13000 Nm/rad
$\bar{\omega}_t$	Normalization of ω_t	100 rad/s
$\bar{\omega}_w$	Normalization of ω_w	10 rad/s
$\Delta\bar{\omega}$	Normalization of $\Delta\omega$	100 rad/s
$\bar{T}_s$	Normalization of T_s	1000 Nm
$\bar{p}_{cb}$	Normalization of p_{cb}	10^5 Pa

T_l is the resistant torque delivered from the tires, F_s denotes the return spring force of clutch B and μ is the friction coefficient of clutch B depending on the speed difference $\Delta\omega$. The definition of the other parameters can be found in Table 2.4.

The turbine torque T_t and resistance torque T_l in (2.61a)–(2.61d) are calculated as follows [7]:

$$T_t = t(\lambda)C(\lambda)\omega_e^2, \tag{2.62a}$$

$$T_l = T_w + C_A R_w^3 \omega_w^2, \tag{2.62b}$$

where $C(\lambda)$ denotes the capacity factor of the torque converter, $t(\lambda)$ is the torque ratio, ω_e is the engine speed and λ is the speed ratio defined as $\lambda = \frac{\omega_t}{\omega_e}$, T_w denotes the rolling resistance moment of tires, R_w is the tire radius, and C_A is a constant coefficient depending on air density, aerodynamic drag coefficient and the front area of the vehicle.

Inertia Phase

In the inertia phase, the pressure of cylinder A is greatly reduced, and the pressure of cylinder B increases so that the speed difference between ring gear and turbine can be reduced to zero, i.e., we have the engagement of clutch B. The dynamic motion of this phase can be described by the following equations if the drive axle shaft compliance is considered:

$$\dot{\omega}_t = C_{11}T_t + C_{13}\mu(\Delta\omega)RN(Ap_{cb} - F_s) + C_{14}\frac{T_s}{i_{df}}, \tag{2.63a}$$

$$\Delta\dot{\omega} = (C_{11} - C_{21})T_t + (C_{13} - C_{23})\mu(\Delta\omega)RN(Ap_{cb} - F_s) + (C_{14} - C_{24})\frac{T_s}{i_{df}}, \tag{2.63b}$$

$$\dot{\omega}_w = C_{34}T_s + C_{35}T_l, \tag{2.63c}$$

$$\dot{T}_s = \frac{K_s}{i_{df}}\left(\omega_t - \frac{1}{1+\gamma}\Delta\omega\right) - K_s\omega_w, \tag{2.63d}$$

$$\dot{p}_{cb} = -\frac{1}{\tau_{cv}}p_{cb} + \frac{K_{cv}}{\tau_{cv}}u, \tag{2.63e}$$

where $\Delta\omega$ is the slip speed of clutch B, i.e., the speed difference between the turbine and the ring gear, C_{ij} are constant coefficients determined by inertia moments of the vehicle and transmission shafts; note that C_{ij} are different from c_{ij} of (2.61a)–(2.61d).

The models in consideration of the drive shaft stiffness are constructed for the observer design. State variables are selected as

$$x_1 = \frac{\omega_t}{\bar{\omega}_t}, \qquad x_2 = \frac{\Delta\omega}{\Delta\bar{\omega}}, \qquad x_3 = \frac{\omega_w}{\bar{\omega}_w}, \qquad x_4 = \frac{T_s}{\bar{T}_s}, \qquad x_5 = \frac{p_{cb}}{\bar{p}_{cb}},$$

so that the variables are normalized to have the same level of magnitude. The driveline motion of the upshift torque phase is then expressed in the following state space form:

$$\dot{x}_1 = \frac{c_{14}\bar{T}_s}{\bar{\omega}_t i_{df}}x_4 + \frac{c_{13}\mu(\bar{\omega}_t x_1)RNA\bar{p}_{cb}}{\bar{\omega}_t}x_5 + \frac{1}{\bar{\omega}_t}f_{t1}(\omega_e, x_1), \tag{2.64a}$$

$$\dot{x}_3 = \frac{c_{34}\bar{T}_s}{\bar{\omega}_w}x_4 + \frac{1}{\bar{\omega}_w}f_{t2}(x_3), \tag{2.64b}$$

$$\dot{x}_4 = \frac{K_s\bar{\omega}_t}{i_{df}i_1\bar{T}_s}x_1 - \frac{K_s\bar{\omega}_w}{\bar{T}_s}x_3, \tag{2.64c}$$

$$\dot{x}_5 = -\frac{1}{\tau_{cv}}x_5 + \frac{K_{cv}}{\tau_{cv}\bar{p}_{cb}}u, \tag{2.64d}$$

where $u = i_b$ is the control input and

$$f_{t1}(\omega_e, x_1) = c_{11}T_t - c_{13}\mu(\Delta\omega)RNF_s, \tag{2.65a}$$

$$f_{t2}(x_3) = c_{35}T_l. \tag{2.65b}$$

Similarly, the inertia phase can also be described in the following state space form with state variables of x_1 to x_5:

$$\dot{x}_1 = \frac{C_{14}\bar{T}_s}{\bar{\omega}_t i_{df}} x_4 + \frac{C_{13}\mu(x_2)RNA\bar{p}_{cb}}{\bar{\omega}_t} x_5 + \frac{1}{\bar{\omega}_t} f_1(\omega_e, x_1), \tag{2.66a}$$

$$\dot{x}_2 = \frac{(C_{14} - C_{24})\bar{T}_s}{\Delta\bar{\omega} i_{df}} x_4 + \frac{C_{13} - C_{23}\mu(x_2)RNA\bar{p}_{cb}}{\Delta\bar{\omega}} x_5 + \frac{1}{\Delta\bar{\omega}} f_2(\omega_e, x_1), \tag{2.66b}$$

$$\dot{x}_3 = \frac{C_{34}\bar{T}_s}{\bar{\omega}_w} x_4 + \frac{1}{\bar{\omega}_w} f_3(x_3), \tag{2.66c}$$

$$\dot{x}_4 = \frac{K_s\bar{\omega}_t}{i_{df}\bar{T}_s} x_1 - \frac{K_s\Delta\bar{\omega}}{i_{df}(1+\gamma)\bar{T}_s} x_2 - \frac{K_s\bar{\omega}_w}{\bar{T}_s} x_3, \tag{2.66d}$$

$$\dot{x}_5 = -\frac{1}{\tau_{cv}} x_5 + \frac{K_{cv}}{\tau_{cv}\bar{p}_{cb}} u, \tag{2.66e}$$

with the nonlinear functions

$$f_1(\omega_e, x_1) = C_{11}T_t - C_{13}\mu(\Delta\omega)RNF_s, \tag{2.67a}$$

$$f_2(\omega_e, x_1) = (C_{11} - C_{21})T_t - (C_{13} - C_{23})\mu(\Delta\omega)RNF_s, \tag{2.67b}$$

$$f_3(x_3) = C_{35}T_l. \tag{2.67c}$$

The problem considered here is to estimate the pressure of clutch B x_4 (drive shaft torque x_5, too) both in the torque and inertia phases, in the presence of model errors, given the measured rotational speeds of transmission x_1, x_2, x_3, ω_e and valve electric current u.

2.4.2 Design of Reduced-Order Nonlinear State Observer

Reduced-Order Nonlinear Observer

Denote the variable to be estimated as z, and rewrite the dynamics of the system for estimating the clutch pressure as follows:

$$\dot{\boldsymbol{y}} = \boldsymbol{F}(\boldsymbol{y}, u) + \boldsymbol{G}(\boldsymbol{y}, u)z + \boldsymbol{H}\boldsymbol{w}(\boldsymbol{y}, u, z), \tag{2.68a}$$

$$\dot{z} = \boldsymbol{A}_{21}\boldsymbol{y} + \boldsymbol{A}_{22}z + \boldsymbol{B}_2(u), \tag{2.68b}$$

where $\boldsymbol{y}$ is the measured outputs, $\boldsymbol{w}(\boldsymbol{y}, u, \boldsymbol{z})$ summarizes model uncertainties which is normalized by $\boldsymbol{H}$ as $\|\boldsymbol{w}\|_\infty \approx 1$, and in particular

$$\boldsymbol{y} = [x_1, x_3]^T, \qquad \boldsymbol{z} = [x_4, x_5]^T, \qquad u = i_b, \tag{2.69a}$$

$$\boldsymbol{F}(\boldsymbol{y}, u) = \begin{pmatrix} \frac{1}{\bar{\omega}_t} f_{t1}(\omega_e, y_1) \\ \frac{1}{\bar{\omega}_w} f_{t2}(y_2) \end{pmatrix}, \tag{2.69b}$$

$$\boldsymbol{G}(\boldsymbol{y}, u) = \begin{pmatrix} \frac{c_{14}\bar{T}_s}{\bar{\omega}_t R_{df}} & \frac{c_{13}\mu(x_2) RNA\bar{p}_{cb}}{\bar{\omega}_t} \\ \frac{c_{34}\bar{T}_s}{\bar{\omega}_w} & 0 \end{pmatrix}, \tag{2.69c}$$

$$\boldsymbol{A}_{21} = \begin{pmatrix} \frac{K_s \bar{\omega}_t}{R_{df} R_1 \bar{T}_s} & -\frac{K_s \bar{\omega}_w}{\bar{T}_s} \\ 0 & 0 \end{pmatrix}, \tag{2.69d}$$

$$\boldsymbol{A}_{22} = \begin{pmatrix} 0 & 0 \\ 0 & -\frac{1}{\tau_{cv}} \end{pmatrix}, \tag{2.69e}$$

$$\boldsymbol{B}_2(u) = \begin{pmatrix} 0 \\ \frac{K_{cv}}{\tau_{cv}\bar{p}_{cb}} \end{pmatrix} u. \tag{2.69f}$$

for the torque phase.

For the inertia phase, $\boldsymbol{y} = [x_1, x_2, x_3]^T$ is the measurement. Hence, (2.69b)–(2.69e) are replaced by

$$\boldsymbol{F}(\boldsymbol{y}, u) = \begin{pmatrix} \frac{1}{\bar{\omega}_t} f_1(\omega_e, y_1) \\ \frac{1}{\Delta\bar{\omega}} f_2(\omega_e, y_1) \\ \frac{1}{\bar{\omega}_w} f_3(y_3) \end{pmatrix}, \tag{2.70a}$$

$$\boldsymbol{G}(\boldsymbol{y}, u) = \begin{pmatrix} \frac{C_{14}\bar{T}_s}{\bar{\omega}_t R_{df}} & \frac{C_{13}\mu(y_2) RNA\bar{p}_{cb}}{\bar{\omega}_t} \\ \frac{(C_{14}-C_{24})\bar{T}_s}{\Delta\bar{\omega} R_{df}} & \frac{C_{13}-C_{23}\mu(y_2) RNA\bar{p}_{cb}}{\Delta\bar{\omega}} \\ \frac{C_{34}\bar{T}_s}{\bar{\omega}_w} & 0 \end{pmatrix}, \tag{2.70b}$$

$$\boldsymbol{A}_{21} = \begin{pmatrix} \frac{K_s \bar{\omega}_t}{R_{df} \bar{T}_s} & -\frac{K_s \Delta\bar{\omega}}{R_{df}(1+\gamma)\bar{T}_s} & -\frac{K_s \bar{\omega}_w}{\bar{T}_s} \\ 0 & 0 & 0 \end{pmatrix}, \tag{2.70c}$$

$$\boldsymbol{A}_{22} = \begin{pmatrix} 0 & 0 \\ 0 & -\frac{1}{\tau_{cv}} \end{pmatrix}. \tag{2.70d}$$

The observer is then designed in the form of

$$\dot{\hat{\boldsymbol{z}}} = \boldsymbol{A}_{21}\boldsymbol{y} + \boldsymbol{A}_{22}\hat{\boldsymbol{z}} + \boldsymbol{B}_2(u) + \boldsymbol{L}\big(\dot{\boldsymbol{y}} - \boldsymbol{F}(\boldsymbol{y}, u) - \boldsymbol{G}(\boldsymbol{y}, u)\hat{\boldsymbol{z}}\big), \tag{2.71}$$

where $\boldsymbol{L} \in \mathbb{R}^{2\times 2}$ ($\boldsymbol{L} \in \mathbb{R}^{2\times 3}$ for the inertia phase) is the constant observer gain to be determined [4].

In order to avoid taking derivatives of the measurements $\boldsymbol{y}$, the following transformation is made. Let

$$\eta = \hat{z} - \boldsymbol{L}\boldsymbol{y}, \tag{2.72}$$

then, we can infer for a time-invariant $\boldsymbol{L}$ that

$$\dot{\eta} = \big(\boldsymbol{A}_{22} - \boldsymbol{L}\boldsymbol{G}(\boldsymbol{y},u)\big)(\eta + \boldsymbol{L}\boldsymbol{y}) + \boldsymbol{A}_{21}\boldsymbol{y} + \boldsymbol{B}_2(u) - \boldsymbol{L}\boldsymbol{F}(\boldsymbol{y},u). \tag{2.73}$$

Equations (2.72) and (2.73) constitute then the reduced-order observer of the clutch pressure for the nonlinear driveline system. Obviously, the nonlinearities of the powertrain system appear in the observer in their original form. Therefore, the characteristics of powertrain mechanical systems, such as characteristics of the engine and the aerodynamic drag, are represented in the form of lookup tables, which is easily processed in computer control.

Then the error dynamics of the designed shaft torque observer is analyzed using the concept of ISS (input-to-state stability) [11, 12, 18]. By defining the observer error as

$$\boldsymbol{e} = z - \hat{z}, \tag{2.74}$$

the error dynamics can then be described by

$$\dot{\boldsymbol{e}} = \big(\boldsymbol{A}_{22} - \boldsymbol{L}\boldsymbol{G}(\boldsymbol{y},u)\big)\boldsymbol{e} - \boldsymbol{L}\boldsymbol{H}\boldsymbol{w}. \tag{2.75}$$

We define $V(\boldsymbol{e}) = \frac{1}{2}\boldsymbol{e}^T\boldsymbol{e}$ and differentiate it along the solution of (2.75) to obtain

$$\dot{V} = \boldsymbol{e}^T\big(\boldsymbol{A}_{22} - \boldsymbol{L}\boldsymbol{G}(\boldsymbol{y},u)\big)\boldsymbol{e} - \boldsymbol{e}^T\boldsymbol{L}\boldsymbol{H}\boldsymbol{w}, \tag{2.76}$$

and then

$$\dot{V} \le \boldsymbol{e}^T\big(\boldsymbol{A}_{22} - \boldsymbol{L}\boldsymbol{G}(\boldsymbol{y},u) + \kappa_1\boldsymbol{I}\big)\boldsymbol{e} + \frac{1}{4\kappa_1}\boldsymbol{w}^T\boldsymbol{H}^T\boldsymbol{L}^T\boldsymbol{L}\boldsymbol{H}\boldsymbol{w}, \tag{2.77}$$

where $\kappa_1 > 0$. We now choose $\boldsymbol{L}$ to satisfy the following matrix inequality:

$$\boldsymbol{A}_{22} - \boldsymbol{L}\boldsymbol{G}(\boldsymbol{y},u) \le -(\kappa_1 + \kappa_2)\boldsymbol{I} \tag{2.78}$$

with $\kappa_2 > 0$, then we arrive at

$$\dot{V} \le -\kappa_2\boldsymbol{e}^T\boldsymbol{e} + \frac{1}{4\kappa_1}\boldsymbol{w}^T\boldsymbol{H}^T\boldsymbol{L}^T\boldsymbol{L}\boldsymbol{H}\boldsymbol{w}, \tag{2.79}$$

which implies that the error dynamics admits the input-to-state stability property if the model error $\boldsymbol{w}$ is supposed to be bounded in amplitude.

It follows from (2.79) that

$$\|\boldsymbol{e}(t)\|^2 \le \|\boldsymbol{e}(0)\|^2 e^{-2\kappa_2 t} + \frac{\|\boldsymbol{w}\|_\infty^2 \sup\limits_{[0,t]} \lambda_{\max}(\boldsymbol{H}^T\boldsymbol{L}^T\boldsymbol{L}\boldsymbol{H})}{2\kappa_1}\int_0^t e^{-2\kappa_2(t-\tau)}\,d\tau, \tag{2.80}$$

which implies that

$$\left\|e(t)\right\|^2 \to \frac{\|w\|_\infty^2 \sup(\lambda_{\max}(H^T L^T L H))}{4\kappa_1\kappa_2} \quad \text{as } t \to \infty. \tag{2.81}$$

For a more detailed deduction, please refer to Sect. 2.3.

Gain Determination

Now we discuss how to choose parameters κ_1, κ_2, and finally, the observer gain L.

κ_1 and κ_2 It follows from (2.79) that κ_2 can be chosen according to the required decay rate of the error. If it is desired that the error converges in 0.05 s, then $\frac{4}{2\kappa_2} = 0.05$, which results in $\kappa_2 = 40$.

According to (2.81), one may choose a larger κ_1 to reduce the offset. From (2.78), however, one should notice that the larger the κ_1, the higher the observer gain.

Optimization of L We now give a solution of (2.78) for constant L through solving a set of linear matrix inequalities (LMIs). If $A_{22}(u)$ and $G(y, u)$ in (2.78) vary in a polytope with r vertices, i.e.,

$$\begin{aligned} &\begin{pmatrix} A_{22} & G(y,u) \end{pmatrix} \\ &\quad \in \mathrm{Co}\left\{\begin{pmatrix} A_{22,1} & G_1 \end{pmatrix}, \begin{pmatrix} A_{22,2} & G_2 \end{pmatrix}, \ldots, \begin{pmatrix} A_{22,r} & G_r \end{pmatrix}\right\}, \end{aligned} \tag{2.82}$$

where $\mathrm{Co}\{\cdot\}$ denotes the convex hull of the polytope. Then, a constant observer gain L satisfying the following Linear Matrix Inequalities (LMIs):

$$A_{22,i} - LG_i \le -(\kappa_1 + \kappa_2)I, \quad i = 1, 2, \ldots, r, \tag{2.83}$$

meets the observer gain condition (2.78).

Moreover, we prefer low observer gains, due to robustness against noises and also the reduction of the estimation error offset estimated as (2.81). Hence, L can be obtained through the following optimization:

$$\min_{\alpha, L} \mathrm{trace}(\alpha) \quad \text{subject to LMIs (2.83) and} \begin{pmatrix} \alpha & LH \\ H^T L^T & I \end{pmatrix} \ge 0, \tag{2.84}$$

where α is a 2×2 positive diagonal matrix for both the torque and inertia phase. Given κ_1 and κ_2, the solution of (2.84) gives then the lowest possible gains.

Solution and Evaluation To calculate the observer gain and the error offset, the bound of the modeling error should be calculated first. It is indeed difficult, if not impossible, to obtain a comprehensive estimate of the modeling error bound. Hence some major uncertainties are taken into consideration to estimate the value of the modeling errors. If the estimation error of the turbine torque T_t is bounded within

15 %, the variation of vehicle mass is ±500 kg and the variation of road slope is ±5°, the modeling error for the torque phase can be calculated as

$$\|\boldsymbol{w}\| = \begin{pmatrix} 1 \\ 1 \end{pmatrix} \tag{2.85}$$

with the normalization matrix

$$\boldsymbol{H} = \begin{pmatrix} 2.38 & 0 \\ 0 & 0.13 \end{pmatrix}. \tag{2.86}$$

Given the above modeling error bound and the system parameters (shown in Table 2.4), (2.84) and (2.81) can be used to calculate $\boldsymbol{L}$ and then check the error offset under the calculated gain. The final tuned results for the torque phase are $\kappa_1 = 15$ and

$$\boldsymbol{L} = \begin{pmatrix} -2.08 & 30.49 \\ -6.46 & -152.48 \end{pmatrix}, \tag{2.87}$$

and the calculated error offset is $\boldsymbol{e}(\infty) \leq 0.51$, which means that the error offset of the clutch pressure p_{cb} is not larger than 0.051 MPa, which is considered acceptable [4].

Similarly, following the procedure given above, the observer gain for the inertia phase can also be calculated, and the result reads

$$\boldsymbol{L} = \begin{pmatrix} 0.023 & 0.22 & 56.9 \\ -1.74 & -0.23 & 1.73 \end{pmatrix}. \tag{2.88}$$

2.4.3 Simulation Results

Besides the continuous simulation, discrete implementations are carried out as well to get an in-vehicle assessment of the proposed observer. The sampling rate is chosen to be 100 Hz in order to test the feasibility of implementing the resulting observer for real applications [9]. The discrete characteristics and random noise of the speed sensor are included as well.

The major concern is put on the power-on 1st-to-2nd gear upshift process. Figure 2.10 gives the simulation results of the shift process with the nominal driving condition, i.e., the condition for the observer design. The continuous and discrete results are listed simultaneously. During the shift process, the engine throttle angle is adjusted to cooperate with the transmission shift. It can be seen that during the torque phase (between 7.74 and 7.94 s) the rotational speeds of the shafts do not change much, whereas during the inertia phase (between 7.94 and 8.24 s), the rotational speeds change extensively because of the clutch slip.

In the torque and inertia phase, the pressure of cylinder B is estimated by the designed observers. After the inertia phase, i.e., after 8.24 s, because of the engagement of the clutch, the observer is not valid any more, and the pressure is estimated

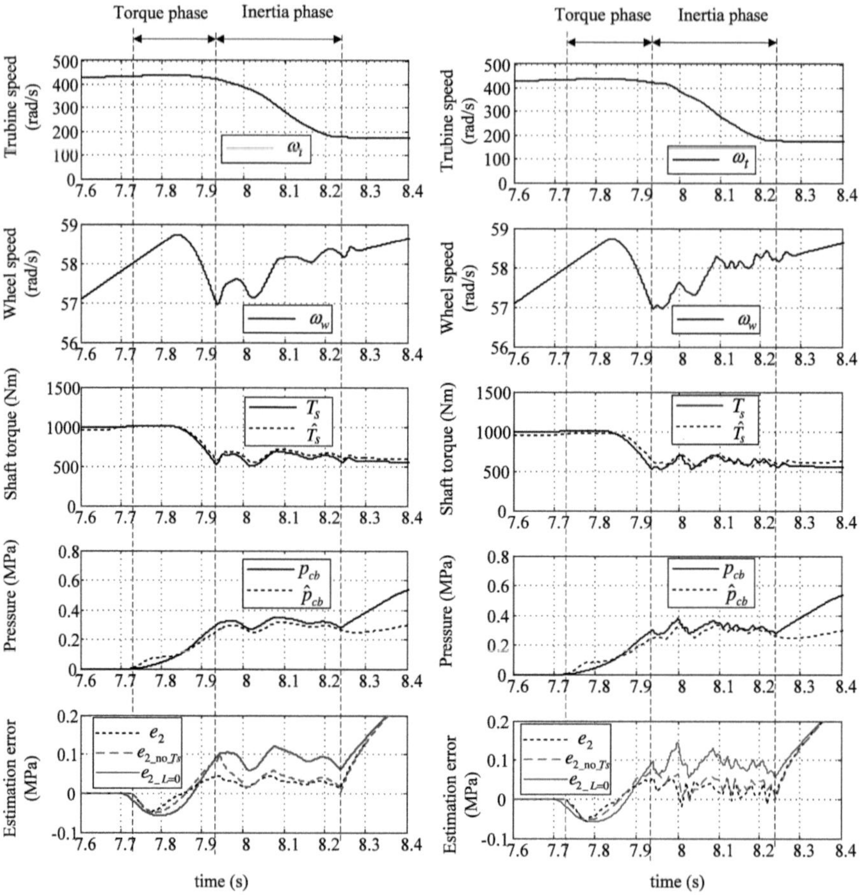

Fig. 2.10 Simulation results of nominal condition (torque characteristics of engine and torque converter: nominal, $m = 1500$ kg, $\theta_g = 0°$, $K_s = 220$ Nm/deg, $I_t = 0.06$ kg m^2). (*Left*) Continuous implementation; (*Right*) Discrete implementation

by the simplified control valve dynamics (2.61a)–(2.61d), i.e., the estimation with observer gains being zero. In order to show the effectiveness of the newly designed observers, the error of estimation with the observer gains being zero during all time periods $e_{2\_L=0}$, and the error of the observer of [4] in which the drive shaft stiffness is not considered during the design procedure $e_{2\_no\_Ts}$ are given as well. It can be seen that the result of the observer designed in this section, e_2, has the best performance. It is also shown that the estimated drive shaft torque $\hat{T}_s$ can track the true values without large errors.

It should be noted that in the discrete implementation, the observer gains have to be reduced in order to restrain oscillations resulting from sampling, and the tuned

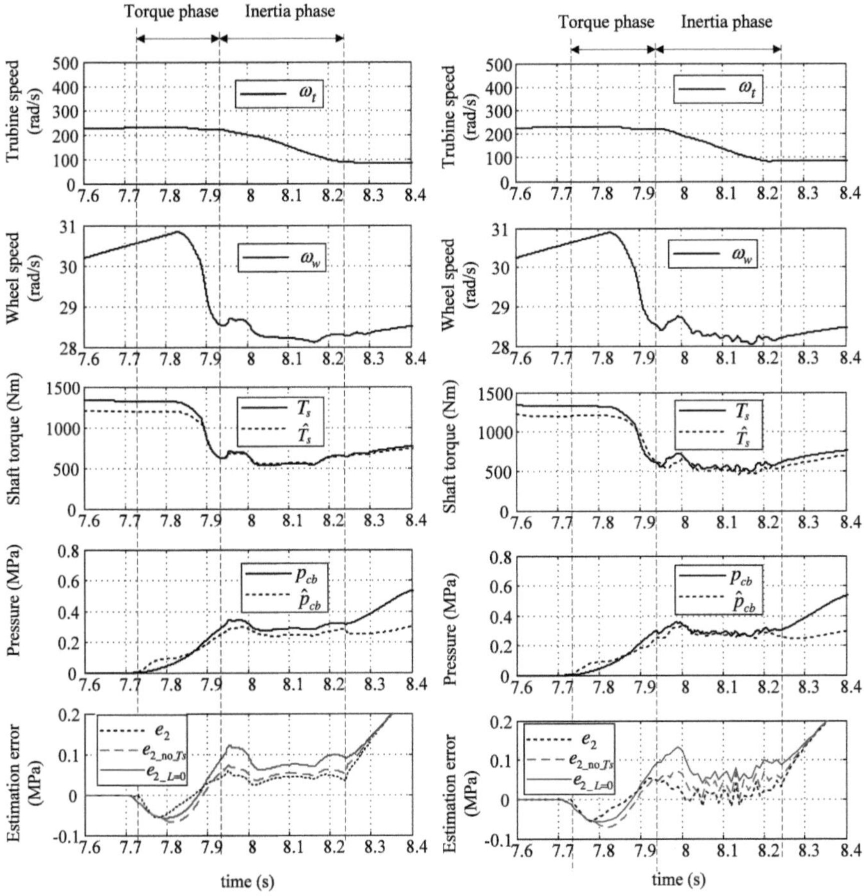

Fig. 2.11 Simulation results of different driving condition (torque characteristics of engine and torque converter are enlarged by 15 %, $m = 1725$ kg, $\theta_g = 5°$, $K_s = 242$ Nm/deg, $I_t = 0.09$ kg m^2). (*Left*) Continuous implementation; (*Right*) Discrete implementation

values are given as

$$L = \begin{pmatrix} -2.0 & 15.0 \\ -6.0 & -20.0 \end{pmatrix} \tag{2.89}$$

for the torque phase, and

$$L = \begin{pmatrix} 0.02 & 0.2 & 15 \\ -1.7 & -0.2 & 1.7 \end{pmatrix} \tag{2.90}$$

for the inertia phase.

Then the proposed observer is tested under driving conditions that deviate from the nominal driving setting. The following items are changed, and the variation bounds are also given as follows:

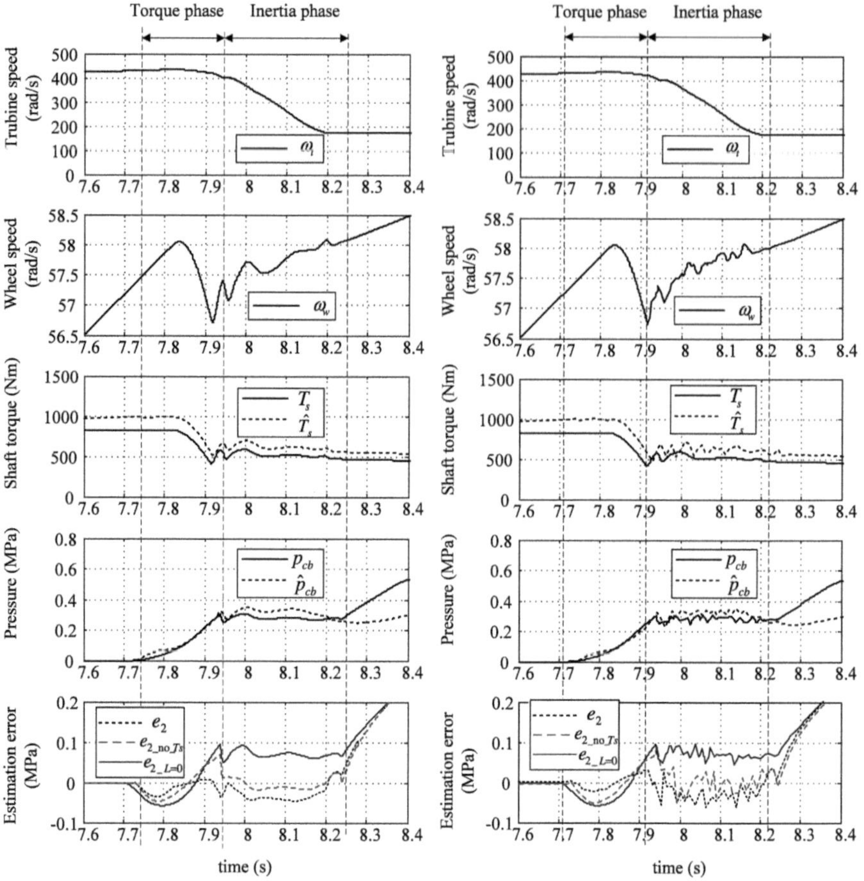

Fig. 2.12 Simulation results of different driving condition (torque characteristics of engine and torque converter are reduced by 15 %, $m = 1250$ kg, $\theta_g = 0°$, $K_s = 198$ Nm/deg, $I_t = 0.03$ kg m^2). (*Left*) Continuous implementation; (*Right*) Discrete implementation

(a) Torque characteristics of engine and torque converter, ±15 %;
(b) Vehicle mass m, ±16 %;
(c) Road grade θ_g, 0–5°;
(d) Drive shaft stiffness K_s, ±10 %;
(e) Turbine shaft inertia I_t, ±50 %.

The results with relatively large estimation errors are shown in Figs. 2.11 and 2.12. It should be noted that because the engine simulation model is based on the torque and friction maps, only a relatively large (±15 %) variation of steady state characteristics of the engine torque is represented. In other words, the steady map of the engine torque is increased or decreased by 15 %, and it is assumed that the bound covers the transient estimation error and the torque variation due to long-term aging. Then the map of the capacity factor of the torque converter $C(\lambda)$ (see (2.62a)) is adjusted accordingly, to make the turbine torque T_t increase or decrease by 15 %.

It can be seen from Fig. 2.11 that, due to large model errors, the pressure estimation error using the observer of the last section, $e_{2\_no\_Ts}$, becomes larger than that in Fig. 2.10, especially in the torque phase. The observer designed in this section, however, can still work with acceptable performance, and the maximum estimation error e_2 is 0.05 MPa, which is about 10 % of the maximum working pressure of the clutch. In the results of Fig. 2.12, although the proposed observer does not outperform the observer of the last section during the inertia phase, it does perform better during the torque phase. At the end of the torque phase, the error of the proposed observer e_2 is less than 0.035 MPa while the error of the observer of the last section, $e_{2\_no\_Ts}$, is 0.07 MPa. Reducing the estimation error at the end of the torque phase is critical because it determines the smoothness of the torque transferring between the two clutches if the estimated pressure is used for closed-loop control of the torque phase.

2.5 Notes and References

In the discrete implementation, the observer gains have to be reduced in order to restrain oscillations resulting from sampling. This is because the inter-sample behavior of the real system is not captured, which may be critical in a number of applications.

The analysis incorporating full time information leads to challenging control problems with a rich mathematical structure, and could be done in the framework of the sampled-data system theory, which is out of the scope of this book. Please refer to [1, 2, 5] for some theoretic discussions.

References

1. Bamieh BA, Pearson JB Jr (1992) A general framework for linear periodic systems with applications to H_∞ sampled-data control. IEEE Trans Autom Control 37(4):418–435
2. Chen T, Bruce F (1995) Optimal sampled-data control systems. Springer, London
3. Drivetrain HS (2007) In: Recent 10 years of automotive engineering. Society of Automotive Engineers of Japan, Tokyo, pp 134–137. In Japanese
4. Gao B-Z, Chen H, Zhao H-Y, Sanada K (2010) A reduced-order nonlinear clutch pressure observer for automatic transmission. IEEE Trans Control Syst Technol 18(2):446–453
5. Gao HJ, Sun WC, Shi P (2010) Robust sampled-data h-infinity control for vehicle active suspension systems. IEEE Trans Control Syst Technol 18(1):238–245
6. Gao B-Z, Chen H, Tian L, Sanada K (2012) A nonlinear clutch pressure observer for automatic transmission: considering drive shaft compliance. ASME J Dyn Syst Meas Control 134(1):1–8
7. Gillespie TD (1992) Fundamentals of vehicle dynamics. Society of Automotive Engineers, New York
8. Goetz M, Levesley MC, Crolla DA (2005) Dynamics and control of gearshifts on twin-clutch transmissions. Proc Inst Mech Eng, Part D, J Automob EngMech 219(8):951–963

9. Hahn JO, Lee KI (2002) Nonlinear robust control of torque converter clutch slip system for passenger vehicles using advanced torque estimation algorithms. Veh Syst Dyn 37(3):175–192
10. Hahn JO, Hur JW, Cho YM, Lee KI (2002) Robust observer-based monitoring of a hydraulic actuator in a vehicle power transmission control system. Control Eng Pract 10(3):327–335
11. Khalil HK (2002) Nonlinear Systems. Prentice Hall, New York
12. Krstić M, Kanellakopoulos I, Kokotović P (1995) Nonlinear and adaptive control design. Wiley, New York
13. Masmoudi RA, Hedrick K (1992) Estimation of vehicle shaft torque using nonlinear observers. ASME J Dyn Syst Meas Control 114:394–400
14. Misawa EA, Hedrick JK (1989) Nonlinear observers—a state-of-the-art survey. ASME J Dyn Syst Meas Control 111:344–352
15. Ogata K (2001) Modern control engineering, 4th edn. Prentice Hall, New York
16. Sanada K, Kitagawa A (1998) A study of two-degree-of-freedom control of rotating speed in an automatic transmission, considering modeling errors of a hydraulic system. Control Eng Pract 6:1125–1132
17. Shin BK, Hahn JO, Lee KI (2000) Development of shift control algorithm using estimated turbine torque. SAE technical paper 2000-01-1150
18. Sontag ED (2005) Input to state stability: basic concepts and results. Lecture notes in mathematics. Springer, Berlin
19. Sun Z, Hebbale K (2005) Challenges and opportunities in automotive transmission control. In: Proceedings of American control conference, vol 5, pp 3284–3289
20. Vahidi A, Stefanopoulou A, Peng H (2005) Recursive least squares with forgetting for online estimation of vehicle mass and road grade: theory and experiments. Veh Syst Dyn 43(1):31–55
21. Watechagit S, Srinivasan K (2003) On-line estimation of operating variables for stepped automatic transmissions. In: IEEE conference on control applications (CCA 2003), Istanbul, Turkey, vol 1, pp 279–284
22. Watechagit S, Srinivasan K (2005) Implementation of on-line clutch pressure estimation for stepped automatic transmissions. In: Proc American control conference, vol 3, pp 1607–1612
23. Yi K, Shin BK, Lee KL (2000) Estimation of turbine torque of automatic transmissions using nonlinear observers. ASME J Dyn Syst Meas Control 122:276–283

Chapter 3
Torque Phase Control of the Clutch-to-Clutch Shift Process

3.1 Introduction

It is well known that the dynamic behavior of the engine and clutch greatly affect the torque oscillation of the driveline, and even the steering system [5, 10]. Hence a smooth and fast clutch-to-clutch shift is necessary. As aforementioned, the clutch-to-clutch shift is usually divided into two phases, the torque phase and the inertia phase, and during the torque phase, the precise timing of releasing and applying of clutches is crucial for the prevention of clutch tie-up and traction interruption.

In the torque phase, the rotational speeds of clutch shafts do not change much. In order to achieve a smooth torque transfer between the two clutches, the off-going clutch is required to mimic the operation of a one-way clutch so that it can be disengaged at the moment when the direction of transmitted torque switches over. In [3], a clutch slip control scheme is suggested to accomplish the function of a one-way clutch, i.e., the off-going clutch is controlled to track a small reference speed (such as 5 rad/s). This control objective can effectively prevent clutch tie-up. However, if the pressure of the off-going clutch is not manipulated well, the stick-slip phenomenon [1] is apt to be caused and results in some powertrain vibration.

As mentioned above, a smooth torque transfer can be assured if the off-going clutch is disengaged when the transmitted torque is reduced to zero. If the knowledge of the transmitted torque of the clutch is available, the pressure of the off-going clutch can be controlled using the torque information. Therefore, this text proposes another control scheme which is based on a clutch pressure/torque observer. The observer designed in Chap. 2 is used, and a closed loop control scheme is proposed for the shift torque phase. The vehicle of interest is still the medium-size passenger car of the last chapter.[1]

[1]This chapter uses the content of [2], with permission from Inderscience Enterprises Ltd.

H. Chen, B. Gao, *Nonlinear Estimation and Control of Automotive Drivetrains*,
DOI 10.1007/978-3-642-41572-2_3,

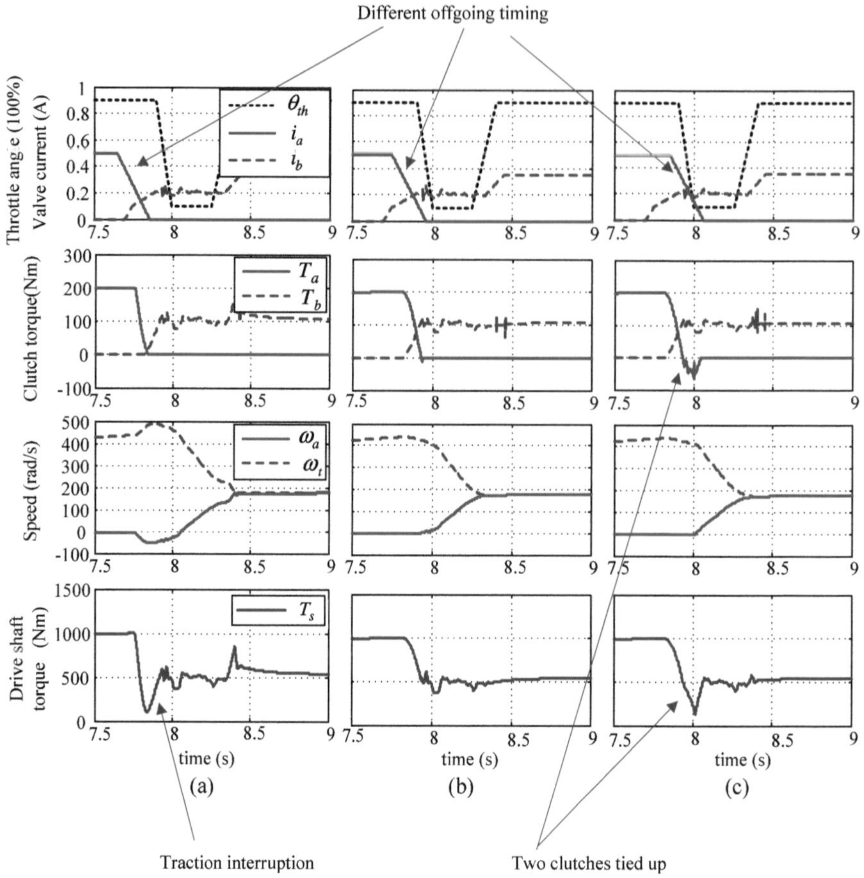

Fig. 3.1 Comparison of the different release times of the off-going clutch: (**a**) 0.1 s ahead of optimal timing, (**b**) optimal timing, (**c**) 0.1 s after optimal timing. (θ_{th}, throttle angle; i_a, current of the off-going clutch; i_b, current of the on-coming clutch; T_a, torque of the off-going clutch; T_b, torque of the on-coming clutch; ω_t, turbine speed; ω_a, speed difference of the off-going clutch; T_s, drive shaft torque)

3.2 Motivation of Clutch Timing Control

During a clutch-to-clutch shift process, if the torque transfer between the two clutches is not well controlled, clutch tie-up or torque interruption may be caused.

Figure 3.1 gives the simulation results of a typical power-on upshift process. During the torque phase, the pressure of the on-coming clutch is ramped up, and the off-going clutch is controlled by three patterns in order to show the effects of the disengagement timing of the off-going clutch. Pattern (b) gives the best result because the off-going clutch is disengaged just when its transmitted torque is reduced to zero at 7.93 s. Pattern (a) releases the off-going clutch 0.1 s earlier than the optimal release time, and pattern (c) gives a postponed disengagement timing. It is

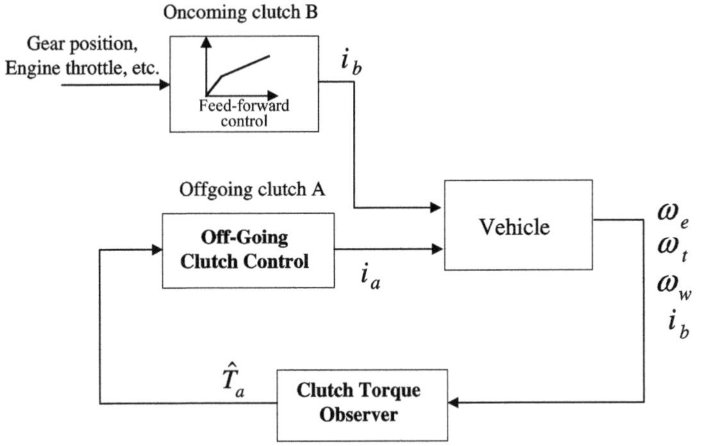

Fig. 3.2 Block diagram of a clutch disengagement system ($\hat{T}_a$, estimated torque of clutch A; i_a, valve current of clutch A; i_b, valve current of clutch B; ω_e, engine speed, ω_t, turbine speed, ω_w, driven wheel speed)

shown that an earlier releasing will cause traction interruption, and the engine speed and the turbine speed will flare up. On the other hand, a postponed timing will lead to clutch tie-up, and consequently, the shift shock will be enlarged and the friction loss will be increased.

From the results of Fig. 3.1, one sees that the precise timing of releasing the clutch is crucial for the shift quality during the shift torque phase. Therefore, a strategy of clutch timing control is proposed, and the block diagram of the proposed system is described in Fig. 3.2.

As shown in Fig. 3.2, during the torque phase, the pressure p_{cb} of clutch B is ramped up to undertake the traction torque T_t of the torque converter. At the same time, the torque T_a delivered to clutch A reduces accordingly. The output torque T_0 of the transmission is determined by the turbine torque T_t and the torque T_b of clutch B, and, in order to make the control strategy easy to be implemented, T_t and T_b are controlled feed-forwardly. The pressure p_{cb} of clutch B is ramped up according to a pre-determined pattern, while the torque T_a delivered to clutch A is estimated by the "Clutch Torque Observer". For a given T_a, there exist threshold values of the clamp force, pressure p_{ca}, and consequently, valve current i_a, which are just big enough to prevent clutch A from slipping. Thus the block of the "Off-going Clutch Control" is designed to calculate these threshold values, and finally gives the command of the valve current i_a, which assures that clutch A does not slip before T_a reaches zero, and after that, the clamp force of clutch A is totally withdrawn to avoid tie-up with clutch B.

3.3 Clutch Control Strategy

Using the estimation results of the pressure of clutch B $\hat{p}_{cb}$ and the drive shaft torque $\hat{T}_s$ shown in Chap. 2, the transmitted torque of clutch A can be calculated according to the following relationship of the planetary gear set:

$$T_t - T_b = \frac{\gamma}{1+\gamma} T_O = \gamma (T_a + T_b), \tag{3.1}$$

where T_a and T_b are the transmitted torque of clutch A and clutch B, respectively, T_O is the transmission output torque, γ is the ratio of the teeth number of the sun gear to that of the ring gear. The transformation of the above equation yields

$$T_a = T_O - T_t, \tag{3.2}$$

or

$$T_a = \frac{1}{\gamma} T_t - \frac{\gamma+1}{\gamma} T_b. \tag{3.3}$$

Then the estimated torque of clutch A can be given as

$$\hat{T}_a = \frac{1}{R_{df}} \hat{T}_s - \hat{T}_t, \tag{3.4}$$

or

$$\hat{T}_a = \frac{1}{\gamma} \hat{T}_t - \frac{\gamma+1}{\gamma} \mu R N (A \hat{p}_{cb} - F_s), \tag{3.5}$$

where $\hat{T}_t$ is the estimated turbine torque

$$\hat{T}_t = t(\lambda) C(\lambda) \omega_e^2. \tag{3.6}$$

In the results, Eq. (3.4) is used to estimate the torque of clutch A because of its simpler form compared to Eq. (3.5).

During the torque phase, the valve current of the on-coming clutch (clutch B) is controlled feed-forwardly to ramp up its pressure, while the off-going clutch (clutch A) is controlled according to the estimated torque $\hat{T}_a$. With the increase of pressure of clutch B, the transmitted torque to clutch A decreases. It is desired that the engagement force of clutch A is controlled to zero when its transmitted torque decreases to zero.

When the clutch is sticking (locked up), the maximally transmittable torque is limited by p_{ca}, i.e.,

$$T_{a\max} = (A_a p_{ca} - F_{sa}) \mu_s R_a N_a, \tag{3.7}$$

where p_{ca} is the pressure of clutch A, F_{sa} is the return spring force, μ_s is the static friction coefficient, A_a, R_a, N_a are the friction area, effective radius and plate number, respectively.

Together with the dynamics of valve A,

$$\dot{p}_{ca} = -\frac{1}{\tau_{cva}} p_{ca} + \frac{K_{cva}}{\tau_{cva}} i_a, \tag{3.8}$$

and using the static relationship of the current i_a and the pressure p_{ca}, we can determine the desired current i_a as

$$i_a = \kappa_{ca} \frac{1}{K_{cva}} \frac{1}{A_a} \left(\frac{\hat{T}_a}{\mu_s R_a N_a} + F_{sa} \right), \tag{3.9}$$

where κ_{ca} is a coefficient larger than 1. If the value is small, unwanted clutch slip may be caused before the transmitted torque reaches zero. On the other hand, if κ_{ca} is too large, the disengagement timing may be delayed. The tuned value is $\kappa_{ca} = 1.3$. It is clear that by such a clutch disengagement strategy, the off-going clutch will be disengaged when the transmitted torque $\hat{T}_a$ approaches zero, and before that the clutch is locked up.

3.4 Simulation Results

3.4.1 Powertrain Simulation Model

In this section, the proposed clutch control strategy (3.4) and (3.9) is evaluated on a powertrain simulation model. The model is established by the commercial simulation software AMESim, which supports the Simulink environment by the S-Function. The constructed model can capture the important transient dynamics during vehicle shift process, such as the drive shaft oscillation and tire slip. Moreover, time-delay and time-varying parameters of the proportional valves [8] are also considered in the simulation model, which are neglected in the controller design. The detailed description can be found in Chaps. 1 and 2.

3.4.2 Simulation Results

In order to get an in-vehicle assessment of the proposed clutch control system, the designed observer is discretized by a sampling rate of 100 Hz [4] with zero-order holder discretization.

Figure 3.3 shows the simulation results of a power-on 1st-to-2nd upshift. During the torque phase, the pressure of clutch B is ramped up, and clutch A is controlled by the proposed feedback control strategy. The driving conditions are the same as the nominal driving conditions, i.e., the conditions for the controller design. The observer gain used in Fig. 3.3 is

$$\boldsymbol{L} = \begin{pmatrix} -2.0 & 15.0 \\ -6.0 & -20.0 \end{pmatrix}, \tag{3.10}$$

which is kept constant in all the simulations.

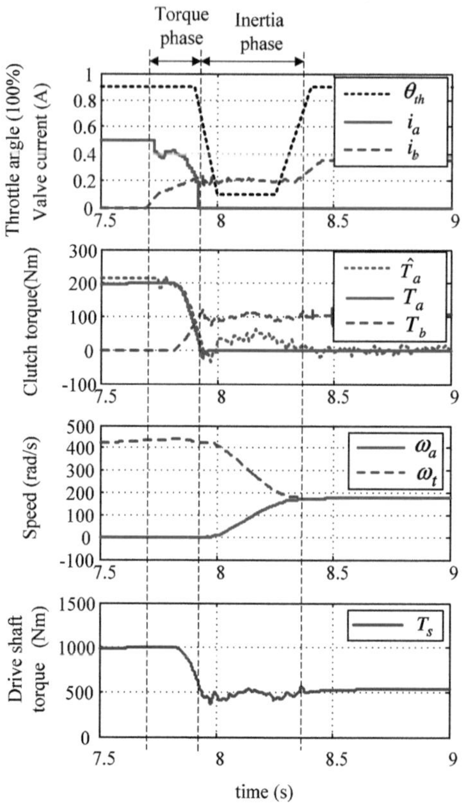

Fig. 3.3 Simulation results under nominal driving conditions (torque characteristics of engine and torque converter are standard; $m = 1500$ kg; $\theta_g = 0°$; $I_t = 0.06$ kg m^2)

Because there are inevitable errors associated with the estimated clutch torque, in order to avoid clutch tie-up, after $\hat{T}_a$ reaches a small value (such as 50 Nm), clutch A is controlled by the following on–off logic:

$$\text{if } \omega_a \leq -5 \text{ rad/s}, \quad \text{then } i_a = 0.3 \text{ A}; \tag{3.11a}$$

$$\text{if } \omega_a > -5 \text{ rad/s}, \quad \text{then } i_a = 0, \tag{3.11b}$$

where ω_a is the speed difference of clutch A, i.e., the speed of the ring gear. ω_a becomes negative when clutch A is released earlier than it should be (see Fig. 3.1(a) for reference). Note that because the transmitted torque is already reduced to a low level, the switching control of the pressure valve will not bring about a large driveline oscillation. However, if the switching logic is triggered from the first beginning of the torque phase, it is demonstrated through simulations that the torque oscillation will become unacceptable.

It is shown that clutch A is fully disengaged at 7.90 s when the estimated torque of clutch A, $\hat{T}_a$, approaches zero. We can see that the turbine speed does not flare up, and there is no clutch tie-up and torque interruption shown in the result of the drive shaft torque.

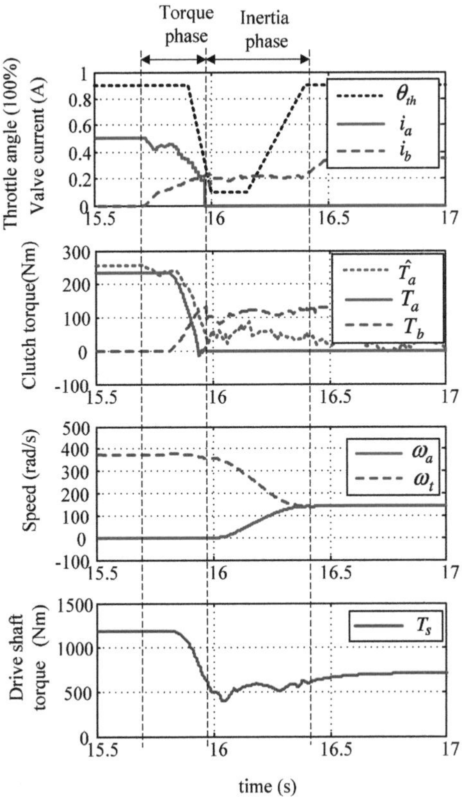

Fig. 3.4 Simulation results under different driving conditions (torque characteristics of engine and torque converter are standard × 115 %; $m = 2000$ kg; $\theta_g = 5°$; $I_t = 0.1$ kg m^2)

In order to examine the robustness of the proposed control strategy, the driving conditions and parameters are changed, and the results are shown in Figs. 3.4 and 3.5. The following items are changed:

- Torque characteristics of the engine and the torque converter;
- Vehicle mass;
- Road slope angle;
- Inertia moment of the torque converter turbine;

because they are highly correlated with the performance of the torque observer, but it is difficult to obtain the true values. We can see that, although the enlarged modeling errors bring about a larger estimation error of $\hat{T}_a$, the timing of release of clutch A is not seriously affected (it is 7.95 s in Fig. 3.4 and 7.92 s in Fig. 3.5) and there is no intensive fluctuation of the drive shaft torque, which shows that the shift quality is still good enough.

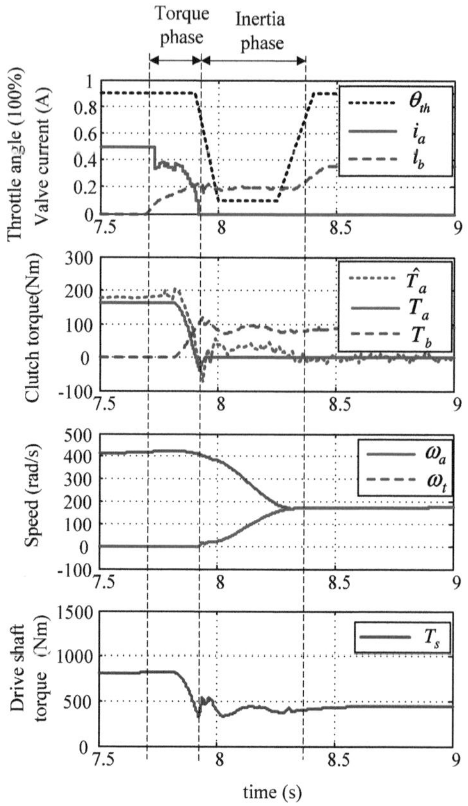

Fig. 3.5 Simulation results under different driving conditions (torque characteristics of engine and torque converter are standard × 85 %; $m = 1275$ kg; $\theta_g = 5°$; $I_t = 0.04$ kg m^2)

3.5 Notes and References

In this chapter, a new observer-based clutch control strategy is proposed for the torque phase of the clutch-to-clutch shift process.

The output torque T_0 of the transmission is determined by the turbine torque T_t and the torque T_b of clutch B, and in order to make the control strategy easy to implement, T_t and T_b are controlled feed-forwardly.

Along with the increase of the pressure of the on-coming clutch, the off-going clutch is fully disengaged when its transmitted torque approaches zero.

An AMESim powertrain simulation model is constructed to test the proposed clutch control strategy. Simulation results show that, by using the estimated clutch torque, the strategy can provide smooth torque transfer in the torque phase without clutch tie-up or traction interruption.

It is also demonstrated that the control strategy is robust to the variations of driving conditions and parameters, such as a change of the engine characteristics, vehicle mass, and the road grade, etc.

Although the control strategy was designed for a hydraulic Automatic Transmission, it is also applicable to the shift control of DCT (Dual Clutch Transmission) due to its similar clutch-to-clutch shift process [3, 6, 7, 9].

References

1. Crowther A, Zhang N, Liu DK, Jeyakumaran JK (2004) Analysis and simulation of clutch engagement judder and stick-slip in automotive powertrain systems. Proc Inst Mech Eng, Part D, J Automob EngMech 218(12):1427–1446
2. Gao B-Z, Chen H, Li J, Tian L, Sanada K (2012) Observer-based feedback control during torque phase of clutch-to-clutch shift process. Int J Veh Des 58(1):93–108
3. Goetz M, Levesley MC, Crolla DA (2005) Dynamics and control of gearshifts on twin-clutch transmissions. Proc Inst Mech Eng, Part D, J Automob EngMech 219(8):951–963
4. Hahn JO, Lee KI (2002) Nonlinear robust control of torque converter clutch slip system for passenger vehicles using advanced torque estimation algorithms. Veh Syst Dyn 37(3):175–192
5. Hohn BR, Pflaum H, Lechner C, Draxl T (2010) Efficient CVT hybrid driveline with improved drivability. Int J Veh Des 53(1/2):70–88
6. Kulkarni M, Shim T, Zhang Y (2007) Shift dynamics and control of dual-clutch transmissions. Mech Mach Theory 42(2):168–182
7. Minowa T, Ochi T, Kuroiwa H, Liu KZ (1999) Smooth gear shift control technology for clutch-to-clutch shifting. SAE technical paper 1999-01-1054
8. Sanada K, Kitagawa A (1998) A study of two-degree-of-freedom control of rotating speed in an automatic transmission, considering modeling errors of a hydraulic system. Control Eng Pract 6:1125–1132
9. Watechagit S (2004) Modeling and estimation for stepped automatic transmission with clutch-to-clutch shift technology. PhD Thesis, The Ohio State University
10. Yao Z, Mousseau C, Kao BG, Nikolaidis E (2008) An efficient powertrain simulation model for vehicle performance. Int J Veh Des 47(1–4):189–214

Chapter 4
Inertia Phase Control of the Clutch-to-Clutch Shift Process

4.1 Introduction

As aforementioned, during the shift inertia phase [18], the applying (on-coming) clutch slips, and the rotational speeds change intensively. The clutch slip control during the inertia phase greatly affects the shift quality (smoothness and efficiency).

The clutch slip control of stepped ratio transmissions has been intensively discussed by many researchers [17]. Because the clutch engagement is expected to satisfy different and sometimes conflicting objectives, e.g., minimizing clutch lock-up time, minimizing the friction losses during the slipping phase, ensuring a smooth acceleration of the vehicle, optimization based algorithms are a potential solution for this problem. For example, Hybrid Model Predictive Control (HMPC) [3] and Linear Quadratic Optimal Control (LQOC) [8, 9, 15, 16] are used to control the engagement of a dry clutch. In [3], Model Predictive Control is used so that the constraints on the control and state variables can be considered in an explicit and optimal way. In [16], it is pointed out that, to overcome the problem of high online computational demand, the time evolution computed off-line under different driving conditions can be fitted by polynomials for online application. Dynamic programming-based optimal control [20] and Sequential Quadratic Programming (SQP) [21] are also used for gear shift operations in automatic transmissions.

Different from optimal algorithms which use penalty functions to formulate multiple control objectives simultaneously, there is another kind of controller design method for clutch slip control, in which the only control objective is to make clutch speed track a pre-designed reference trajectory. Especially for gear shift operation during driving, where the duration is much shorter than that of the start-up scenario, the speed tracking control method is widely used [6, 11, 30, 36, 37]. Toward the highly nonlinear powertrain dynamics, such as the characteristics of the engine and torque converter, and the uncertainties, such as the parameter variation of hydraulic systems and the perturbations of road resistance torque, sliding mode control [6, 36], μ synthesis [30], and two-degree-of-freedom control [11, 30] are used to ensure consistent control performance. Although during the shift operations, speed tracking control does not consider the multiple control objectives directly, the required

H. Chen, B. Gao, *Nonlinear Estimation and Control of Automotive Drivetrains*,
DOI 10.1007/978-3-642-41572-2_4,

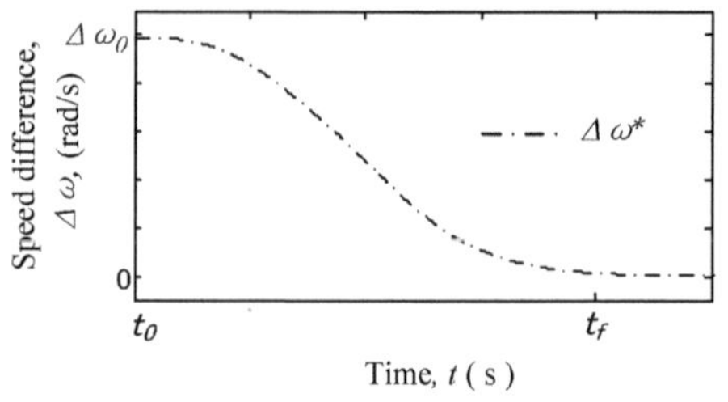

Fig. 4.1 Reference trajectory of clutch speed difference

shift time and shift comfort can be reached by selecting proper reference trajectory; the friction losses can also be reduced by suitable engine torque coordination during the shift process [11].

This chapter will focus on the latter method, i.e., on carrying out speed tracking control of the wet clutch of a stepped ratio automatic transmission during the shift inertia phase. During the inertia phase, the pressure of clutch A (see Chaps. 2 and 3) has already been reduced to a very low level, and the dynamics of the clutch system can be described by the following equations:

$$\Delta\dot{\omega} = (C_{13} - C_{23})\mu(\Delta\omega)R \cdot N \cdot Ap_{cb} + f(\Delta\omega, \omega_t, \omega_e), \tag{4.1a}$$

$$\dot{p}_{cb} = -\frac{1}{\tau_{cv}}p_{cb} + \frac{K_{cv}}{\tau_{cv}}i_b \tag{4.1b}$$

with

$$\begin{aligned} f(\Delta\omega, \omega_t, \omega_e) = {} & (C_{11} - C_{21})T_t(\omega_t, \omega_e) + (C_{14} - C_{24})T_{ve}(\Delta\omega, \omega_t) \\ & - (C_{13} - C_{23})\mu(\Delta\omega)R \cdot N \cdot F_s, \end{aligned} \tag{4.2}$$

where $\Delta\omega$ is the speed difference of clutch B, i.e., the speed difference between the sun gear and the ring gear; p_{cb} is the pressure of cylinder B; i_b is the current of valve B; T_t is the turbine torque; T_{ve} is the equivalent resistant torque delivered from the tire to the drive shaft; C_{ij} is the constant coefficients determined by the inertia moments of vehicle and transmission shafts; F_s denotes the return spring force of clutch B.

In this study, the engine control, T_t, is regarded as a non-controlled input (it is decided by an open-loop algorithm based on shift timing), and the electric current of valve B, i_b, is manipulated to make the speed difference of clutch B, $\Delta\omega$, track a reference trajectory. The shift process should assure driving comfort and minimize the dissipated friction energy. In general, if shift duration is limited to a suitably short time, there will not be too much dissipated energy. As for the driving comfort, because the lock-up of the clutch tends to cause a sudden change of the transmission output torque, the clutch engagement should satisfy the so-called no-lurch condition [3, 8, 20], i.e., the rotational acceleration of the clutch input shaft should be equal to that of the output shaft at the synchronization point. Therefore, the desired trajectory shown in Fig. 4.1 should satisfy the following requirements:

- $t_f - t_0$ does not exceed the required shift time;

- The change rate of the trajectory at t_f is zero;
- Moreover, in order to avoid control saturation, the change rate of the clutch speed at t_0 should be a small value.

Three different control methodologies will be adopted, and the results are given as well to show their different characteristics. The three methodologies used are the two-degree-of-freedom linear control scheme, backstepping approach and nonlinear feedback-feedforward control scheme, which will be respectively addressed below.[1]

4.2 Two-Degree-of-Freedom Linear Controller

The two-degree-of-freedom control scheme is suitable to many automotive control systems for it can show good tracking performance and robustness simultaneously [34]. This section, therefore, uses the two-degree-of-freedom controller design method to carry out the clutch slip control of an Automatic Transmission with proportional pressure control valves. The clutch cylinder pressure, which is necessary for state feedback control is estimated by the reduced-order pressure observer of Chap. 2. The feedback gain is calculated by robust pole assignment methods while the feedforward compensator aims to improve the system response [11].

4.2.1 Controller Design

We rewrite the dynamics system for clutch slip control as follows:

$$\dot{x}_1 = (C_{13} - C_{23})\mu(x_1)RNAx_2 + f(\omega_e, \omega_t, x_1), \tag{4.3a}$$

$$\dot{x}_2 = -\frac{1}{\tau_{cv}}x_2 + \frac{K_{cv}}{\tau_{cv}}u, \tag{4.3b}$$

where $x_1 = \Delta\omega$, $x_2 = p_{cb}$, and

$$\begin{aligned} f(\omega_e, \omega_t, x_1) = {} & (C_{11} - C_{21})T_t(\omega_e, \omega_t) + (C_{14} - C_{24})T_{ve}(\omega_t, x_1) \\ & - (C_{13} - C_{23})\mu(x_1)RNF_s. \end{aligned} \tag{4.4}$$

The two-degree-of-freedom controller is a control system with a forward compensator besides the feedback controller. Model matching controller is a type of extensively used two-degree-of-freedom controller [29]. Its block diagram is shown in Fig. 4.2.

According to the diagram, we have

$$Y(s) = P(s)\big(P^{-1}(s)M(s)R(s) + K_b(s)M(s)R(s) - Y(s)\big). \tag{4.5}$$

---

[1]This chapter uses the content of [14], with permission from IEEE.

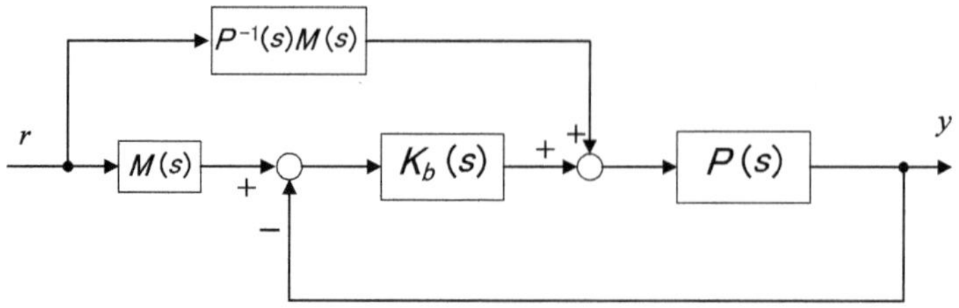

Fig. 4.2 Block diagram of a 2 DOF controller

If the to-be-controlled plant is modeled accurately enough, the transfer function from input to output turns to be

$$\frac{Y(s)}{R(s)} = M(s), \tag{4.6}$$

which means that the closed-loop system transfer function depends only on the dynamics of $M(s)$. Therefore, the quality of the output response can be improved by giving suitable $M(s)$. While, on the other hand, the feedback controller $K_b(s)$ can be designed for stability and robustness.

Ignoring the nonlinearities of the friction coefficient, and assuming it to be constant $\mu_0 = 0.13$, the system equation can be rewritten in the matrix form

$$\dot{\boldsymbol{x}} = \boldsymbol{A}\boldsymbol{x} + \boldsymbol{B}u + \boldsymbol{E}d \tag{4.7}$$

and the output equation is given by

$$y = \boldsymbol{C}\boldsymbol{x}, \tag{4.8}$$

where

$$\boldsymbol{x} = (\Delta\omega p_{cb})^T, \qquad y = \Delta\omega, \tag{4.9a}$$

$$\boldsymbol{A} = \begin{pmatrix} 0 & (C_{13} - C_{23})\mu_0 RNA \\ 0 & -\frac{1}{\tau_{cv}} \end{pmatrix}, \qquad \boldsymbol{B} = \begin{pmatrix} 0 & \dfrac{K_{cv}}{\tau_{cv}} \end{pmatrix}^T, \tag{4.9b}$$

$$\boldsymbol{E} = (1 \quad 0)^T, \qquad \boldsymbol{C} = (1 \quad 0), \tag{4.9c}$$

$$u = i_b, \qquad d = f(\omega_e, \omega_t, x_1). \tag{4.9d}$$

Based on the above linear state equations, the two-degree-of-freedom clutch slip controller is designed. The block diagram is given in Fig. 4.3, where $\Delta\omega^*$ is the initial speed reference and $\Delta\omega^{*'}$ is the modified reference as

$$\Delta\omega^{*'}(s) = \Delta\omega^*(s)M(s). \tag{4.10}$$

The difference of $\Delta\omega$ and $\Delta\omega^*$ depends on $M(s)$, which can be seen from the following results in Figs. 4.4 and 4.5.

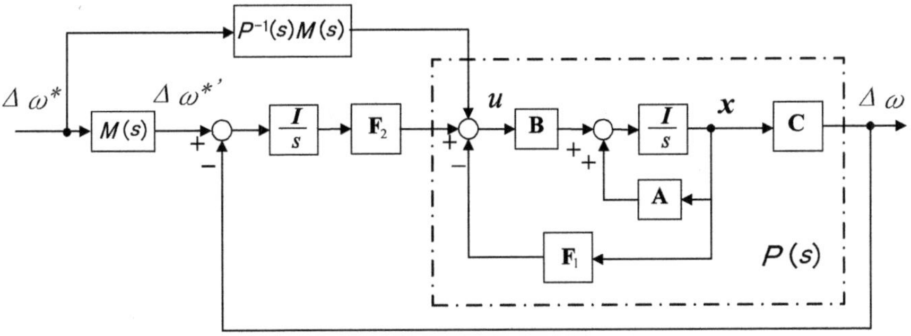

Fig. 4.3 2 DOF clutch slip controller

Fig. 4.4 Simulation results with $p_0 = 100$ without engine control

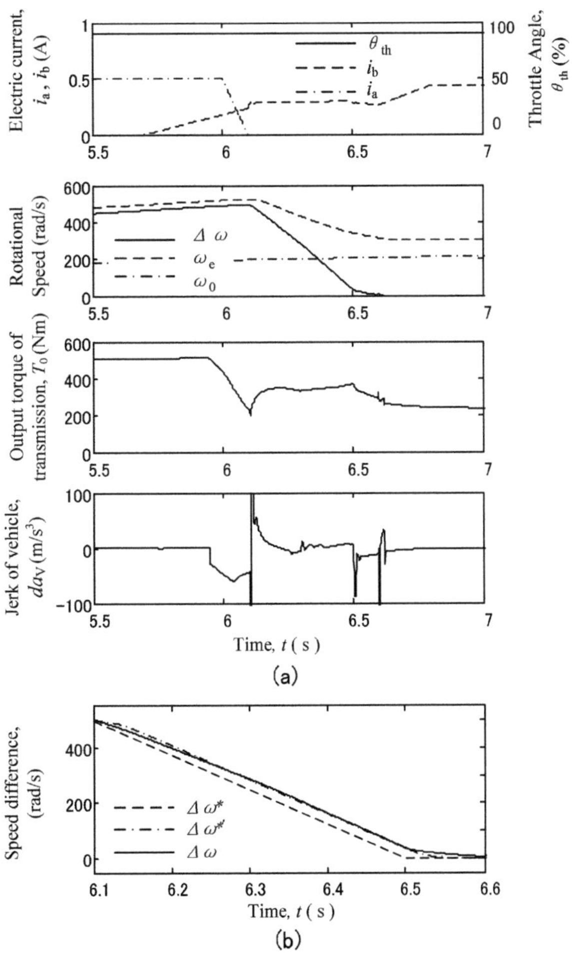

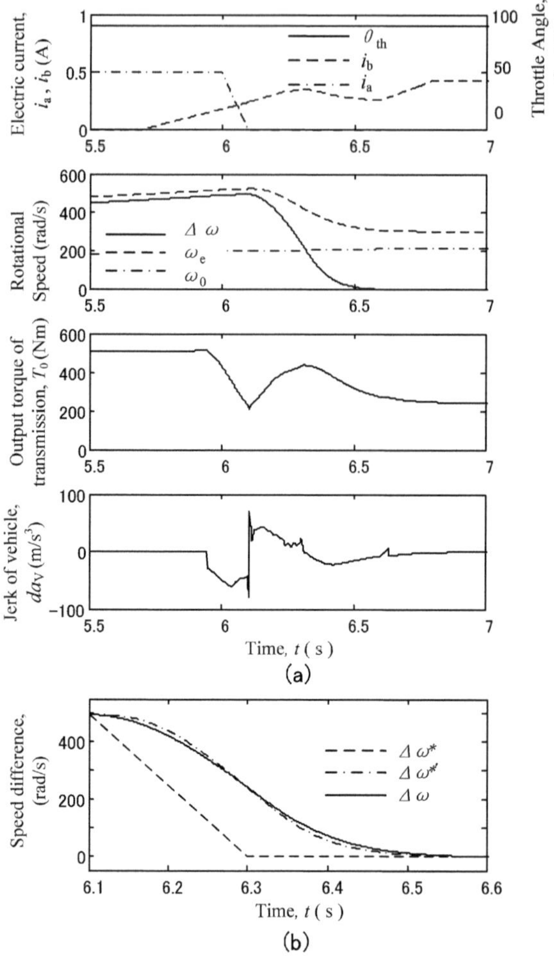

Fig. 4.5 Simulation results with $p_0 = 30$ without engine control

If no feedforward compensator is included, the gains $\boldsymbol{F}_1$ and $\boldsymbol{F}_2$ turn to be the commonly used linear servo system for the output tracking control [29, p. 847]. The robust pole assignment method proposed by [24] is used here to calculate $\boldsymbol{F}_1$ and $\boldsymbol{F}_2$, which is also the algorithm of the command "place" in the control toolbox of MATLAB.

After determining the feedback gains $\boldsymbol{F}_1$ and $\boldsymbol{F}_2$, the forward compensator can be derived. First, the part circled by the dashed line is labeled as $P(s)$. Being different from the $P(s)$ of Fig. 4.2, $P(s)$ defined here includes the state feedback besides the plant to be controlled. This treatment allows for convenient design of the feedforward compensator, and the later simulation results show its validity. Thus, $P(s)$ can be calculated as

$$P(s) = \boldsymbol{C}\left(s\boldsymbol{I} - \boldsymbol{A}'\right)^{-1}\boldsymbol{B} = \frac{P_n(s)}{P_d(s)} \tag{4.11}$$

with

$$A' = A - BF_1, \tag{4.12}$$

where $P_n(s)$ is a constant and $P_d(s)$ is a second-order polynomial of the Laplace variable s.

Because $P^{-1}(s)M(s)$ must be a proper transfer function, $M(s)$ is set as a third-order transfer function of the following form:

$$M(s) = \frac{p_0^3}{(s + p_0)^3}. \tag{4.13}$$

After getting $P(s)$ and $M(s)$, the feedforward compensator can be calculated as

$$P^{-1}(s)M(s) = \frac{P_d(s)M(s)}{P_n(s)}. \tag{4.14}$$

4.2.2 Simulation Results

The AMESim model constructed in the previous two chapters (please, refer to Sect. 2.2) is used to verify the designed controller. The 1st-to-2nd gear upshift is simulated, and during the inertia phase, the designed controller works to make the clutch slip speed track the desired trajectory shown in Fig. 4.1. The feedback gain used here is

$$F_1 = \begin{bmatrix} -7.8 \times 10^{-3} & 1.9 \times 10^{-6} \end{bmatrix}, \tag{4.15a}$$

$$F_2 = [-0.081]. \tag{4.15b}$$

Figure 4.4 shows the results with $p_0 = 100$, and no engine control is involved, i.e., the engine throttle is 90 % open, constantly during the gear shift. The gear shift process consists of three parts: before 6.1 s, the 1st gear torque phase; after 6.5 s, the 2nd gear torque phase, and between 6.1 and 6.5 s, the inertia phase. During the torque phases, the rotational speeds of shafts do not change much, while during the inertia phase, the rotational speeds change intensively because of the clutch slip.

The desired time of the inertia phase is set to be 0.4 s. The simulation results of the speed difference are shown in Fig. 4.4(b), and $\Delta\omega^*$ and $\Delta\omega^{*'}$ are also given. It can be seen that the slip speed $\Delta\omega$ can track reference value $\Delta\omega^{*'}$ with little error (refer to Fig. 4.1). The output torque of transmission T_0 and the jerk of vehicle da_v, i.e., the rate of the change of vehicle longitudinal acceleration, are shown to examine the shift shock. At the times the inertia phase begins and ends, there is an intensive change of the output torque and it results in a large jerk of the vehicle. The jerk during the 1st gear torque phase reached—60 m/s^3, the reason can be considered to be the large gear ratio difference between the 1st and 2nd gear.

Simulation results with $p_0 = 30$ are given in Fig. 4.5. The response of the reference model $M(s)$ is slow when $p_0 = 30$. Hence, the desired time for $\Delta\omega^*$ is set to

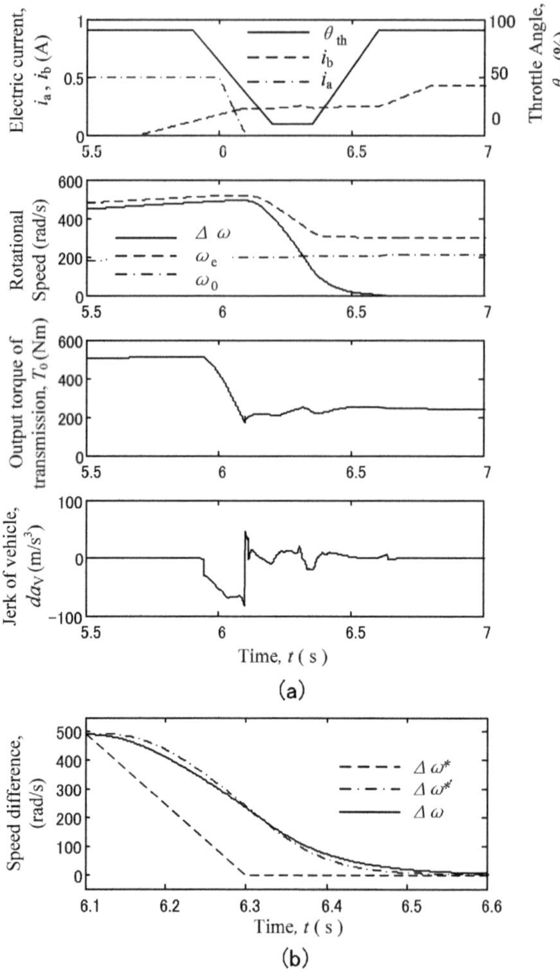

Fig. 4.6 Simulation results with $p_0 = 30$ with engine control

be of 0.2 s in order to finish the inertia phase in about 0.4 s. It can be seen that at the times the inertia phase begins and ends, there is no sharp change in the electric current of valve B, which results in a smoother output torque of the transmission and thus lesser jerk of the vehicle. Especially before 6.6 s, the time clutch B is locked up, the electric current of valve B decreases for a while, which makes the lock up of the clutch smooth.

Finally, the integrated control of the engine and transmission is used extensively on newly developed vehicles, for it can reduce clutch load, shorten shift time and improve fuel economy. Figure 4.6 is the simulated gear shift with the engine control. Only the throttle angle is controlled to cooperate with the shift process of the transmission. It can be seen that the speed difference can track reference value with enough precision despite the fluctuation of the engine throttle angle, which shows the robustness of the designed controller. In addition, the output torque of the trans-

Table 4.1 Friction work of clutch B, W_b, (in J)

Case in Fig. 4.4	Case in Fig. 4.5	Case in Fig. 4.6
27400	25400	17800

mission shows to have even less fluctuation compared to Fig. 4.5, thus even lesser jerk is obtained during the inertia phase.

Moreover, the friction work of clutch B during the gear shift is also calculated as

$$W_b = \int_{t0}^{tf} T_{cb} \Delta\omega \, dt \tag{4.16}$$

for the simulation cases discussed above. The result is shown in Table 4.1. One can see that the engine control greatly reduces the friction work of the clutch.

4.3 Nonlinear Feedback–Feedforward Controller

In recent years, differential flatness [10] was widely used for trajectory planning and tracking control [7, 23]. For a differentially flat system, if the trajectory for the flat outputs is given, the desired states and inputs can be derived as functions of the flat outputs and their derivatives. The advantages of flatness-based control include at least computational efficiency and avoidance of control saturation [7, 23]. *Moreover, the flatness-based control can improve the performance of an existing linear feedback control system by introducing a feedforward compensator, which is suitable for a large amount of presently produced automotive systems.*

This section will construct a nonlinear feedforward–feedback control where the feedforward control is designed based on differential flatness with the flat output being the clutch speed. In order to accommodate the model errors and the disturbances, a linear feedback controller is added. The feedback control is calculated through Linear Matrix Inequalities (LMIs) and convex optimization such that the control system is robust against the parameter uncertainties. The vehicle of interest is still the mid-size passenger vehicle equipped with an automatic transmission (AT).

The power-on 1st-to-2nd upshift and 2nd-to-1st downshift are considered here. During the upshift, the gear shift process is divided into the torque phase where the turbine torque is transferred from clutch A to clutch B and the inertia phase where clutch B is synchronized [18]. In the case of the downshift, the shift process starts from the inertia phase where clutch A is synchronized (realized through the disengagement of clutch B), followed by the torque phase where the torque is transferred from clutch B to clutch A. The downshift can be approximately regarded as a reverse process of the upshift. The focus is put on the clutch slip control during shift inertia phase. The reference trajectories of both maneuvers are shown in Fig. 4.7 [13].

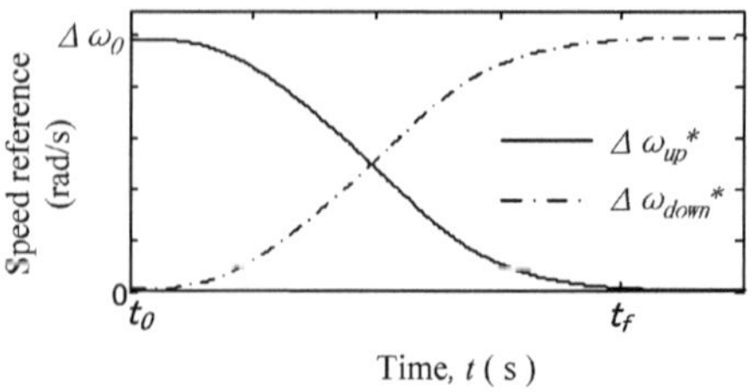

Fig. 4.7 Reference trajectories of the clutch speed in both upshift and downshift

Table 4.2 Parameters for controller design

C_{13}	Coefficient of clutch torque	$-25.85\ \frac{1}{\text{kg m}^2}$
C_{23}	Coefficient of clutch torque	$17.38\ \frac{1}{\text{kg m}^2}$
R	Effective radius of plates of clutch B	0.13 m
N	Plate number of clutch B	3
A	Piston area of clutch B	0.01 m^2
τ_{cv}	Time constant of valve B	0.04 s
K_{cv}	Gain of valve B	1.0 MPa/A
$\mu_{\min}$	Minimum friction coefficient	0.10
$\mu_{\max}$	Maximum friction coefficient	0.16

4.3.1 Clutch Slip Controller

In this section we will derive a nonlinear controller for the problem stated in Sect. 4.1. To do this, we rewrite the dynamics of the system (4.1a), (4.1b) for clutch slip control as follows:

$$\dot{x}_1 = a_1 \mu(x_1) x_2 + f_2(\omega_e, \omega_t, \omega_0), \tag{4.17a}$$

$$\dot{x}_2 = a_2 x_2 + b_{22} u, \tag{4.17b}$$

where $x_1 = \Delta\omega$; $x_2 = p_{cb2}/1000$ so that x_1 and x_2 are of the same order of magnitude. Moreover,

$$a_1 = (C_{13} - C_{23}) RNA \times 1000, \tag{4.18a}$$

$$a_2 = -\frac{1}{\tau_{cv}}, \tag{4.18b}$$

$$b_{22} = \frac{K_{cv}}{\tau_{cv} \times 1000}. \tag{4.18c}$$

The electric current of valve B is chosen as system input, i.e.,

$$u = i_b. \tag{4.19}$$

The parameter values are given in Table 4.2.

Nonlinear Feedforward Controller

The control objective is to track a given smooth trajectory of x_1, denoted as x_{1d}. To derive the feedforward control law,

$$y = x_1 \tag{4.20}$$

is chosen as the output. Differentiating (4.20) and inserting the state equations (4.17a), (4.17b) gives

$$\dot{y} = a_1\mu(x_1)x_2 + f_2(\omega_e, \omega_t, \omega_0), \tag{4.21a}$$

$$\ddot{y} = a_1\dot{\mu}(x_1)x_2 + a_1\mu(x_1)(a_2x_2 + b_{22}u) + \dot{f}_2(\omega_e, \omega_t, \omega_0). \tag{4.21b}$$

The relative degree of the system equals the system order, which implies that the clutch system is flat and $y = x_1$ is a flat output.

Hence, the state variables and the system input can be expressed by the following functions of the system output y and a finite number of its time derivatives:

$$x_1 = y, \tag{4.22a}$$

$$x_2 = \frac{\dot{y} - f_2(\omega_e, \omega_t, \omega_0)}{a_1\mu(y)}, \tag{4.22b}$$

$$u = \frac{\left(\frac{\dot{y} - f_2(y, \omega_e, \omega_t)}{a_1\mu(y)}\right)' - a_2\left(\frac{\dot{y} - f_2(y, \omega_e, \omega_t)}{a_1\mu(y)}\right)}{b_{22}}. \tag{4.22c}$$

Inserting the desired system output $y_d = x_{1d}$ and its time derivatives yields the nonlinear feedforward control

$$x_{2d} = \frac{\dot{y}_d - f_2(\omega_e, \omega_t, \omega_0)}{a_1\mu(y_d)}, \tag{4.23a}$$

$$u_f = \frac{\dot{x}_{2d} - a_2x_{2d}}{b_{22}}. \tag{4.23b}$$

Since $\mu(y_d)$ and $f_2(\omega_e, \omega_t, \omega_0)$ are given as lookup tables, it is impossible to obtain the explicit form of $\dot{x}_{2d}$. Hence, we apply the input shaping technique [5], which is occasionally named as Dynamic Surface Control (DSC) in the backstepping literature [33]. The result of (4.23a) is labeled as $\bar{x}_2$, and passed through a first order filter,

$$\tau\dot{x}_{2d} + x_{2d} = \bar{x}_2, x_{2d}(0) = \bar{x}_2(0), \tag{4.24}$$

which yields

$$\dot{x}_{2d} = \frac{\bar{x}_2 - x_{2d}}{\tau}. \tag{4.25}$$

Together with filter (4.24), Eqs. (4.23a), (4.23b) constitute the proposed nonlinear feedforward control scheme for the clutch slip control problem.

Linear Feedback Controller

Flatness-based control allows using simple linear feedback part in a two-degree-of-freedom control structure. Here, a P controller is adopted and it is designed such that the control system is robust against the parameter uncertainties.

Substituting the nonlinear feedforward control law (4.23a), (4.23b) into the state equations (4.17a), (4.17b), and assuming that $\mu(x_1) \approx \mu(x_{1d})$, we have

$$\dot{e}_1 = a_1 \mu(x_1) e_2, \tag{4.26a}$$

$$\dot{e}_2 = a_2 e_2 + b_{22} \Delta u, \tag{4.26b}$$

where $e_1 = x_1 - x_{1d}$, $e_2 = x_2 - x_{2d}$, and Δu is the linear feedback control law to be determined.

By considering modeling errors, including the torque and pressure computation error as additive input, we rewrite (4.26a), (4.26b) in state space form as

$$\dot{\boldsymbol{e}} = \boldsymbol{A}(x_1)\boldsymbol{e} + \boldsymbol{B}\Delta u + \boldsymbol{w}. \tag{4.27}$$

Matrix $\boldsymbol{A} = \left(\begin{smallmatrix} 0 & a_1\mu(x_1) \\ 0 & a_2 \end{smallmatrix}\right)$ varies in a convex envelope of a set of LTI models

$$(\boldsymbol{A}) = \mathrm{Co}\big\{(\boldsymbol{A}_1), (\boldsymbol{A}_2), \ldots, (\boldsymbol{A}_r)\big\}. \tag{4.28}$$

In this study, $r = 2$, i.e., there are two vertices, and they are determined by $\mu = \mu_{\min}$ and $\mu = \mu_{\max}$.

The question of the simultaneous stabilization amounts to finding a state feedback law $\Delta u = \boldsymbol{K}_p \boldsymbol{e}$ with $\boldsymbol{K}_p \in \mathbb{R}^{1\times 2}$ such that the eigenvalues $\lambda(\boldsymbol{A}_i - \boldsymbol{B}\boldsymbol{K}_p)$ belong to the left-half complex plane for both $i = 1, 2$. The problem has solutions if and only if there exists a matrix $\boldsymbol{X}_i \in \mathbb{R}^{2\times 2}$ such that the following matrix inequalities are feasible [4]:

$$\boldsymbol{X}_i > 0, \tag{4.29a}$$

$$(\boldsymbol{A}_i - \boldsymbol{B}\boldsymbol{K}_p)^T \boldsymbol{X}_i + \boldsymbol{X}_i(\boldsymbol{A}_i - \boldsymbol{B}\boldsymbol{K}_p) < 0. \tag{4.29b}$$

Because this is not a system of linear matrix inequalities (LMIs) in the variables $\boldsymbol{X}_i$ and $\boldsymbol{K}_p$, we assume that there exists a joint Lyapunov function $\boldsymbol{X}$ and introduce new variables $\boldsymbol{Y} = \boldsymbol{X}^{-1}$ and $\boldsymbol{K}_y = \boldsymbol{K}_p \boldsymbol{Y}$ [4, p. 100]. Moreover, we prefer low gains $\boldsymbol{K}_p$ due to robustness against noises, hence we restrict $\boldsymbol{Y}$ to be larger than a certain positive value and calculate a result as small as possible for $\boldsymbol{K}_y$. Moreover, for a rapid enough response, we define $\boldsymbol{A}_{2i} = \boldsymbol{A}_i + p_0 \boldsymbol{I}$, where $p_0 = 40$ for a settling time of less than 0.1 s [29, p. 221], to make the eigenvalues $\lambda(\boldsymbol{A}_i - \boldsymbol{B}\boldsymbol{K}_p)$ belong to the left of $s = -40$ in the complex plane. Then (4.29a), (4.29b) reads

$$\boldsymbol{Y} > \beta \boldsymbol{I}, \tag{4.30a}$$

$$\boldsymbol{A}_{2i}\boldsymbol{Y} + \boldsymbol{Y}\boldsymbol{A}_{2i}^T - \boldsymbol{B}\boldsymbol{K}_y - \boldsymbol{K}_y^T \boldsymbol{B}^T - 2p_0 Y < 0, \tag{4.30b}$$

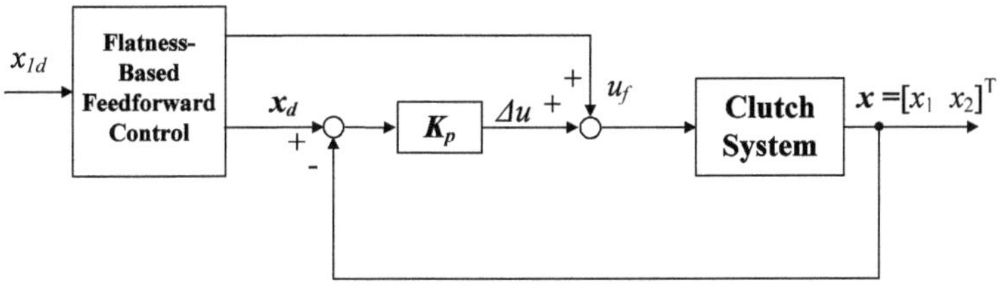

Fig. 4.8 Block diagram of the designed nonlinear feedforward–feedback control

with β being a small positive value. The result of $\boldsymbol{K}_y$ is then obtained through the following convex optimization

$$\min_{\alpha,\boldsymbol{K}_y,\boldsymbol{Y}} \alpha \quad \text{subject to LMIs (4.30a), (4.30b) and} \begin{pmatrix} \alpha & \boldsymbol{K}_y \\ \boldsymbol{K}_y^T & \boldsymbol{I} \end{pmatrix} \geq 0. \tag{4.31}$$

Consequently, the gain of the feedback controller is obtained as

$$\boldsymbol{K}_p = \boldsymbol{K}_y \boldsymbol{Y}^{-1}. \tag{4.32}$$

According to the parameter values shown in Table 4.2 and using $\beta = 0.01$, the convex optimization problem (4.31) can be solved, and the final result is

$$\boldsymbol{K}_p = (-0.0074, 0.0024), \tag{4.33}$$

which assures the stability of the controller under the variation of $\mu(\Delta\omega)$. The stability of the closed-loop system with the linear feedback can be analyzed in the framework of input-to-state stability. Following [12], it can be easily shown that the closed-loop system is input-to-state stable in the presence of bounded modeling and estimation errors.

The final control law is a combination of the feedforward and the feedback control

$$u = u_f + \Delta u, \tag{4.34}$$

and the structure of the complete controller is shown in Fig. 4.8.

Note that the clutch slip controller is built assuming that the clutch pressure p_{cb} is available. On production vehicles, however, not all transmissions are equipped with pressure sensors. Fortunately, a pressure observer can be constructed using the measured speed information. Please refer to Chap. 2 for detailed discussions.

4.3.2 Simulation Results

The powertrain simulation model is established in the environment of the commercial simulation software AMESim. The parameters used here represent a typical front-wheel-drive mid-size passenger car equipped with a 2000 cc injection gasoline engine. Please refer to Sect. 2.2 for details.

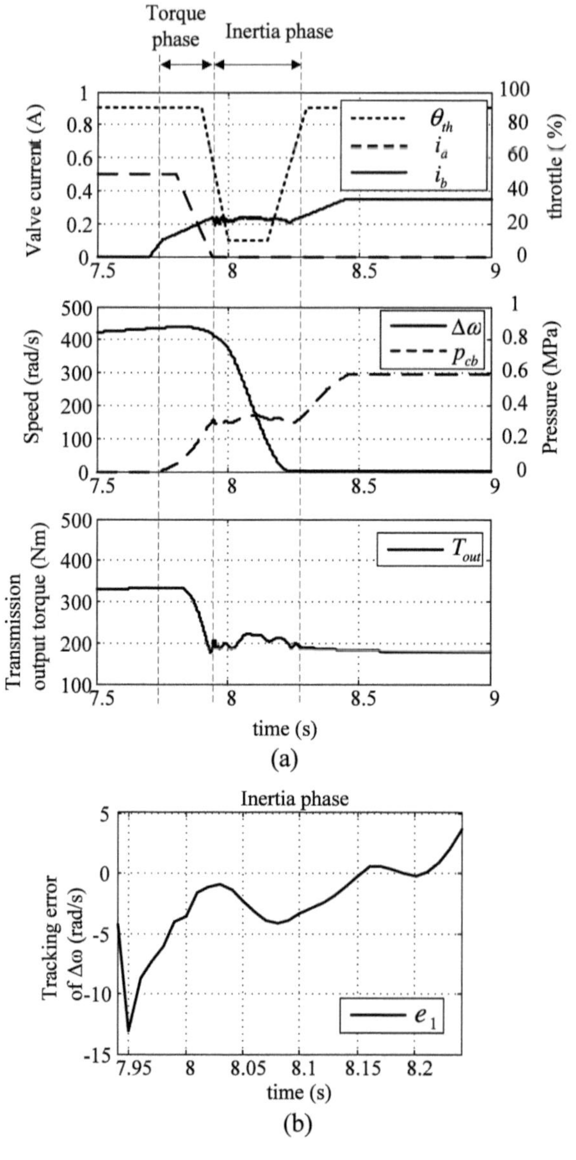

Fig. 4.9 Simulation results under the driving conditions used for controller design (torque characteristics of the engine and torque converter are standard; $m = 1500$ kg; $\theta_g = 0°$; $I_t = 0.06\ \text{kg m}^2$)

Gear Upshift

First, the major concern is put on the power-on 1st-to-2nd gear upshift process. Figure 4.9 gives the simulation results of the shift process under the driving conditions used in controller design. During the torque phase (between 7.7 and 7.94 s), we assume that the timing of releasing and applying the clutches has been well set, and refer to Chap. 3 for this part of work. In the inertia phase (between 7.94 and 8.24 s), clutch B is controlled by the designed controller. The parameters of the desired trajectory are $\Delta\omega_0 = 420$ rad/s and $(t_f - t_0) = 0.3$ s (see the solid line of Fig. 4.7). After the inertia phase (8.24 s), when the clutch speed difference is small enough, the valve current is increased by a pre-determined pattern to make the clutch lock up reliably.

The tracking error of the clutch slip speed during the inertia phase is plotted in Fig. 4.9(b). The maximum error is 13 rad/s, which is small enough for the vehicle clutch system. The transmission output torque T_{out} is given as well to examine the shift shock. We can see that there is no sharp change of T_{out}, which implies a smooth gear shift.

Then, in order to get an in-vehicle assessment of the proposed clutch slip control system, the designed controller is discretized by a sampling rate of 100 Hz [19] with zero-order hold discretization. The results are shown in Fig. 4.10. It should be noted that in the discrete implementations, the gain of the controller $\boldsymbol{K}_p$ has to be reduced in order to restrain the chatters caused by the sample time of 10 ms, i.e.,

$$\boldsymbol{K}_p = (-0.0025, 0.0015). \tag{4.35}$$

It can be seen that the sampling rate causes some small-magnitude chattering of the responses and a relatively large tracking error at the beginning of the inertia phase. The tracking error, however, decays rapidly. Moreover, because the chattering magnitude of the transmission output torque T_{out} is not large and the frequency is high, it is reasonable to believe that it will not noticeably affect driving comfort.

Finally, the controller is tested under different driving conditions. The results are shown in Fig. 4.11. We can see that, although there exist large modeling errors, the maximum tracking error is still about 20 rad/s and there is no large shift shock. The tracking error is no larger than that of Fig. 4.10 because the initial speed of the clutch is much less than that of Fig. 4.10. The initial speeds are different because in Fig. 4.11 the vehicle is fully loaded and is driving on a slope, which results in a different shift point.

Gear Downshift

In contrast with the upshift process, the downshift starts with the inertia phase. The results of the power-on 2nd-to-1st downshift are shown in Fig. 4.12, where the driving conditions are the same as those of Fig. 4.11.

During the inertia phase (from 9 to 9.3 s), the engine throttle follows the driver command, which is fixed at 90 % of the full throttle angle, and clutch B is controlled to track the reference (see the dashed line of Fig. 4.7) trajectory by the proposed controller. When clutch A is synchronized along with the disengagement of clutch B, the inertia phase finishes and the torque phase (from 9.3 to 9.6 s) begins. During the torque phase, it is assumed that the timing of releasing and applying the clutches has been well set.

Although the tracking error reaches 37 rad/s, which is larger than that of the upshift maneuver (because of the steady, large throttle opening angle), the shift process is finished in the required time, and there is no sharp oscillation of the output torque. It is considered as acceptable for the maneuver of the power-on downshift.

Fig. 4.10 Discrete implementation under the driving conditions used for controller design (torque characteristics of the engine and torque converter are standard; $m = 1500$ kg; $\theta_g = 0°$; $I_t = 0.06$ kg m^2)

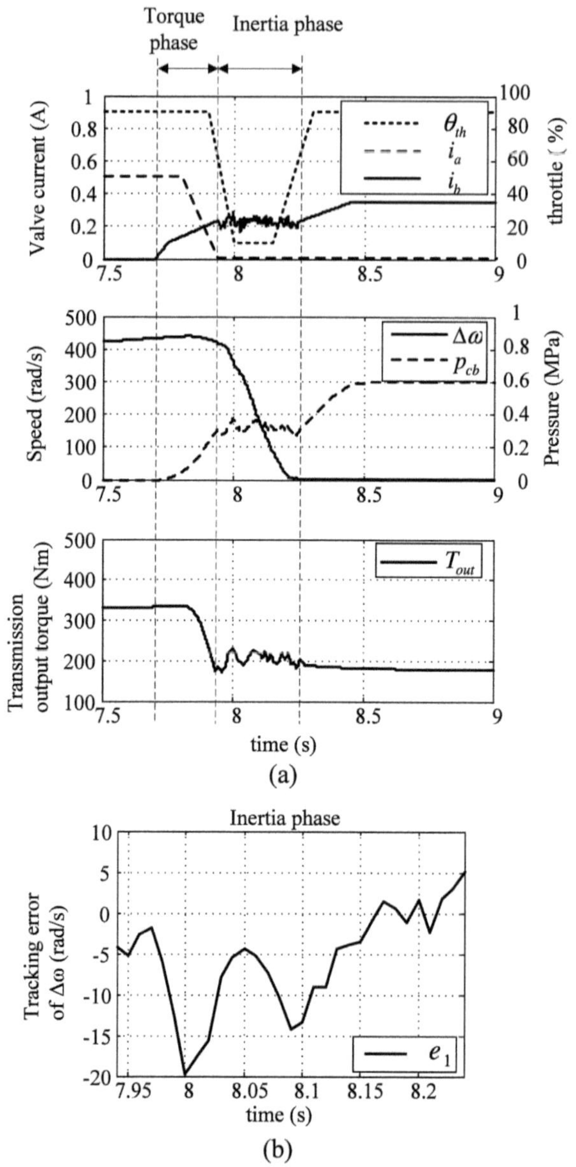

4.4 Backstepping Controller

In the above two sections, the controllers were designed for the clutch slip control. In this section, in order to improve the control performance, a backstepping nonlinear controller will be proposed, which is able to explicitly deal with the system nonlinearities [14].

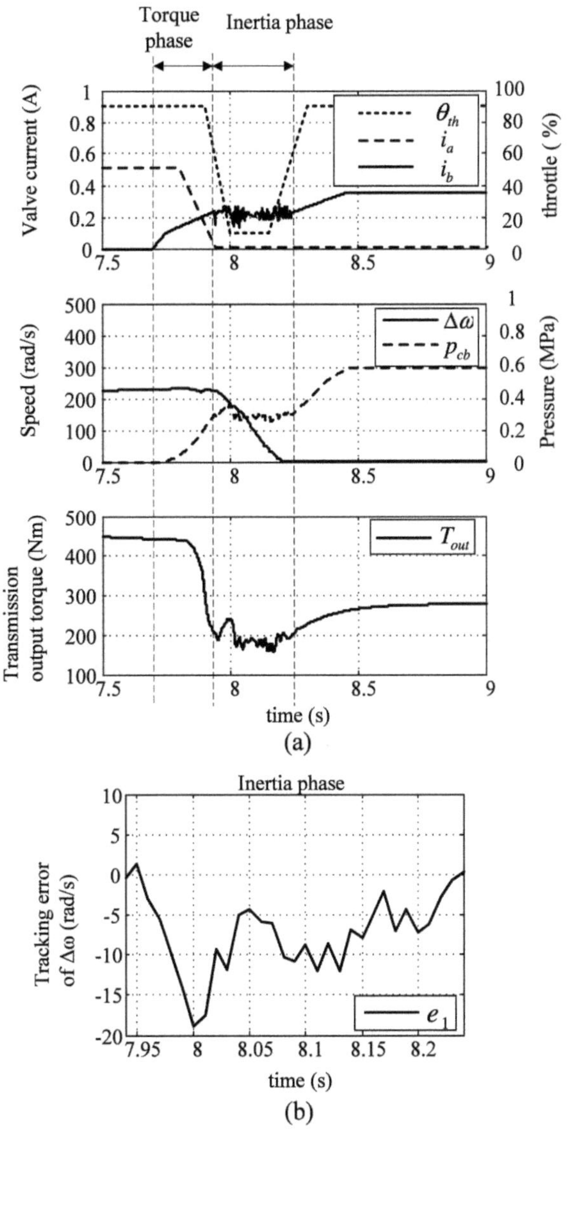

Fig. 4.11 Discrete implementation under different driving conditions (torque characteristics of the engine and torque converter are standard × 115 %; $m = 2000$ kg; $\theta_g = 5°$; $I_t = 0.1$ kg m^2)

4.4.1 Nonlinear Controller with ISS Property

In this section, we will make use of the backstepping technique to derive a nonlinear controller for the problem stated above. The robustness of the designed controller with respect to model errors is achieved in the sense of ISS property. To do this, we

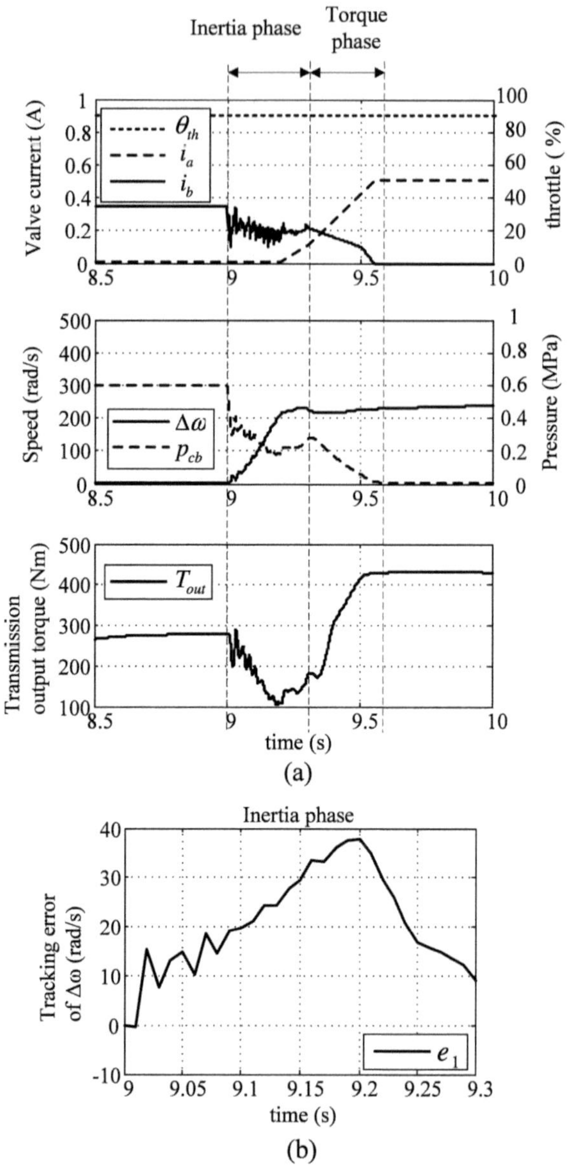

Fig. 4.12 Discrete implementation of power-on downshift (torque characteristics of the engine and torque converter are standard × 115 %; $m = 2000$ kg; road slope $= 5°$; $I_t = 0.1$ kg m^2)

rewrite the dynamical system (4.1a), (4.1b) for clutch slip control as follows:

$$\dot{x}_1 = a_1\mu(x_1)x_2 + f(x_1, \omega_e, \omega_t) + b_{11}w_1, \tag{4.36a}$$

$$\dot{x}_2 = a_2x_2 + b_{22}u + b_{21}w_2, \tag{4.36b}$$

where $x_1 = \Delta\omega$, $x_2 = p_{cb}/1000$ so that x_1 and x_2 are of the same order of magnitude, w_1 and w_2 summarize model uncertainties and b_{11} and b_{21} are known scaling factors, $f(x_1, \omega_e, \omega_t)$ is given in (4.2). Moreover,

$$a_1 = (C_{13} - C_{23})RNA \times 1000, \tag{4.37a}$$

$$a_2 = -\frac{1}{\tau_{cv}}, \tag{4.37b}$$

$$b_{22} = \frac{K_{cv}}{\tau_{cv} \times 1000}. \tag{4.37c}$$

The electric current of valve B is chosen as system input

$$u = i_b. \tag{4.38}$$

The control objective is to track a given smooth trajectory of x_1, denoted as x_{1d}. To do this, we define the tracking error as $e_1 = x_1 - x_{1d}$. We first consider x_2 as a virtual control input and determine a control law of x_{2d} such that the tracking error dynamics is input-to-state stable with respect to the disturbance w_1. Since $x_2 \neq x_{2d}$ indeed, we define $e_2 = x_2 - x_{2d}$ and rewrite the first equation of (4.36a), (4.36b) as follows:

$$\dot{x}_1 = a_1\mu(x_1)(e_2 + x_{2d}) + f(x_1, \omega_e, \omega_t) + b_{11}w_1. \tag{4.39}$$

Letting $V_1 = \frac{1}{2}e_1^2$ and differentiating it along (4.39), we infer

$$\dot{V}_1 = e_1\dot{e}_1 = e_1\big[a_1\mu(x_1)(e_2 + x_{2d}) + f + b_{11}w_1 - \dot{x}_{1d}\big]. \tag{4.40}$$

Using Young's Inequality [25], the above equality becomes

$$\begin{aligned}
\dot{V}_1 &\le e_1\big[a_1\mu(x_1)x_{2d} + f - \dot{x}_{1d}\big] + |b_{11}|\kappa_1 e_1^2 + \frac{|b_{11}|}{4\kappa_1}w_1^2 \\
&\quad + |a_1\mu(x_1)|\kappa_2 e_1^2 + \frac{|a_1\mu(x_1)|}{4\kappa_2}e_2^2 \\
&= e_1\big[a_1\mu(x_1)x_{2d} + f - \dot{x}_{1d} + |b_{11}|\kappa_1 e_1 + \big|a_1\mu(x_1)\big|\kappa_2 e_1\big] \\
&\quad + \frac{|b_{11}|}{4\kappa_1}w_1^2 + \frac{|a_1\mu(x_1)|}{4\kappa_2}e_2^2,
\end{aligned} \tag{4.41}$$

where $\kappa_1 > 0$ and $\kappa_2 > 0$. Note that $\mu(x_1) \neq 0$ in our case and, if the clutch slip controller is well designed, the sign of x_1 does not change during one independent shift inertia phase. Hence, we can choose

$$x_{2d} = \frac{-\kappa_3 e_1 - f + \dot{x}_{1d} - |b_{11}|\kappa_1 e_1 - |a_1\mu(x_1)|\kappa_2 e_1}{a_1\mu(x_1)} \tag{4.42}$$

with $\kappa_3 > 0$ to guarantee

$$\dot{V}_1 \le -\kappa_3 e_1^2 + \frac{|b_{11}|}{4\kappa_1}w_1^2 + \frac{|a_1\mu(x_1)|}{4\kappa_2}e_2^2. \tag{4.43}$$

Hence, we conclude that the control law (4.42) renders the subsystem (4.39) input-to-state stable with respect to w_1 and e_2.

Since the second Eq. (4.36b) is of a linearly parametrized form, we can then use

$$\dot{e}_2 = a_2 e_2 + b_{22} v + b_{21} w_2 \tag{4.44}$$

to design the control law, where $v = u - u_d$ with

$$u_d = \frac{\dot{x}_{2d} - a_2 x_{2d}}{b_{22}} \tag{4.45}$$

for $b_{22} \neq 0$. Because K_{cv} and τ_{cv} are simplified as positive constants for the controller design (see Sect. 4.4.3), $b_{22} \neq 0$ is satisfied in our case.

Let $V_2 = \frac{1}{2} e_2^2$ and infer

$$\begin{aligned}\dot{V}_2 = e_2 \dot{e}_2 &\leq a_2 e_2^2 + b_{22} e_2 v + |b_{21}|\kappa_4 e_2^2 + \frac{|b_{21}|}{4\kappa_4} w_2^2 \\ &= e_2(a_2 e_2 + b_{22} v + |b_{21}|\kappa_4 e_2) + \frac{|b_{21}|}{4\kappa_4} w_2^2,\end{aligned} \tag{4.46}$$

where $\kappa_4 > 0$. If the control law is chosen as

$$v = -\frac{\kappa_5 + a_2 + |b_{21}|\kappa_4}{b_{22}} e_2 \tag{4.47}$$

and hence

$$u = u_d - \frac{\kappa_5 + a_2 + |b_{21}|\kappa_4}{b_{22}} e_2, \tag{4.48}$$

then (4.46) becomes

$$\dot{V}_2 \leq -\kappa_5 e_2^2 + \frac{|b_{21}|}{4\kappa_4} w_2^2, \tag{4.49}$$

where $\kappa_5 > 0$.

Finally, with the control law given by (4.42), (4.45) and (4.48), the total closed-loop error system can be written as

$$\dot{e}_1 = -\big(\kappa_3 + |b_{11}|\kappa_1 + \big|a_1\mu(x_1)\big|\kappa_2\big)e_1 + a_1\mu(x_1)e_2 + b_{11}w_1, \tag{4.50a}$$

$$\dot{e}_2 = -\big(\kappa_5 + |b_{21}|\kappa_4\big)e_2 + b_{21}w_2, \tag{4.50b}$$

for which we define $V = V_1 + V_2$. Then, by exploiting (4.43) and (4.49), we have

$$\dot{V} = \dot{V}_1 + \dot{V}_2 \leq -\kappa_3 e_1^2 + \left(\frac{|a_1\mu(x_1)|}{4\kappa_2} - \kappa_5\right)e_2^2 + \frac{|b_{11}|}{4\kappa_1} w_1^2 + \frac{|b_{21}|}{4\kappa_4} w_2^2. \tag{4.51}$$

Hence, we conclude the following results for the property of the error dynamics of the designed controller.

Theorem 4.1 *Suppose that*

- $\kappa_i > 0, i = 1, \ldots, 5$;
- $\frac{|a_1|\mu(x_1)}{4\kappa_2} - \kappa_5 < 0$.

Then, the tracking error dynamics of the system under controller (4.42), (4.45) *and* (4.48) *is input-to-state stable if* w_1 *and* w_2 *are bounded in amplitude, i.e.,* $w_1, w_2 \in \mathcal{L}_\infty$.

Proof It follows from (4.51) that

$$\dot{V} \leq -\min\left\{\kappa_3, \kappa_5 - \frac{|a_1\mu(x_1)|}{4\kappa_2}\right\}\|\boldsymbol{e}\|_2^2 + \max\left\{\frac{|b_{11}|}{4\kappa_1}, \frac{|b_{21}|}{4\kappa_4}\right\}\|\boldsymbol{w}\|_2^2, \tag{4.52}$$

where $\boldsymbol{e} = [e_1, e_2]^T$ and $\boldsymbol{w} = [w_1, e_2]^T$, which shows that the error dynamics admits the input-to-state stability property [25, p. 503] if the model error $\boldsymbol{w}$ is supposed to be bounded in amplitude. □

Remark 4.1 It follows from (4.49) that

$$\dot{V}_2 \leq -2\kappa_5 V_2 + \frac{|b_{21}|}{4\kappa_4} w_2^2. \tag{4.53}$$

Upon multiplication of (4.53) by $e^{2\kappa_5 t}$, it becomes

$$\frac{d}{dt}\left(V_2 e^{2\kappa_5 t}\right) \leq \frac{|b_{21}|}{4\kappa_4} w_2^2 e^{2\kappa_5 t}. \tag{4.54}$$

Integrating it over $[0, t]$ leads to

$$V_2(t) \leq V_2(0)e^{-2\kappa_5 t} + \frac{|b_{21}|}{4\kappa_4}\int_0^t e^{-2\kappa_5(t-\tau)} w_2(\tau)^2\, d\tau, \tag{4.55}$$

and hence

$$\left\|e_2(t)\right\|^2 \leq \left\|e_2(0)\right\|^2 e^{-2\kappa_5 t} + \frac{|b_{21}|}{2\kappa_4}\int_0^t e^{-2\kappa_5(t-\tau)} w_2(\tau)^2\, d\tau. \tag{4.56}$$

If w_2 is bounded in amplitude, i.e., $w_2 \in \mathcal{L}_\infty$, (4.55) becomes

$$\left\|e_2(t)\right\|^2 \leq \left\|e_2(0)\right\|^2 e^{-2\kappa_5 t} + \frac{|b_{21}|\|w_2\|_\infty^2}{2\kappa_4}\int_0^t e^{-2\kappa_5(t-\tau)}\, d\tau, \tag{4.57}$$

which implies that the tracking error e_2 is bounded as

$$\left\|e_2(t)\right\|^2 \leq \frac{|b_{21}|\|w_2\|_\infty^2}{4\kappa_4\kappa_5} \quad \text{as } t \to \infty. \tag{4.58}$$

Applying a similar procedure for the tracking error e_1, we have

$$\|e_1(t)\|^2 \le \|e_1(0)\|^2 e^{-2\kappa_3 t} + \frac{|b_{11}|}{2\kappa_1}\int_0^t e^{-2\kappa_3(t-\tau)} w_1(\tau)^2\,d\tau + \frac{|a_1|}{2\kappa_2}\int_0^t e^{-2\kappa_3(t-\tau)} e_2(\tau)^2 \big|\mu\big(x_1(\tau)\big)\big|\,d\tau \tag{4.59}$$

and

$$\|e_1(t)\|^2 \le \frac{|b_{11}|\|w_1\|_\infty^2}{4\kappa_1\kappa_3} + \frac{|a_1\mu_{\max}|\|e_2\|_\infty^2}{4\kappa_2\kappa_3} \quad \text{as } t \to \infty, \tag{4.60}$$

since w_1 is bounded in amplitude.

Therefore, it follows from (4.56) and (4.59) that κ_3 and κ_5 can be chosen according to the required decay rate of the errors. And from (4.58) and (4.60), one may choose larger κ_1, κ_2 and κ_4 to reduce tracking offsets. However, one should notice that the larger these tuning parameters, the higher the controller gain. That is, the choice of $\kappa_i, i = 1, 2, 4$, requires the trade-off between the tracking offset and the controller gain.

Remark 4.2 We stress that (4.58) and (4.60) give just upper bounds of the tracking offsets, if the bound of the model error is given. The real offset could be much smaller, due to the multiple use of inequalities in the above derivation. Moreover, we note that (4.50b) is linear time-invariant. Hence, we can compute the tracking offset for e_2 by the use of the final-value theorem [29]

$$e_2(\infty) = \lim_{s\to 0} s \cdot \frac{b_{21}}{s + \kappa_5 + |b_{21}|\kappa_4} w_2(s), \tag{4.61}$$

which implies $e_2(\infty) = 0$ when w_2 is an impulse signal and

$$e_2(\infty) = \frac{b_{21}}{\kappa_5 + |b_{21}|\kappa_4}\bar{w}_2 \tag{4.62}$$

when w_2 is a step signal, where $\bar{w}_2$ is the step magnitude.

According to Theorem 4.1, Remarks 4.1 and 4.2, we now give the following procedure to design the clutch slip controller in the form of (4.42), (4.45) and (4.48):

Step 1: Choose tuning parameters $\kappa_3 > 0$ and $\kappa_5 > 0$ according to the required decay rate of the tracking error for e_1 (Eq. (4.56)) and e_2 (Eq. (4.59)), respectively;

Step 2: Assume w_2 to be a step signal, determine $\kappa_4 > 0$ according to (4.62) and the required offset of e_2;

Step 3: For given bounds of w_1 and e_2, choose $\kappa_1 > 0$ and $\kappa_2 > \frac{|a_1\mu(x_1)|}{4\kappa_5}$ such that the upper bound of the tracking offset $e_1(\infty)$ computed by (4.60) is acceptable.

4.4.2 Implementation Issues

Stability Analysis Considering Pressure Observer

Until now, the clutch slip controller was built assuming that the clutch pressure x_2 was available. When the pressure observer is involved, the stability of the controller should be guaranteed under the interaction between the observer and the controller. With the estimated clutch pressure, denoted as $\hat{x}_2$, the implemented control law (4.48) is given by

$$u_{im} = u_d - \frac{\kappa_5 + a_2 + |b_{21}|\kappa_4}{b_{22}}(\hat{x}_2 - x_{2d}). \tag{4.63}$$

By defining the estimated error by $e_3 = x_2 - \hat{x}_2$, the above control law becomes

$$u_{im} = u_d - \frac{\kappa_5 + a_2 + |b_{21}|\kappa_4}{b_{22}}(e_2 - e_3). \tag{4.64}$$

Consequently, the error dynamics (4.50a), (4.50b) should be rewritten as

$$\dot{e}_1 = -\big(\kappa_3 + |b_{11}|\kappa_1 + |a_1|\mu(x_1)\kappa_2\big)e_1 + a_1\mu(x_1)e_2 + b_{11}w_1, \tag{4.65a}$$

$$\dot{e}_2 = -\big(\kappa_5 + |b_{21}|\kappa_4\big)e_2 + \big(\kappa_5 + a_2 + |b_{21}|\kappa_4\big)e_3 + b_{21}w_2. \tag{4.65b}$$

If we define $w_2' = w_2 + \frac{\kappa_5 + a_2 + |b_{21}|\kappa_4}{b_{21}} e_3$, we get

$$\dot{e}_2 = -\big(\kappa_5 + |b_{21}|\kappa_4\big)e_2 + b_{21}w_2', \tag{4.66}$$

and the derivative of the Lyapunov function, i.e., Eq. (4.51), becomes

$$\dot{V} = \dot{V}_1 + \dot{V}_2 \le -\kappa_3 e_1^2 + \left(\frac{|a_1|\mu(x_1)}{4\kappa_2} - \kappa_5\right)e_2^2 + \frac{|b_{11}|}{4\kappa_1}w_1^2 + \frac{|b_{21}|}{4\kappa_4}w_2'^2. \tag{4.67}$$

From Chap. 2, the pressure observer is designed as

$$\eta = \hat{x}_2 - Lx_1, \tag{4.68a}$$

$$\dot{\eta} = \big(a_2 - La_1\mu(x_1)\big)(\eta + Lx_1) + b_{22}u - Lf, \tag{4.68b}$$

where L is the observer gain, and the estimation error is derived as

$$\dot{e}_3 = \big(a_2 - La_1\mu(x_1)\big)e_3 + b_{21}w_2 - Lb_{11}w_1. \tag{4.69}$$

It is shown in Theorem 2.1 that e_3 is bounded when w_1 and w_2 are bounded. Consequently, w_2' is also bounded. Therefore, with (4.67) and (4.69), the whole system is proved to be input-to-state stable according to Theorem B.1 in Appendix B, wherein

$$\alpha(\boldsymbol{e}) = -\kappa_3 e_1^2 + \left(\frac{|a_1|\mu(x_1)}{4\kappa_2} - \kappa_5\right)e_2^2, \tag{4.70a}$$

$$\gamma(\boldsymbol{w}) = \frac{|b_{11}|}{4\kappa_1}w_1^2 + \frac{|b_{21}|}{4\kappa_4}w_2'^2. \tag{4.70b}$$

Reference Trajectory

There are many methods to define the reference trajectory shown in Fig. 4.1, such as transfer functions or polynomials. Here, the following 3rd-order polynomial

$$x_{1d}(t) = \frac{2\Delta\omega_0}{(t_f - t_0)^3}(t - t_0)^3 - \frac{3\Delta\omega_0}{(t_f - t_0)^2}(t - t_0)^2 + \Delta\omega_0 \tag{4.71}$$

is adopted to meet the requirements shown in Fig. 4.1, i.e., $x_{1d}(t_0) = \Delta\omega_0$, $x_{1d}(t_f - t_0) = 0$ and $\dot{x}_{1d}(t_0) = \dot{x}_{1d}(t_f - t_0) = 0$. Please refer to Fig. 4.1 for the definitions of $\Delta\omega_0$, t_0 and t_f. Then, $\dot{x}_{1d}$ can be calculated as

$$\dot{x}_{1d}(t) = \frac{6\Delta\omega_0}{(t_f - t_0)^3}(t - t_0)^2 - \frac{6\Delta\omega_0}{(t_f - t_0)^2}(t - t_0). \tag{4.72}$$

Moreover, $\dot{x}_{2d}$ is needed for the implementation of (4.45). Since $\mu(x_1)$ and $f(x_1, \omega_e, \omega_t)$ are given as maps, it is impossible to obtain the explicit form of $\dot{x}_{2d}$ by differentiating (4.42). Hence, we apply the input shaping technique [5], which is occasionally named as Dynamic Surface Control (DSC) in the backstepping literature [33]. The result of (4.42) is labeled as $\bar{x}_2$, and passed through a first order filter,

$$\tau_2\dot{x}_{2d} + x_{2d} = \bar{x}_2, \quad x_{2d}(0) = \bar{x}_2(0), \tag{4.73}$$

which yields

$$\dot{x}_{2d} = \frac{\bar{x}_2 - x_{2d}}{\tau_2}. \tag{4.74}$$

4.4.3 Controller of the Considered Clutch System

Together with (4.72) and (4.73), Eqs. (4.42), (4.45) and (4.48) constitute the proposed clutch slip controller. Now we present the concrete controller with all physical and tuning parameters. It is for simplicity assumed that (τ_{cv}, K_{cv}) are constant, and the scaling factors b_{11} and b_{21} are set to be 1.

Nonlinear functions $f(\Delta\omega, \omega_e, \omega_t)$ and $\mu(\Delta\omega)$ are given as lookup tables in the controller. The map of μ is shown in Fig. 4.13, while f is given by 3rd-order maps, and examples when $\omega_e = 200$ rad/s and $\omega_e = 500$ rad/s are shown in Fig. 4.14.

Following the procedure given in Sect. 4.4.2, we first choose κ_3 and κ_5 to meet the requirement for the desired decay rate of the tracking errors e_1 and e_2, respectively. It is desired that the error converges in 0.1 s, and we consider the settling time as 4 time constants [29], which implies $\frac{4}{\kappa_{3,5}} = 0.1$ and results in

$$\kappa_3 = 40 \quad \text{and} \quad \kappa_5 = 40. \tag{4.75}$$

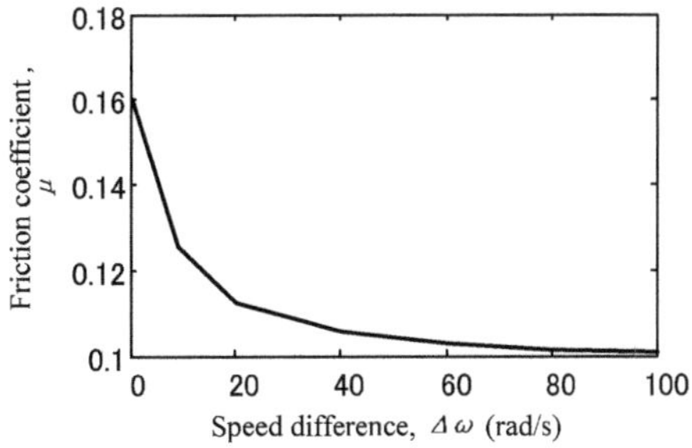

Fig. 4.13 Friction characteristics of clutch plates

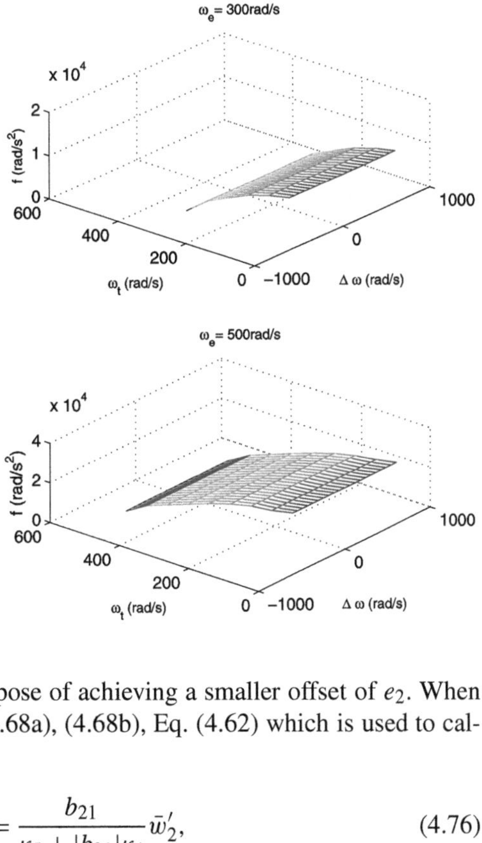

Fig. 4.14 Maps of f when $\omega_e = 300$ and 500 rad/s

Then, we choose κ_4 with the purpose of achieving a smaller offset of e_2. When considering the pressure observer (4.68a), (4.68b), Eq. (4.62) which is used to calculate $e_2(\infty)$ should be rewritten as

$$e_2(\infty) = \frac{b_{21}}{\kappa_5 + |b_{21}|\kappa_4} \bar{w}'_2, \tag{4.76}$$

where $\bar{w}'_2$ is the magnitude of $w'_2 = w_2 + \frac{\kappa_5 + a_2 + |b_{21}|\kappa_4}{b_{21}} e_3$ (see (4.66) for reference). It is assumed here that the model error w'_2 is mainly caused by the pressure estimation error e_3. According to Chap. 2, the maximum value of the pressure estimation error is about 0.06 MPa. Therefore, following (4.76), we have

$$e_2(\infty) = \frac{60(\kappa_5 + a_2 + \kappa_4)}{\kappa_5 + |b_{21}|\kappa_4}. \tag{4.77}$$

Choosing

$$\kappa_4 = 10 \tag{4.78}$$

results in

$$\|e_2(\infty)\| \leq 30(\times 1000 \text{ Pa}), \tag{4.79}$$

which is less than 0.06 MPa, and it is regarded as acceptable for the considered uncertainty.

In order to choose the values of κ_1 and κ_2, we now roughly calculate the bound of the modeling error w_1. Since powertrain systems admit highly nonlinear, complex dynamics and various uncertainties, it is indeed very difficult, if not impossible, to obtain a comprehensive estimate of the modeling error bound. Hence, we only consider some major uncertainties as an example to estimate the value of w_1. The major uncertainties considered here are the estimation error of the turbine torque T_t, and the variations of the road grade θ_g and the vehicle mass m, which affect the driving resistance T_{ve}.

The estimation precision of turbine torque T_t relies on the torque converter modeling [19, 32, 35]. Here the estimation error of the turbine torque is assumed to be $\|\tilde{T}_t\|_\infty = 20$ Nm, which is about ± 12 % of the maximum engine torque. Moreover, it is assumed that the vehicle mass is increased from 1500 kg, the nominal value used for controller design, to $m_{\text{full}} = 2000$ kg, the fully loaded mass, and the road grade angle is increased from 0 to 5 degrees. Hence, $\|w_1\|_\infty$ can be approximately estimated as

$$\|w_1\|_\infty \leq |C_{11} - C_{21}| \|\tilde{T}_t\|_\infty + |C_{14} - C_{24}| \frac{m_{\text{full}} g \sin(\theta_g) R_w}{R_{df}} = 543, \tag{4.80}$$

where $C_{11} - C_{21} = 25.85$, $C_{14} - C_{24} = 0.0911$, and R_w is the tire radius.

Then, according to (4.60), the bound of $e_1(\infty)$ can be estimated as

$$\begin{aligned} \|e_1(\infty)\|^2 &\leq \frac{|b_{11}| \|w_1\|_\infty^2}{4\kappa_1\kappa_3} + \frac{|a_1 \mu_{\max}| \|e_2\|_\infty^2}{4\kappa_2\kappa_3} \\ &= \frac{1816}{\kappa_1} + \frac{152}{\kappa_2}, \end{aligned} \tag{4.81}$$

where we use the value of $\|e_2(\infty)\|$ to replace $\|e_2\|_\infty$, since by design the initial error of e_2 decays rapidly and exponentially.

On the other hand, κ_2 should satisfy $\frac{|a_1|\mu(x_1)}{4\kappa_2} - \kappa_5 < 0$, i.e.,

$$\kappa_2 > \frac{|a_1 \mu_{\max}|}{4\kappa_5} = 0.17. \tag{4.82}$$

If the acceptable control offset is set to be

$$\|e_1(\infty)\| \leq 15 \text{ rad/s}, \tag{4.83}$$

which is precise enough for this application, κ_1 and κ_2 can be chosen as

$$\kappa_1 = 20 \quad \text{and} \quad \kappa_2 = 2, \tag{4.84}$$

and the resulting control offset satisfies

$$\left\| e_1(\infty) \right\| \leq 12.9 \text{ rad/s}. \tag{4.85}$$

4.4.4 Simulation Results

Continuous Implementation

Figure 4.15 gives the simulation results of the shift process under the driving conditions used for controller design. During the shift process, the engine throttle angle is adjusted to cooperate with the transmission shift. Note that in the real world, when the throttle is changed, the engine dynamics is much more complex. The complex engine torque control loop is not included and it is assumed that the engine torque is already well controlled to track the desired value. It can be seen that during the torque phase (between 7.7 and 7.94 s) the rotational speeds of the shafts do not change much, whereas during the inertia phase (between 7.94 and 8.34 s), the rotational speeds change intensively because of the clutch slip.

During the torque phase, we assume that the timing of releasing and applying the clutches has been well set, and this part of work is omitted here. In the inertia phase, clutch B is controlled by the designed controller. The parameters of the desired trajectory are $\Delta\omega_0 = 420$ rad/s and $(t_f - t_0) = 0.4$ s (see Fig. 4.1). After the inertia phase, when the clutch speed difference is small enough, the valve current is increased by a pre-determined pattern to make the clutch lock up reliably. In order to examine the shift shock, the transmission output torque and vehicle acceleration are given as well.

The tracking error of the clutch slip speed during the inertia phase is plotted in Fig. 4.15(b), too. The maximum value of the error reads 7.5 rad/s, which is small enough for the vehicle clutch system. It should be pointed out that the shift process operates under the same driving conditions of controller design, but the stiffness of the drive shaft and a tire model with longitudinal slip are considered in the simulation model, while these are ignored in the model for designing the controller. Moreover, the time-delay in control and time-varying parameters are also considered in the simulation model of the proportional valve.

Furthermore, the proposed controller is tested under different driving conditions. The results are shown in Figs. 4.16 and 4.17, where the driving condition settings are as follows:

- (Figure 4.16) The torque characteristic of the engine is enlarged by 15 %, and consequently, the capacity of the torque converter is also enlarged; the vehicle mass is increased from 1500 to 2000 kg, and the road grade angle is changed from 0 to 5 degrees; the inertia of the turbine shaft is changed from 0.06 to 0.1 kg m^2;

Fig. 4.15 Simulation results under the driving conditions used for controller design (torque characteristics of engine and torque converter are standard; $m = 1500$ kg; $\theta_g = 0°$; $I_t = 0.06$ kg m^2)

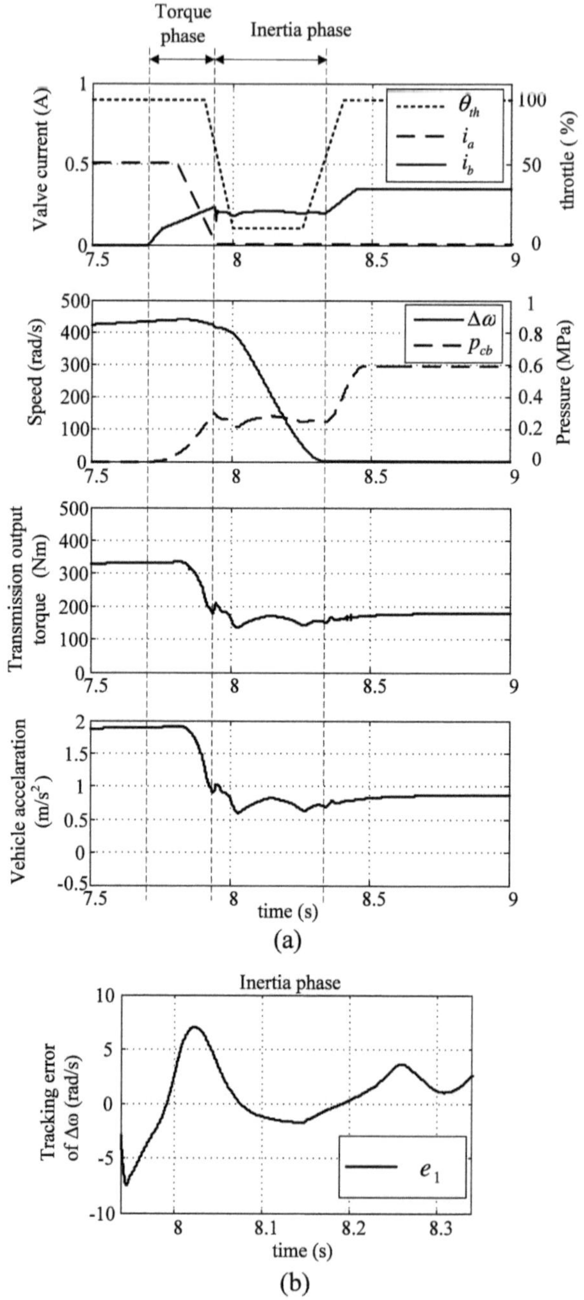

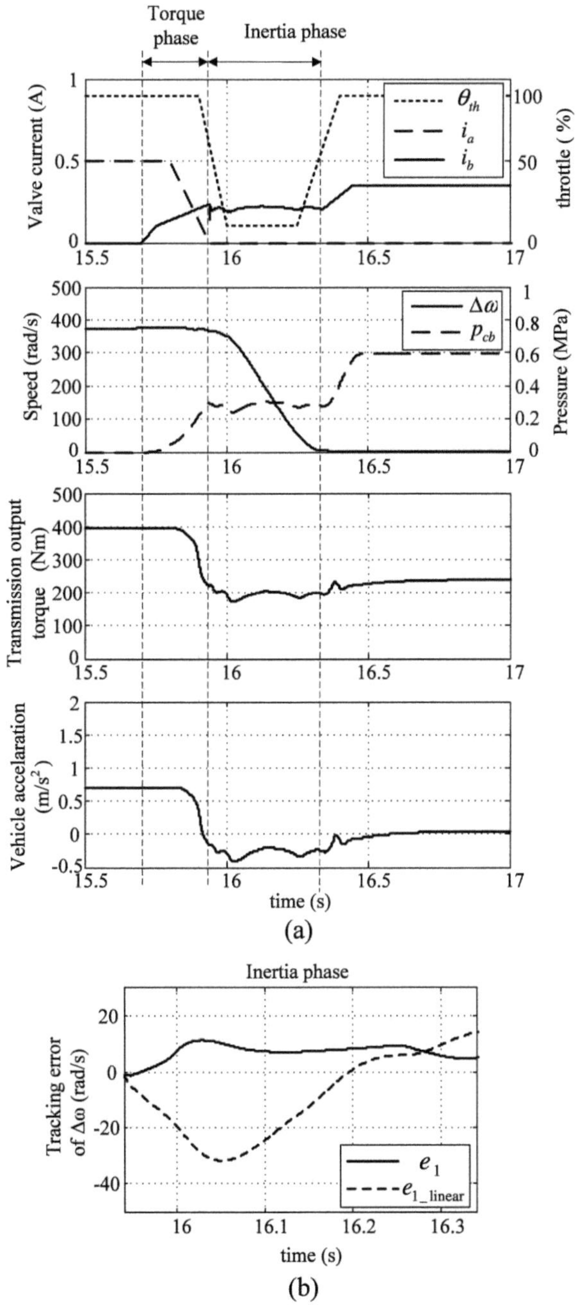

Fig. 4.16 Simulation results under different driving conditions (torque characteristics of engine and torque converter are standard × 115 %; $m = 2000$ kg; $\theta_g = 5°$; $I_t = 0.1$ kg m^2)

Fig. 4.17 Simulation results under different driving conditions (torque characteristics of engine and torque converter are standard × 85 %; $m = 1200$ kg; $\theta_g = 5°$; $I_t = 0.03$ kg m^2)

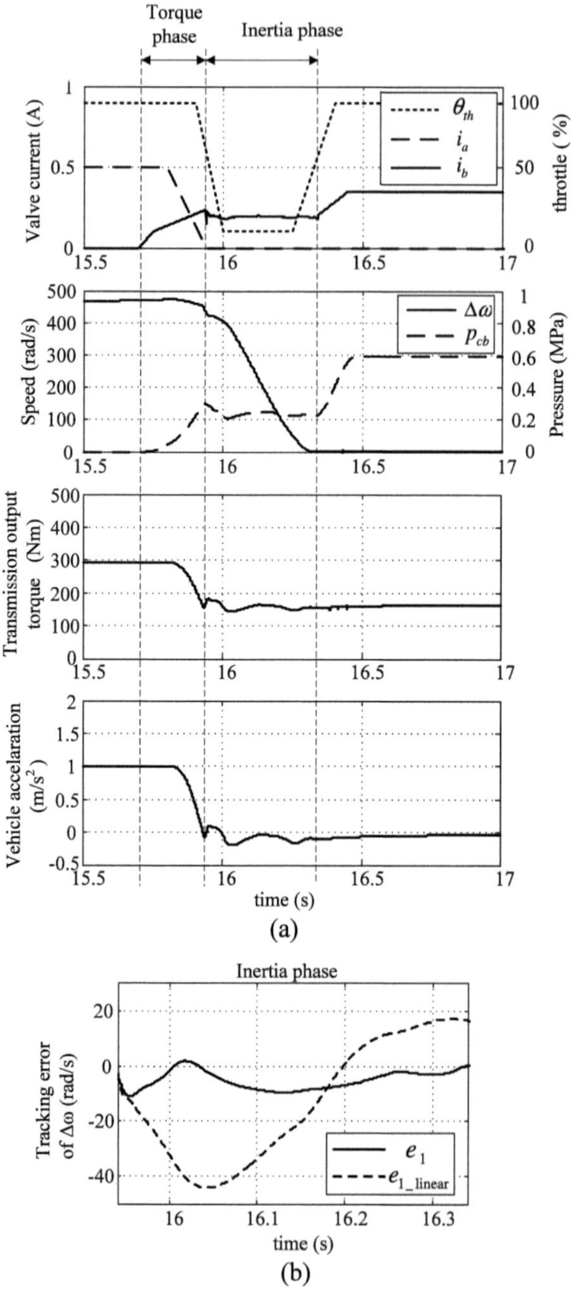

- (Figure 4.17) The torque characteristic of the engine is reduced by 15 %, and consequently, the capacity of the torque converter is also reduced; the vehicle mass is reduced from 1500 to 1200 kg, and the road grade angle is changed from 0 to 5 degrees; the inertia of the turbine shaft is changed from 0.06 to 0.03 kg m^2.

Note that although the driving conditions are changed, because the shift maneuvers are all power-on upshift when the engine load is 90 %, the engine throttle control patterns of Figs. 4.16 and 4.17 are the same as that of Fig. 4.15. It also should be noted that we focus here on the shift transient control and do not consider the determination of the optimal shift point.

As a comparison, the results of the two-degree-of-freedom linear controller from the last section are given as dashed lines of Figs. 4.16(b) and 4.17(b). In order to give a somehow fair comparison, the solid lines in Figs. 4.16(b) and 4.17(b) show the results of the proposed controller, where the same method as in Sect. 4.2 (a 3rd filter) is adopted for the reference trajectory generation. Although there exist large modeling errors, the maximum tracking error of the proposed controller is about 12 rad/s, which is still considered acceptable. Moreover, the comparison with the linear controller verifies the potential benefits of the proposed nonlinear controller in achieving smaller tracking errors.

Discrete Implementation

In order to get an in-vehicle assessment of the proposed clutch slip control system, the designed controller is discretized by a sampling rate of 100 Hz [19] with zero-order hold discretization. Furthermore, the discrete speed sensor models [27, 31] are used to give the clutch speed ω_c and the wheel speed ω_w. The speed sensors are assumed to have 48 teeth, and the time interval corresponding to 3 teeth is recorded to calculate the speeds. A relative tolerance of teeth location of 0.169 % [27] and a trigger (to convert the analog signal into the square-wave signal) randomness of 1.5 % are considered. Note that in the case of discrete implementation, which is a sampled-data system, there are some relatively large chatters because of the system discretization. The gains have to be adjusted to restrain the chatters, and the tuned results are

$$\begin{aligned} &\kappa_1 = 12; \qquad \kappa_2 = 1.2; \qquad \kappa_3 = 24; \\ &\kappa_4 = 7; \qquad \kappa_5 = 28. \end{aligned} \tag{4.86}$$

The time constant τ_2 of filter (4.74) is chosen as

$$\tau_2 = 0.05 \text{ s}. \tag{4.87}$$

The discrete implementation results are shown in Fig. 4.18, where the driving conditions are the same as those of Fig. 4.16. The sampling rate causes some small-magnitude chattering of the responses and a relatively large tracking error at the beginning of the inertia phase. The tracking errors, however, decay to the bound of

Fig. 4.18 Simulation results of discrete implementation (torque characteristics of the engine and torque converter are standard × 115 %; $m = 2000$ kg; road slope is 5°; $I_t = 0.1$ kg m^2)

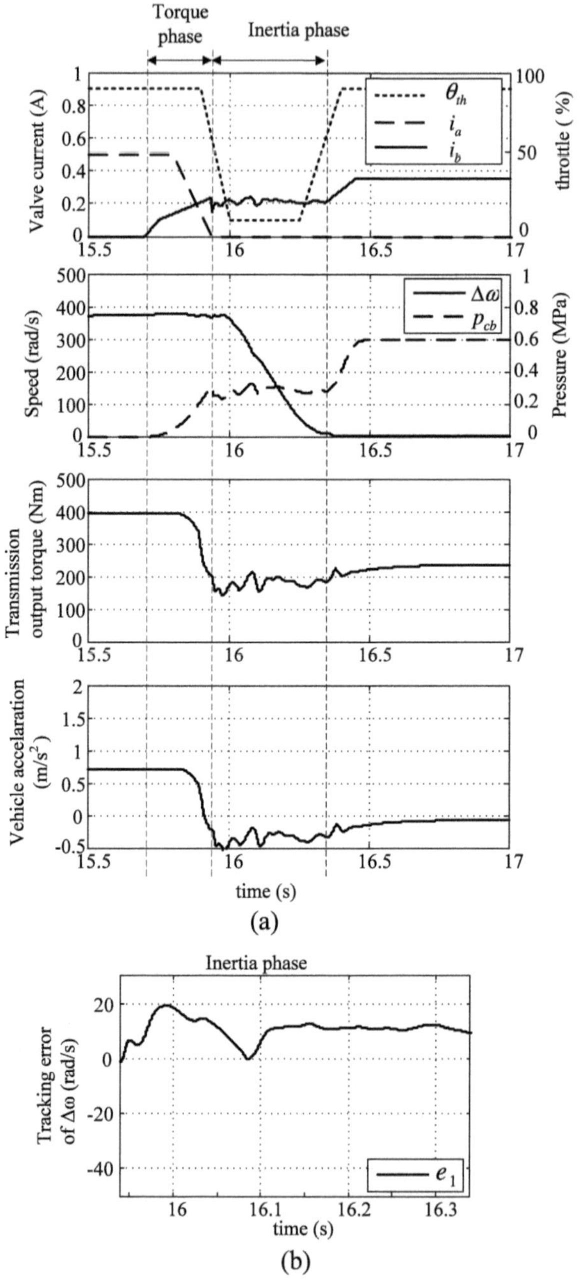

±12 rad/s rapidly. It can also be seen that the fluctuation magnitude of the transmission output torque is less than 50 Nm, which is considered acceptable for a fully loaded mid-size car shifting on a slope.

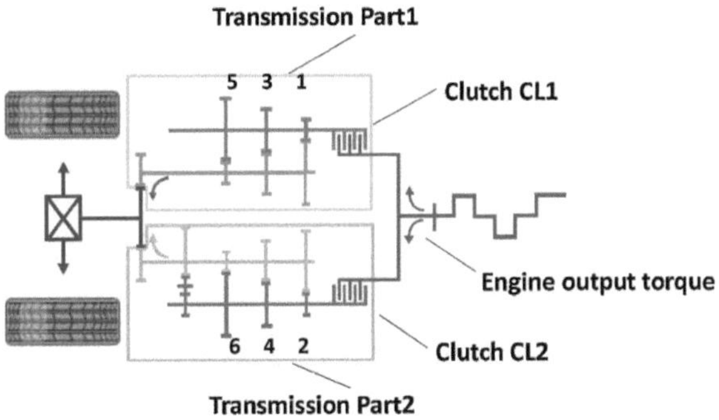

Fig. 4.19 Schematic diagram of a considered DCT

4.5 Backstepping Controller for DCTs

Both ATs (Automatic Transmissions) and DCTs (Dual Clutch Transmissions) adopt the clutch-to-clutch shift technique. Originally marketed by Volkswagen as DSG and by Audi as S-Tronic, the potential of DCT is tremendous because of its ability to shift gears very quickly and to have the same driving characteristics of a manual transmission with the convenience of an automatic [28]. For DCTs, the design differences compared with ATs (such as the absence of one-way clutches and torque converter) make achieving good shift quality in various operating conditions more difficult.

It has been shown in the last section that the methodology of backstepping is able to provide good control performance for the shift inertia phase of ATs. In this section, backstepping can be adopted to deal with the challenging control task during the inertia phase of the DCT shift.

4.5.1 System Modeling and Problem Statement

Here a 6-speed wet DCT is considered, and the system diagram is shown in Fig. 4.19. The two clutches, used as the actuators, are connected to two separate sets of gears. The first gear set is connected to clutch CL1 and the second gear set to clutch CL2. The two clutches, which are controlled by two proportional pressure valves, can operate independently.

The engine torque T_e is modeled as a function of the engine speed ω_e and throttle angle θ_{th} in the form of a look-up table (map)

$$T_e = T_e(\omega_e, \theta_{th}). \tag{4.88}$$

The two clutches are modeled as Coulomb friction elements which transfer torque between the engine and the driving unit. If the return spring force is treated as a constant, the torque T_c of the clutch transmitted in the slipping state depends on the cylinder pressure p_c and is described by the following equation:

$$T_c = \mu RN(Ap_c - F_s), \tag{4.89}$$

where μ is the friction coefficient, N is the clutch disc number, A is the piston area of the clutch, F_s is the return spring force of the clutch, R is the effective radius of the friction disc.

The cylinder pressure is determined by the input current of the proportional pressure control valve. For simplicity, the proportional pressure control valve is modeled as a first-order system

$$\tau_{cv}\dot{p}_c = -p_c + K_{cv}i, \tag{4.90}$$

where τ_{cv} and K_{cv} are the time constant and the gain of the valve, and i is the electric current of the valve.

Moreover, the vehicle road load is modeled taking the rolling resistance F_r and aerodynamic resistance F_A into account as

$$F_{\text{load}} = F_A + F_r = \frac{1}{2}\rho C_D A_A v^2 + F_r, \tag{4.91}$$

where C_D is the aerodynamic drag coefficient, A_A is the front area of a vehicle, ρ is the air density, v is the vehicle velocity, F_r is the rolling resistance considered as a constant.

For instance, when the transmission is shifted from the 1st gear to the 2nd gear, clutch CL1 is disengaged and CL2 is engaged, and before the clutch-to-clutch shift motion is carried out, gear 2 has already been pre-engaged, which prevents traction interruption. The following assumptions are made in the development of the inertia phase model of the 1st-to-2nd gear shift:

- During the inertia phase, the pressure of clutch CL1 has already been reduced to a very low level and so can be ignored;
- Gears have no backlash;
- Temperature effects of the powertrain are not taken into account.

Then the fundamental equations for the inertia phase of 1st-to-2nd gear shift can be derived by using Newton's second law as follows:

$$I_e\dot{\omega}_e = T_e - T_{c2}, \tag{4.92a}$$

$$I_t\dot{\omega}_{c2} = T_{c2} - b_t\omega_{c2} - \frac{F_{\text{load}} r_w}{R_{t2}R_{df}}, \tag{4.92b}$$

with

$$I_t = I_{t2} + \frac{I_{df} + I_w + m r_w^2}{R_{t2}^2 R_{df}^2}, \tag{4.93a}$$

$$b_t = b_{t2} + \frac{b_{df} + b_w}{R_{t2}^2 R_{df}^2}, \tag{4.93b}$$

where T_{c2} is the torque delivered at clutch CL2; ω_{c2} is the speed of clutch CL2; I_{t2}, I_{df}, I_w are inertias of transmission, differential box and wheels, respectively;

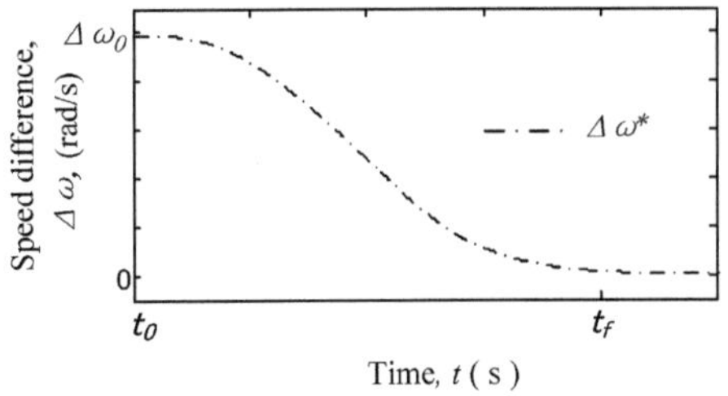

Fig. 4.20 Reference trajectory of the clutch speed difference

m is the vehicle mass; r_w is the wheel radius; R_{t2} is the gear ratio of the 2nd gear; R_{df} is the gear ratio of the differential box; b_{t2}, b_{df}, b_w are the damping coefficients of transmission, differential box, and wheels, respectively.

When considering the clutch torque equation (4.89) and valve dynamic equation (4.90), the dynamics of the 1st-to-2nd upshift inertia phase is described as

$$\Delta\dot{\omega} = -\left(\frac{1}{I_e} + \frac{1}{I_t}\right)\mu RNAp_{c2} + f(\omega_e, \omega_{c2}), \tag{4.94a}$$

$$\dot{p}_{c2} = -\frac{1}{\tau_{cv}}p_{c2} + \frac{K_{cv}}{\tau_{cv}}i_{c2}, \tag{4.94b}$$

with

$$f(\omega_e, \omega_c) = \frac{T_e}{I_e} + \left(\frac{1}{I_e} + \frac{1}{I_t}\right)\mu RNF_s + \frac{b_t\omega_{c2} + \frac{F_{\text{load}}r_w}{R_{t2}R_{df}}}{I_t}, \tag{4.95}$$

where p_{c2} is the pressure of clutch CL2, i_{c2} is the valve current of clutch CL2, and

$$\Delta\omega = \omega_e - \omega_{c2}. \tag{4.96}$$

Our goal is to design a controller to make the speed difference of clutch CL2 track a reference trajectory, where the current of valve CL2 is considered as control input. As for the driving comfort, the clutch engagement should satisfy the no-lurch condition, imposing a zero time derivative of the clutch sliding speed at synchronization. Then a desired trajectory is programmed in Fig. 4.20, which has the same pattern as in Fig. 4.1.

4.5.2 Controller Design

Choosing $\Delta\omega$ and p_{c2} as system states x_1 and x_2, respectively, and i_{c2} as control input u, the state space equation can be given as

$$\dot{x}_1 = a_1\mu(x_1)x_2 + f(\omega_e, \omega_{c2}) + b_{11}w_1, \tag{4.97a}$$

$$\dot{x}_2 = a_2x_2 + b_{22}u + b_{21}w_2, \tag{4.97b}$$

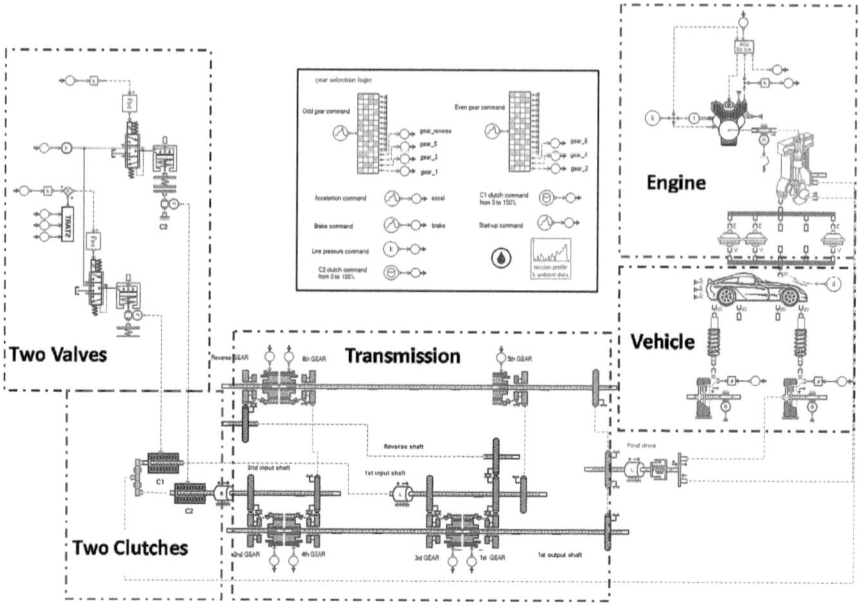

Fig. 4.21 DCT simulation model

where b_{11} and b_{21} are known scaling factors, and

$$a_1 = -\left(\frac{1}{I_e} + \frac{1}{I_t}\right) RNA, \tag{4.98a}$$

$$a_2 = -\frac{1}{\tau_{cv}}, \tag{4.98b}$$

$$b_{22} = \frac{K_{cv}}{\tau_{cv}}. \tag{4.98c}$$

It is clear that the dynamics equation (4.97a), (4.97b) has the same form as (4.36a), (4.36b), the dynamics equation of AT.

Then the same manipulation used in Sect. 4.4 is applied, and the control law for the considered DCT is obtained as follows:

$$x_{2d} = \frac{-\kappa_3 e_1 - f + \dot{x}_{1d} - |a_1 \mu(x_1)| \kappa_2 e_1}{a_1 \mu(x_1)}, \tag{4.99a}$$

$$u = \frac{\dot{x}_{2d} - a_2 x_{2d}}{b_{22}} - \frac{\kappa_5 + a_2 + |b_{21}| \kappa_4}{b_{22}} e_2, \tag{4.99b}$$

with

$$e_1 = x_1 - x_{1d}, \tag{4.100a}$$

$$e_2 = x_2 - x_{2d}. \tag{4.100b}$$

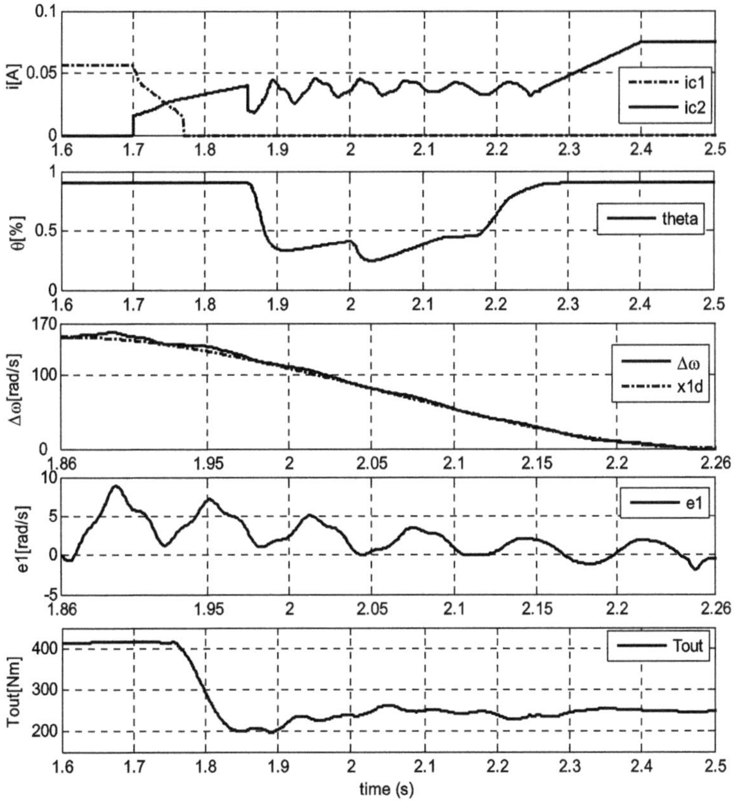

Fig. 4.22 Simulation results of PID controller (vehicle mass is 1200 kg, road slope angle is 0°)

It is shown using Theorem 4.1 of Sect. 4.4 that if $\kappa_i, i = 1, \ldots, 5$ are chosen suitably, the system dynamics will be input-to-state stable.

4.5.3 Simulation Results

The DCT simulation model is created in the environment of AMESim, which is shown in Fig. 4.21. The model is used to verify the performance of the proposed controller. It takes into account the important transient dynamics during the vehicle shift process, such as dual-mass fly-wheel, drive shaft oscillation and tire slip. These dynamics are crucial for shift dynamic quality during the controller design; however, they are not considered in controller design in order to obtain a practically applicable controller. The designed controller is simulated by MATLAB/Simulink and connected to the aforementioned AMESim simulation model through the cosimulation technique.

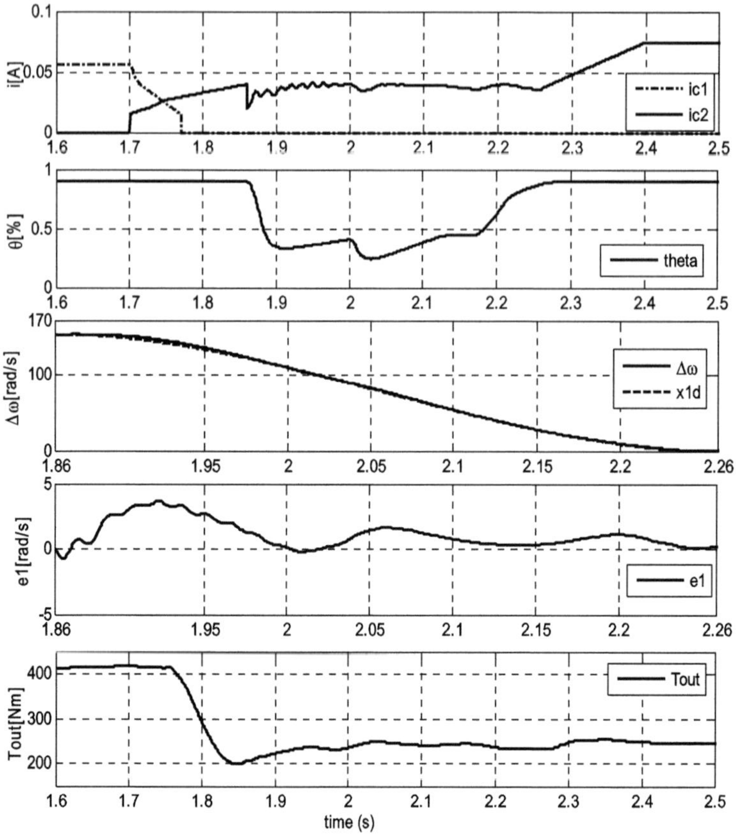

Fig. 4.23 Simulation results of a backstepping controller (vehicle mass is 1200 kg, road slope angle is 0°)

For comparison, the results of a well-tuned PID controller are given first in Fig. 4.22, wherein the driving conditions are as same as those used for the controller design, i.e., the vehicle mass is 1200 kg and road slope angle is 0 degrees. The graph depicts the variables including the currents profiles i at two clutches, the rotational speed $\Delta\omega$ of clutch CL2, the tracking error e_1 of clutch CL2, and the throttle angle θ_{th}, transmission output torque T_{out}. In can be seen that during the inertia phase, between 1.86 and 2.26 s, although there is no large fluctuation in T_{out}, the transmission output torque and the valve current i_{c2} vibrate frequently, which results in some vibration of the tracking error e_1.

Figure 4.23 gives the results of the designed backstepping controller. It can be seen that the backstepping controller achieves much less vibration in valve current i_{c2}, tracking error e_1, and output torque of the transmission T_{out}, when compared with the PID controller (refer to Fig. 4.24 for a direct comparison). This justifies the potential benefit of the nonlinear design method of backstepping.

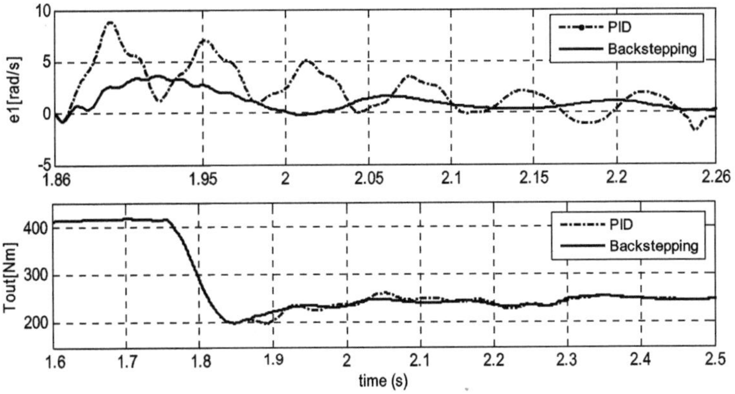

Fig. 4.24 Backstepping controller vs. PID controller

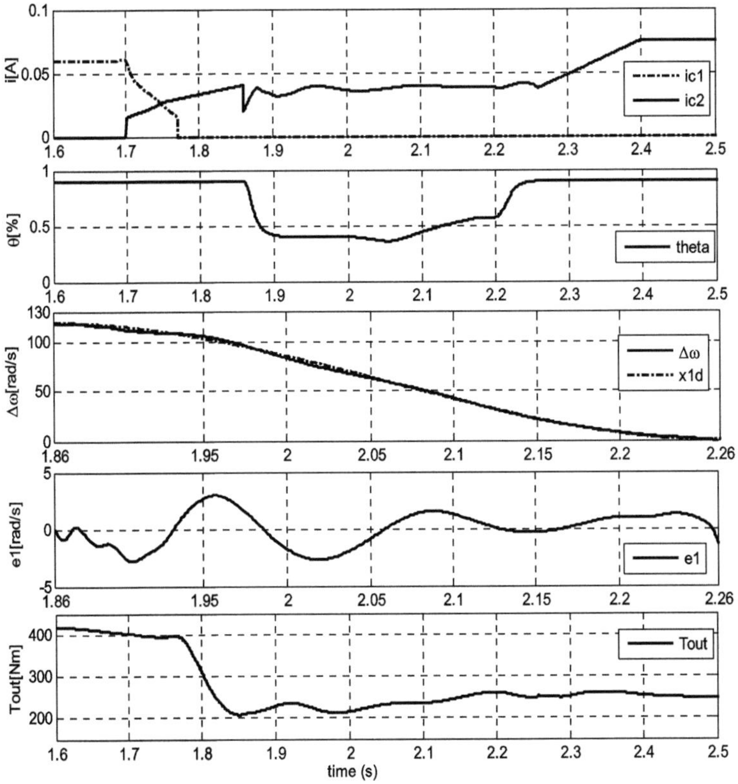

Fig. 4.25 Simulation results of a backstepping controller (vehicle mass is 1500 kg, road slope angle is 5°)

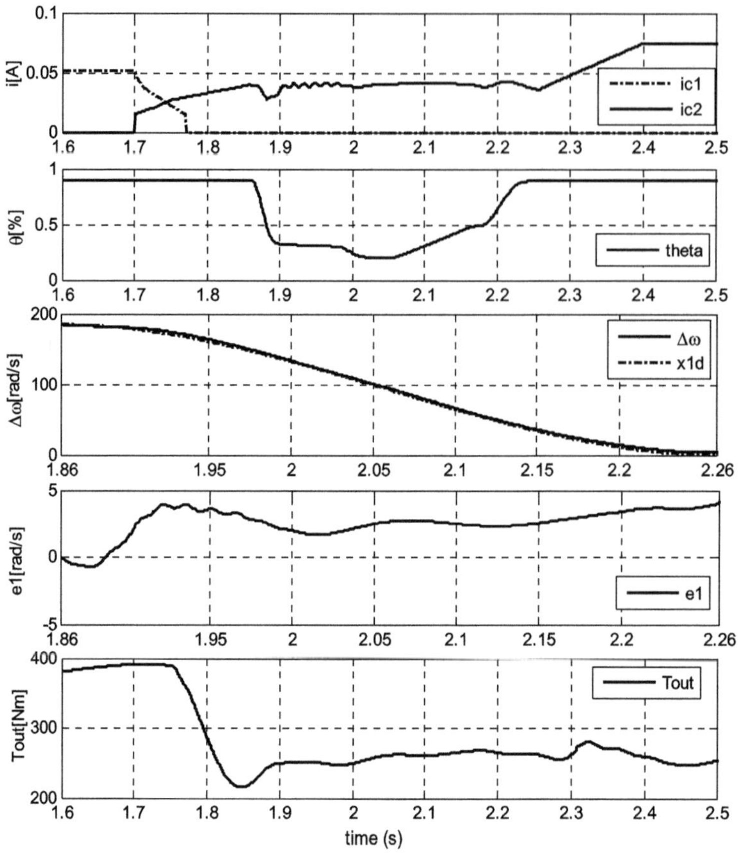

Fig. 4.26 Simulation results of a backstepping controller (vehicle mass is 1000 kg, road slope angle is 5°)

The designed clutch slip controller is then tested under driving conditions of full load and empty load, and the results are given in Figs. 4.25 and 4.26, respectively. It is shown that although there are large modeling errors, the output torque is still smooth enough and the tracking error is less than 5 rad/s.

4.6 Notes and References

For ATs and DCTs, which use proportional pressure control valves to control the clutches directly, three different control methodologies, including the linear 2 DOF control, nonlinear feedforward–feedback control, and the backstepping technique, are used for designing the inertia phase controller of the gear shift, wherein the control objective is to make the clutch speed track a given reference trajectory.

In Chap. 3 and in this chapter, the torque phase and the inertia phase of clutch-to-clutch shift are controlled. Actually, from the point of view of hybrid control theory, the transition from the shift torque phase to the shift inertia phase is a state switching along with the stick-slipping of the clutch, and hybrid control [1–3, 22, 26] is a possible solution to deal with this problem under a uniform framework explicitly so that extensive control system calibration could be avoided.

References

1. Balluchi A, Benvenuti L, Ferrari A, Sangiovanni-Vincentelli AL (2006) Hybrid systems in automotive electronics design. Int J Control 79(5):375–394
2. Bemporad A, Morari M (1999) Control of systems integrating logic, dynamics, and constraints. Automatica 35:407–427
3. Bemporad A, Borrelli F, Glielmo L, Vasca F (2001) Hybrid control of dry clutch engagement. In: Proceedings of the European control conference, Porto, Portugal
4. Boyd S, El Ghaoui L, Feron E, Balakishnan V (1994) Linear matrix inequalities in system and control theory. SIAM, Philadelphia
5. Chang PH, Park HS (2005) Time-varying input shaping technique applied to vibration reduction of an industrial robot. Control Eng Pract 13(1):121–130
6. Cho D (1987) Nonlinear control methods for automotive powertrain systems. PhD Thesis, MIT
7. Chung SK, Koch CR, Lynch AF (2007) Flatness-based feedback control of an automotive solenoid valve. IEEE Trans Control Syst Technol 15(2):394–401
8. Dolcini P, Béchart H (2005) Observer-based optimal control of dry clutch engagement. In: Proceedings of the 44th IEEE conference on decision and control, Seville, Spain, pp 440–445
9. Dolcini P, Wit CC, Béchart H (2008) Lurch avoidance strategy and its implementation in amt vehicles. Mechatronics 18(5–6):289–300
10. Fliess M, Lévine J, Martin P, Rouchon P (1995) Flatness and defect of nonlinear systems: introductory theory and examples. Int J Control 61:1327–1361
11. Gao B-Z, Chen H, Sanada K (2008) Two-degree-of-freedom controller design for clutch slip control of automatic transmission. SAE technical paper 2008-01-0537
12. Gao B-Z, Chen H, Zhao H-Y, Sanada K (2010) A reduced-order nonlinear clutch pressure observer for automatic transmission. IEEE Trans Control Syst Technol 18(2):446–453
13. Gao B-Z, Chen H, Hu YF, Sanada K (2011) Nonlinear feedforward-feedback control of clutch-to-clutch shift technique. Veh Syst Dyn 49(12):1895–1911
14. Gao B-Z, Chen H, Sanada K, Hu Y-F (2011) Design of clutch slip controller for automatic transmission using backstepping. IEEE/ASME Trans Mechatron 16(3):498–508
15. Garofalo F, Glielmo L, Iannelli L, Vasca F (2002) Optimal tracking for automotive dry clutch engagement. In: Proceedings of the 15th IFAC Congress, Barcelona, Spain
16. Glielmo L, Vasca F (2000) Optimal control of dry clutch engagement. SAE technical paper 2000-01-0837
17. Glielmo L, Iannelli L, Vacca V, Vasca F (2006) Gearshift control for automated manual transmissions. IEEE/ASME Trans Mechatron 11(1):17–26
18. Goetz M, Levesley MC, Crolla DA (2005) Dynamics and control of gearshifts on twin-clutch transmissions. Proc Inst Mech Eng, Part D, J Automob EngMech 219(8):951–963
19. Hahn JO, Lee KI (2002) Nonlinear robust control of torque converter clutch slip system for passenger vehicles using advanced torque estimation algorithms. Veh Syst Dyn 37(3):175–192
20. Haj-Fraj A, Pfeiffer F (2001) Optimal control of gear shift operations in automatic transmissions. J Franklin Inst 338(2–3):371–390

21. Haj-Fraj A, Pfeiffer F (2002) A model based approach for the optimisation of gearshifting in automatic transmissions. Int J Veh Des 28(1–3):171–188
22. Heijden ACVD, Serrarens AFA, Camlibel MK, Nijmeijer H (2007) Hybrid optimal control of dry clutch engagement. Int J Control 80(11):1717–1728
23. Horn J, Bamberger J, Michau P, Pindl S (2003) Flatness-based clutch control for automated manual transmissions. Control Eng Pract 11(12):1353–1359
24. Kautsky J, Nichols NK (1985) Robust pole assignment in linear state feedback. Int J Control 41:1129–1155
25. Krstić M, Kanellakopoulos I, Kokotović P (1995) Nonlinear and adaptive control design. Wiley, New York
26. Liberzon D (2003) Switching in systems and control. Birkhäuser, Boston
27. Masmoudi RA, Hedrick K (1992) Estimation of vehicle shaft torque using nonlinear observers. ASME J Dyn Syst Meas Control 114:394–400
28. Matthes B, Guenter F (2005) Dual clutch transmissions—lessons learned and future potential. SAE technical paper 2005-01-1021
29. Ogata K (2001) Modern control engineering, 4th edn. Prentice Hall, New York
30. Sanada K, Kitagawa A (1998) A study of two-degree-of-freedom control of rotating speed in an automatic transmission, considering modeling errors of a hydraulic system. Control Eng Pract 6:1125–1132
31. Schwarz R, Nelles O, Scheerer P, Isermann R (1997) Increasing signal accuracy of automotive wheel-speed sensors by on-line learning. In: Proceedings of American control conference, Albuquerque, NM, pp 1131–1135
32. Shin BK, Hahn JO, Lee KI (2000) Development of shift control algorithm using estimated turbine torque. SAE technical paper 2000-01-1150
33. Swaroop D, Hedrick JK, Yip PP, Gerdes JC (2000) Dynamic surface control for a class of nonlinear systems. IEEE Trans Autom Control 45(10):1893–1899
34. Tsutsumi J, Higashimata A (2005) Application of advanced control technologies to the vehicle control. J Soc Automot Eng Jpn 59(5):10–15. In Japanese
35. Yi K, Shin BK, Lee KL (2000) Estimation of turbine torque of automatic transmissions using nonlinear observers. ASME J Dyn Syst Meas Control 122:276–283
36. Yokoyama M (2008) Sliding mode control for automatic transmission systems. J Jpn Fluid Power Syst Soc 39(1):34–38. In Japanese
37. Zheng Q, Srinivasan K, Rizzoni G (1999) Transmission shift controller design based on a dynamic model of transmission response. Control Eng Pract 7(8):1007–1014

Chapter 5
Torque Estimation of the Vehicle Drive Shaft

Until now, the estimation and control problems involved in ATs or DCTs were addressed. From now on, the estimation and control problems of AMT will be discussed, and at first, in this chapter, the estimation of the axle drive shaft torque will be analyzed because it is the basis for later chapters.

5.1 Introduction

Mechanical resonance of vehicle drivelines may occur due to the elasticity of the driveline parts, such as clutch spring, propeller shaft and drive axle shaft. Driveline oscillations are a kind of disturbance to the driver. They also lead to overlarge mechanical stress and affect the dynamic performance of the drivelines [6, 24]. The question of how to avoid or reduce the oscillations of the driveline is an important issue, especially for heavy duty vehicles which have relatively large driveline torsion.

There is some literature on active damping of vehicle drivelines published in recent years [3, 13]. The engine torque is controlled actively to damp the driveline oscillations during transient maneuvers, such as when pressing and releasing the accelerator pedal. Because the drive axle shaft is the main component of the driveline, driving performance can be improved by controlling the axle shaft torsion. In order to design the longitudinal speed controller handling the drive shaft torsion, it is often necessary to know the angle/torque of the axle shaft [1, 22, 30].

It is also well known that the gear shift quality can be improved if an accurate measurement of the axle shaft torque is available [17, 23, 29]. One example is the shift process of Automated Manual Transmissions (AMTs) [16], which are widely adopted to offer easy drive and fuel efficiency for trucks. At the beginning of the gear shift of AMTs, the torque transmitted by the transmission is decreased and then cut off by active engine control and clutch disengagement. If the timing of the clutch disengagement is not well controlled, the potential energy of the driveline will lead to unwanted driveline and vehicle oscillations [7, 23]. Knowing the axle shaft

H. Chen, B. Gao, *Nonlinear Estimation and Control of Automotive Drivetrains*,
DOI 10.1007/978-3-642-41572-2_5,

torque helps to determine the most optimal time point to disengage the clutch (or directly engage the neutral gear). On the other hand, at the end of the shift process, when the clutch is engaged and the engine torque level is recovered, closed-loop shift control algorithms could greatly benefit if the measurement of the axle shaft torque is available.

Although the knowledge of the axle shaft torque is necessary for improving the longitudinal speed control performance of vehicles, shaft torque sensors [27] or high precision encoders [18] (the drive shaft torque could be calculated if the twist angle measurement is available) are seldom used in production vehicles because of the cost and durability. Hence, it is required to estimate the axle shaft torque. Luenberger observer [1, 31] and Kalman Filter [23, 24] have been used to estimate the drive axle shaft torque. Although automotive powertrains contain complex nonlinearities, these observers are designed based on the linearized models. The sliding mode observer [19] has also been designed to estimate the axle shaft torque in [17]. A sliding mode observer offers a way to ensure robustness to modeling errors and parameter uncertainties if the uncertainties are limited in their assumed bounds [17]. Kalman filtering [15] and recursive least squares method with multiple forgetting factors [28] are used for simultaneous estimation of the road grade and the vehicle mass, which helps to improve the estimation accuracy of the driving load.

In Chap. 2, a nonlinear clutch pressure observer is proposed for automatic transmissions, where robustness is guaranteed in the sense of input-to-state stability (ISS). The order of the designed observer is reduced to one, nevertheless complex nonlinear characteristics of powertrain systems are included and appear in their usual form of maps. A comparison with the existing sliding mode observer verifies the potential benefits of the proposed observer in eliminating chatters and in achieving satisfactory estimation performance.

In this chapter, therefore, the methodology of Chap. 2 is extended, and an axle shaft torque observer is discussed for trucks with a stepped ratio transmission. The observer is designed for all gear positions and the error dynamics is input-to-state stable, where modeling errors and external disturbances are considered as input. Compared with passenger cars, the truck mass varies greatly, hence a small road grade seriously increases the load. These properties are taken into account by the proposed observer, and the observer gains obtained by convex optimization are robust against large variations of driving conditions.[1]

5.2 Driveline Modeling and Problem Statement

5.2.1 Driveline Modeling

We consider the powertrain in a medium-duty truck with an AMT, which contains a dry clutch and a 6-speed manual transmission. The powertrain is schematically shown in Fig. 5.1.

---

[1]This chapter uses the content of [9], with permission from Elsevier.

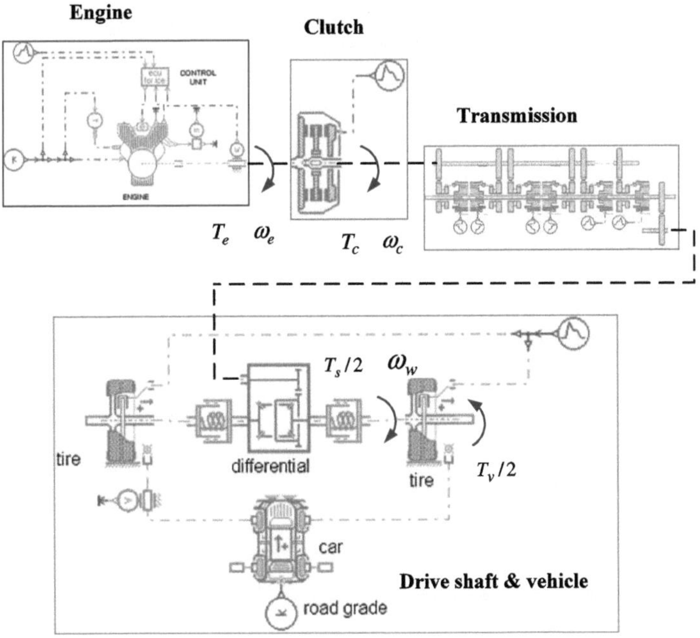

Fig. 5.1 Schematic graph of a medium-duty truck

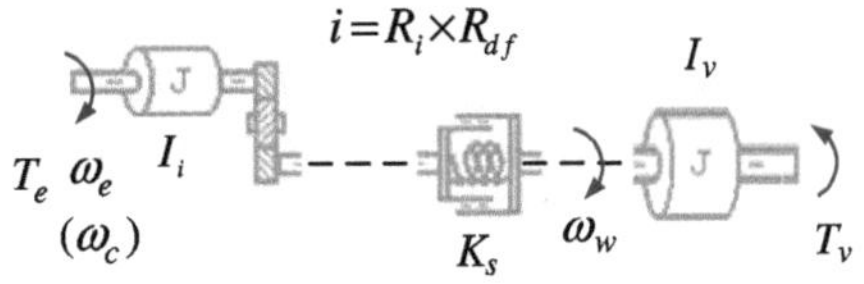

Fig. 5.2 Simplified driveline model

When the vehicle runs in a certain gear position (no clutch operation), the driveline is simplified as a spring–mass system shown in Fig. 5.2. The motion of the driveline is described by the following equations:

$$\dot{\omega}_c = \frac{1}{I_i}\left(T_e - \frac{1}{R_i R_{df}} T_s\right), \tag{5.1a}$$

$$\dot{\omega}_w = \frac{1}{I_v}(T_s - T_v), \tag{5.1b}$$

$$\dot{T}_s = K_s\left(\frac{1}{R_i R_{df}}\omega_c - \omega_w\right), \tag{5.1c}$$

where ω_c is the output speed of the clutch; ω_w is the wheel speed; T_s is the axle shaft torque; I_i denotes the equivalent inertia moment from the engine to the axle shaft at the ith gear position, $i = 1, 2, \ldots, 6$; I_v is the equivalent inertia of the vehicle; T_e is the engine torque, and T_v is the driving resistance torque. R_i denotes the gear ratio of the ith gear position, and R_{df} is the ratio of the differential gear box; K_s is the stiffness of the axle shaft. The nominal value of the damping coefficient is set to zero in these dynamical equations, because the damping torque changes greatly along with temperature variation and it is indeed very difficult to determine a constant damping coefficient. It should be noted that if a nominal value of the damping coefficient is valid, the design method shown in the following is still applicable for deriving the observer.

The engine torque T_e is described by the torque map. The inputs of the map are the engine rotational speed ω_e and engine "throttle angle" θ_{th}. Because diesel engines do not have a butterfly valve throttle, here θ_{th} represents the load requested by the engine control unit. When the vehicle is driven in a certain gear position and there is no clutch slip, we have

$$T_e = T_e(\omega_e, \theta_{th}) \approx T_e(\omega_c, \theta_{th}). \tag{5.2}$$

If the tire slip and road grade are ignored, the resistance torque from the tire to the drive axle shaft is calculated as

$$T_v = T_w + C_A R_w^3 \omega_w^2, \tag{5.3}$$

where T_w denotes the rolling resistant moment of tires; R_w is the tire radius; C_A is a constant coefficient depending on air density, aerodynamic drag coefficient and the front area of the vehicle.

5.2.2 Estimation Problem Statement

The state variables are selected as $x_1 = \omega_c$, $x_2 = \omega_w R_i R_{df}$ and $x_3 = \frac{T_s}{\bar{T}_s}$, so that x_1 and x_2 are of the same order of magnitude, and x_3, the variable to be estimated, is normalized into a level of ± 1 through $\bar{T}_s$, the nominal value of the drive shaft torque T_s. Note that T_s may take negative values because of shaft vibration or engine braking.

The driveline motion is then expressed in the following state space form:

$$\dot{x}_1 = \frac{-\bar{T}_s}{I_i R_i R_{df}} x_3 + f_1(x_1, u), \tag{5.4a}$$

$$\dot{x}_2 = \frac{R_i R_{df} \bar{T}_s}{I_v} x_3 + R_i R_{df} f_2(x_2), \tag{5.4b}$$

$$\dot{x}_3 = \frac{K_s}{R_i R_{df} \bar{T}_s} (x_1 - x_2), \tag{5.4c}$$

where $u = \theta_{th}$ is the throttle angle and

$$f_1(x_1, u) = \frac{1}{I_i} T_e(x_1, u), \tag{5.5a}$$

$$f_2(x_2) = \frac{-1}{I_v} T_v(x_2). \tag{5.5b}$$

In order to estimate the drive shaft torque x_3, the rotational speeds x_1, x_2 are used as the measurable outputs, i.e.,

$$\boldsymbol{y} = [x_1 \quad x_2]^T. \tag{5.6}$$

The nonlinear functions in (5.5a), (5.5b) are in general given as lookup tables (i.e., maps), which are obtained by a series of steady state experiments and inherently contain errors. Other modeling uncertainties include uncertain parameters, such as the vehicle mass, the road grade and the damping coefficient of shafts. The approximation of (5.2) may bring about modeling error as well.

Hence, the problem considered here is to design an observer of the axle shaft torque for all gear positions. The observer estimates the shaft torque in the presence of model errors, given the engine throttle input and the measured rotational speeds of the transmission.

5.3 Reduced-Order Nonlinear Shaft Torque Observer

5.3.1 Structure of the Observer

In this section, the special structure of the driveline system is exploited, and the methodology in Chap. 2 (or [8]) is extended to derive a reduced-order observer of the axle shaft torque. The robustness of the observer with respect to model errors is achieved in the sense of input-to-state (ISS) property. To do this, we denote the variable to be estimated as z, and rewrite the system dynamics as follows:

$$\dot{\boldsymbol{y}} = \boldsymbol{F}(\boldsymbol{y}, u) + \boldsymbol{G} z + \boldsymbol{H} \boldsymbol{w}(\boldsymbol{y}, u, z), \tag{5.7a}$$

$$\dot{z} = \boldsymbol{A} \boldsymbol{y}, \tag{5.7b}$$

where $\boldsymbol{y}$ is the measured outputs, $\boldsymbol{w}(\boldsymbol{y}, u, z)$ summarizes model uncertainties which is normalized as $\|\boldsymbol{w}\|_\infty \approx 1$. In particular, $\boldsymbol{H}$ is the matrix for the normalization of $\boldsymbol{w}$ and

$$\boldsymbol{F}(\boldsymbol{y}, u) = \begin{pmatrix} f_1(x_1, u) \\ R_i R_{df} f_2(x_2) \end{pmatrix}, \tag{5.8a}$$

$$\boldsymbol{G} = \begin{pmatrix} \frac{-\bar{T}_s}{I_i R_i R_{df}} \\ \frac{R_i R_{df} \bar{T}_s}{I_v} \end{pmatrix}, \tag{5.8b}$$

$$\boldsymbol{A} = \left(\frac{K_s}{R_i R_{df} \bar{T}_s}, -\frac{K_s}{R_i R_{df} \bar{T}_s} \right). \tag{5.8c}$$

Because the shaft torque directly affects the related shaft accelerations, the difference between the true accelerations $\dot{y}$ and the estimated values $\boldsymbol{F}(y,u)+\boldsymbol{G}\hat{z}$ is used to constitute the correction term. The observer is then designed in the form of

$$\dot{\hat{z}} = \boldsymbol{A}y + \boldsymbol{L}\big(\dot{y} - \boldsymbol{F}(y,u) - \boldsymbol{G}\hat{z}\big), \tag{5.9}$$

where $\boldsymbol{L} \in \mathbb{R}^{1\times 2}$ is the time-invariant (constant) observer gain to be determined.

In order to avoid taking derivatives of the measurements $\boldsymbol{y}$, the following transformation is made. Let

$$\eta = \hat{z} - \boldsymbol{L}y, \tag{5.10}$$

then, we can infer for a time-invariant $\boldsymbol{L}$ that

$$\dot{\eta} = \boldsymbol{A}y - \boldsymbol{L}\boldsymbol{G}(\eta + \boldsymbol{L}y) - \boldsymbol{L}\boldsymbol{F}(y,u). \tag{5.11}$$

Equations (5.10) and (5.11) constitute then the reduced-order observer of the drive axle shaft torque for the nonlinear driveline system. Obviously, the nonlinearities of the powertrain system appear in the observer in their original form. Therefore, the characteristics of powertrain mechanical systems, such as the characteristics of the engine and the aerodynamic drag, can be represented in the form of lookup tables, which are easily processed in computer control.

5.3.2 Properties of the Error Dynamics

In this section, the error dynamics of the designed shaft torque observer is analyzed using the concept of ISS (input-to-state stability) (Appendix B). By defining the observer error as

$$e = z - \hat{z}, \tag{5.12}$$

the error dynamics can then be described by

$$\dot{e} = -\boldsymbol{L}\boldsymbol{G}e - \boldsymbol{L}\boldsymbol{H}\boldsymbol{w}. \tag{5.13}$$

We define $V(e) = \frac{1}{2}e^T e$ and differentiate it along the solution of (5.13) to obtain

$$\dot{V} = -e^T\boldsymbol{L}\boldsymbol{G}e - e^T\boldsymbol{L}\boldsymbol{H}\boldsymbol{w}. \tag{5.14}$$

Using Young's Inequality [12], the above equality becomes

$$\dot{V} \le e^T(-\boldsymbol{L}\boldsymbol{G} + \kappa_1)e + \frac{1}{4\kappa_1}\boldsymbol{w}^T\boldsymbol{H}^T\boldsymbol{L}^T\boldsymbol{L}\boldsymbol{H}\boldsymbol{w}, \tag{5.15}$$

where $\kappa_1 > 0$. We now choose $\boldsymbol{L}$ to satisfy the following inequality:

$$-\boldsymbol{L}\boldsymbol{G} + \kappa_1 \le -\kappa_2 \tag{5.16}$$

with $\kappa_2 > 0$, then we arrive at

$$\dot{V} \le -\kappa_2 e^T e + \frac{1}{4\kappa_1} \boldsymbol{w}^T \boldsymbol{H}^T \boldsymbol{L}^T \boldsymbol{L} \boldsymbol{H} \boldsymbol{w} \tag{5.17}$$

and furthermore,

$$\dot{V} \le -\kappa_2 \|e\|^2 + \frac{1}{4\kappa_1} \lambda_{\max}\big(\boldsymbol{H}^T \boldsymbol{L}^T \boldsymbol{L} \boldsymbol{H}\big) \|\boldsymbol{w}\|_\infty^2 .$$

According to Theorem B.1 in Appendix B, this shows that the error dynamics of the observer (5.9) is input-to-state stable, where the $\mathcal{K}_\infty$ functions are $\alpha(x) = \kappa_2 x^2$ and $\gamma(x) = \frac{1}{4\kappa_1} \lambda_{\max}(\boldsymbol{H}^T \boldsymbol{L}^T \boldsymbol{L} \boldsymbol{H}) x^2$.

Moreover, it follows from (5.17) that

$$\dot{V} \le -2\kappa_2 V + \frac{1}{4\kappa_1} \boldsymbol{w}^T \boldsymbol{H}^T \boldsymbol{L}^T \boldsymbol{L} \boldsymbol{H} \boldsymbol{w}. \tag{5.18}$$

Upon multiplication of (5.18) by $e^{2\kappa_2 t}$, it becomes

$$\frac{d}{dt}\big(V e^{2\kappa_2 t}\big) \le \frac{1}{4\kappa_1} \boldsymbol{w}^T \boldsymbol{H}^T \boldsymbol{L}^T \boldsymbol{L} \boldsymbol{H} \boldsymbol{w} e^{2\kappa_2 t}. \tag{5.19}$$

Integrating it over $[0, t]$ leads to

$$V(t) \le V(0) e^{-2\kappa_2 t} + \frac{1}{4\kappa_1} \int_0^t e^{-2\kappa_2 (t-\tau)} \boldsymbol{w}(\tau)^T \boldsymbol{H}^T \boldsymbol{L}^T \boldsymbol{L} \boldsymbol{H} \boldsymbol{w}(\tau)\, d\tau, \tag{5.20}$$

and furthermore,

$$\big\|e(t)\big\|^2 \le \big\|e(0)\big\|^2 e^{-2\kappa_2 t} + \frac{\|\boldsymbol{w}\|_\infty^2 \lambda_{\max}(\boldsymbol{H}^T \boldsymbol{L}^T \boldsymbol{L} \boldsymbol{H})}{2\kappa_1} \int_0^t e^{-2\kappa_2 (t-\tau)}\, d\tau. \tag{5.21}$$

Hence, we can interpret the ISS property of the designed observer as follows:

(a) The initial estimation error decays exponentially with κ_2;
(b) If a bound of the modeling errors is given, an upper bound of the estimation offset can be computed as

$$\big\|e(\infty)\big\|^2 \le \frac{\|\boldsymbol{w}\|_\infty^2 \lambda_{\max}(\boldsymbol{H}^T \boldsymbol{L}^T \boldsymbol{L} \boldsymbol{H})}{4\kappa_1 \kappa_2}. \tag{5.22}$$

Remark 5.1 We stress that (5.22) gives just an upper bound of the estimation error offset, if a bound of the model error is given. The real offset could be much smaller, due to the multiple use of inequalities in the above derivation. For a fixed gear position, $\boldsymbol{G}$ and $\boldsymbol{H}$ are constant matrices, which implies that the error dynamics (5.13) is time-invariant. We denote them as $\boldsymbol{G}_i$ and $\boldsymbol{H}_i$, $i = 1, 2, \ldots, 6$. Hence, we can compute the estimation error offset by using the final-value theorem [21]

$$e_1(\infty) = \lim_{s \to 0} s \cdot \frac{-a_1}{s + \boldsymbol{L}\boldsymbol{G}_i} w_1(s), \tag{5.23a}$$

$$e_2(\infty) = \lim_{s \to 0} s \cdot \frac{-a_2}{s + \boldsymbol{L}\boldsymbol{G}_i} w_2(s), \tag{5.23b}$$

which implies

$$e_1(\infty) = 0, \tag{5.24a}$$

$$e_2(\infty) = 0, \tag{5.24b}$$

when w_1 and w_2 are impulse signals, and

$$e_1(\infty) = \frac{-a_1}{\boldsymbol{L}\boldsymbol{G}_i}, \tag{5.25a}$$

$$e_2(\infty) = \frac{-a_2}{\boldsymbol{L}\boldsymbol{G}_i}. \tag{5.25b}$$

when w_1 and w_2 are step signals. In the above, a_j is the jth element of $\boldsymbol{L}\boldsymbol{H}_i$ and e_j is the offset resulting from the jth disturbance w_j, $j = 1, 2$.

Remark 5.2 In practice, the gear position changes among $i = 1, 2, \ldots, 6$, which constitutes a switching system. If a constant observer gain $\boldsymbol{L}$ is available for all the gears, it helps to simplify the real world implementation of the designed observer. Therefore, we solve the following LMIs for a constant $\boldsymbol{L}$

$$-\boldsymbol{L}\boldsymbol{G}_i + \kappa_1 \leq -\kappa_2, \quad i = 1, 2, \ldots, 6, \text{ with } \boldsymbol{G}_i = \begin{pmatrix} \frac{-\bar{T}_s}{I_i R_i R_{df}} \\ \frac{R_i R_{df} \bar{T}_s}{I_v} \end{pmatrix}. \tag{5.26}$$

Then, if there exists a constant $\boldsymbol{L}$ satisfying LMIs (5.26), $V(e)$ is a common Lyapunov function to show the observer achieves properties (a) and (b) for any gear position. Moreover, if the gear position were fixed, the decay rate of the error dynamics could be given by $(\boldsymbol{L}\boldsymbol{G}_i - \kappa_1)$.

5.3.3 Guideline of Choosing Tuning Parameters

The above discussion highlights that the observer gain should satisfy (5.26), in order to guarantee the ISS property. In (5.26), $\kappa_1 \geq 0$ and $\kappa_2 \geq 0$ are the tuning parameters. Now we give some guidelines for choosing these tuning parameters.

It follows clearly from the property (a) that κ_2 should be chosen according to the required decay rate of the estimate. According to (b), one may choose a larger κ_1 to reduce the offset. From (5.26), however, one should notice that the larger the κ_1, the higher the observer gain.

Hence, we can give the following systematic procedure to determine the tuning parameters κ_1 and κ_2 of the reduced-order nonlinear drive shaft torque observer in the forms of (5.10) and (5.11):

Table 5.1 Parameters for observer design

I_1–I_6	Inertia from engine to axle shaft	0.6967 kg m^2, 0.7021 kg m^2 0.7135 kg m^2, 0.7399 kg m^2 0.7992kg m^2, 0.9325 kg m^2
R_1–R_6	Gear ratio	7.57, 5.00, 3.38, 2.25, 1.50, 1.00
I_v	Equivalent inertia of vehicle	1560.6 kg m^2
R_{df}	Ratio of differential gear	5
K_s	Axle shaft stiffness	900 Nm/deg
$\bar{T}_s$	Maximum value of axle shaft torque	10^4 Nm

Step 1: Choose the parameter κ_2 according to the required decay rate of the estimation error;
Step 2: Choose the parameter κ_1, where it is suggested to start from some smaller values;
Step 3: Determine the observer gain $\boldsymbol{L}$ such that (5.26) is satisfied;
Step 4: When w_1 and w_2 are step signals, use (5.25a), (5.25b) to compute the estimation error offset for $i = 1, 2, \ldots, 6$ (use (5.22) if $\|\boldsymbol{w}\|_\infty^2$ is available), and check if the offset is acceptable for each gear position;
Step 5: If the offset is acceptable, end the design procedure. If not, go to Step 2.

In order to reduce the offset and to achieve lower observer gains for robustness against noises, $\boldsymbol{L}$ can be obtained through the following convex optimization:

$$\min_{\alpha, \boldsymbol{L}} \alpha \quad \text{subject to LMI (5.26) and} \tag{5.27a}$$

$$\begin{pmatrix} \alpha & \boldsymbol{L}\boldsymbol{H}_i \\ \boldsymbol{H}_i^T \boldsymbol{L}^T & \boldsymbol{I} \end{pmatrix} \geq 0, \quad i = 1, 2, \ldots, 6. \tag{5.27b}$$

Given κ_1 and κ_2, the solution of (5.27a), (5.27b) gives then a constant observer gain with the lowest possible gains satisfying the condition (5.26).

5.3.4 Observer Design for Considered Vehicle

Now the proposed method is applied to design an axle shaft torque observer for the considered vehicle. The parameters required for the observer are listed in Table 5.1. The values of these parameters are derived from the nominal setting of an AMESim simulation model of a medium-duty truck, which is shown in Fig. 5.1 and will be discussed later in Sect. 5.4.1.

Nonlinear functions f_1, f_2 are given as lookup tables for the observer. The maps of f_1, f_2 are shown in Fig. 5.3. These maps are also derived from the steady state characteristics of the AMESim powertrain model described in Sect. 5.4.1.

Following the procedure given in Sect. 5.3.3, we first choose κ_2 to meet the requirement for the desired decay rate of the estimation error. It is desired that the

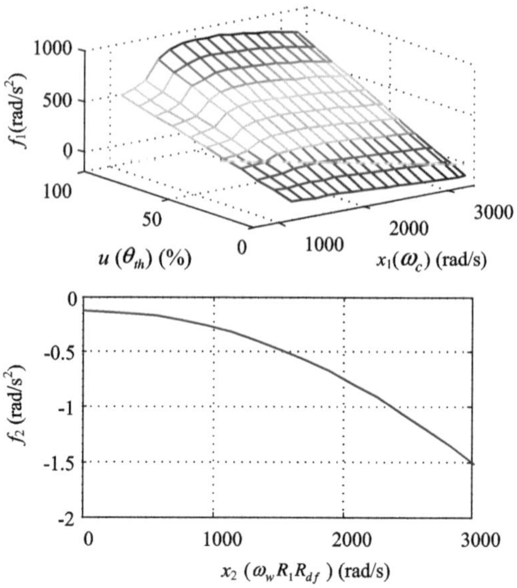

Fig. 5.3 MAPs of nonlinear functions f_1, f_2

error converges in 0.1 s, and we consider the settling time as 4 time constants [21], which implies $\frac{4}{\kappa_2} = 0.1$ and results in $\kappa_2 = 40$.

Then κ_1 is chosen with the purpose of achieving an acceptable offset of the estimation error. To do this, we need first to determine $\boldsymbol{H}_i, i = 1, 2, \ldots, 6$ by the bounds of modeling errors for different gear positions.

Since powertrain systems admit highly nonlinear complex dynamics and various uncertainties, it is indeed very difficult, if not impossible, to obtain a comprehensive estimation of the modeling error bound. Hence, we only consider some major uncertainties as an example to estimate the elements of $\boldsymbol{H}\boldsymbol{w}$. The major uncertainties considered here are the estimation error of the engine torque T_e, the variations of road grade θ_g and vehicle mass m, which affect the driving resistance T_v and the inertia I_v in (5.4a)–(5.4c). The estimation error of the engine torque is assumed to be bounded within $\pm 10\ \%$ of the true value. Hence, $\|h_{i1} w_1\|_\infty$ is estimated by

$$\|h_{i1} w_1\|_\infty = 10\ \% \times \frac{T_{e\max}}{I_i}, \tag{5.28}$$

where $T_{e\max}$ is the maximum value of the engine torque with $T_{e\max} = 620$ Nm. It is assumed here that the error bound ($\pm 10\ \%$) covers the transient estimation error and the torque variation due to long-term aging. It should be noted, however, that the error bound is somewhat conservative for the calculation of the estimation offset.

Then we change the settings of the road grade θ_g and vehicle mass m to calculate $\|h_{i2} w_2\|_\infty$. The un- and full-laden masses of the truck are 4000 and 8000 kg, respectively, hence the nominal mass used for the observer design is set to be 6000 kg. When the vehicle mass is increased from 6000 to 8000 kg, and the road grade angle is increased from 0 to 5 degrees, the modeling error $\|h_{i2} w_2\|_\infty$ under full throttle operation is calculated as

$$\|h_{i2}w_2\|_\infty = \left(\frac{T_{e\max} R_i R_{df}}{I_{v0}} - \frac{T_{e\max} R_i R_{df} - m_1 g R_w \sin(5^\circ)}{I_{v1}}\right) R_i R_{df}, \quad (5.29)$$

where I_{v0} is the nominal value of the vehicle inertia, m_1 is the fully loaded mass, I_{v1} is the vehicle inertia when fully loaded, R_w is the tire radius. The results for the 1st gear position read

$$\|h_{11}w_1\|_\infty = 89 \text{ rad/s}^2, \quad (5.30a)$$

$$\|h_{12}w_2\|_\infty = 205 \text{ rad/s}^2. \quad (5.30b)$$

Hence $\boldsymbol{H}_1$ is set to be

$$\boldsymbol{H}_1 = \begin{pmatrix} 89 & 0 \\ 0 & 205 \end{pmatrix}. \quad (5.31)$$

Similarly, $\boldsymbol{H}_i$ is computed for $i = 2, \ldots, 6$, and reads

$$\boldsymbol{H}_2 = \begin{pmatrix} 88 & 0 \\ 0 & 104 \end{pmatrix},$$

$$\boldsymbol{H}_3 = \begin{pmatrix} 87 & 0 \\ 0 & 57 \end{pmatrix},$$

$$\boldsymbol{H}_4 = \begin{pmatrix} 84 & 0 \\ 0 & 31 \end{pmatrix},$$

$$\boldsymbol{H}_5 = \begin{pmatrix} 78 & 0 \\ 0 & 18 \end{pmatrix},$$

$$\boldsymbol{H}_6 = \begin{pmatrix} 66 & 0 \\ 0 & 11 \end{pmatrix}.$$

Iterate Step 2–Step 5 of the procedure given in Sect. 5.3.3 to determine a suitable κ_1. The result reads $\kappa_1 = 30$ with the observer gain of

$$\boldsymbol{L} = \begin{pmatrix} -0.1714 & 0.0207 \end{pmatrix}. \quad (5.32)$$

According to (5.25a), (5.25b), the corresponding offset of the 1st gear is

$$e_1(\infty) = -0.217, \quad (5.33a)$$

$$e_2(\infty) = 0.060. \quad (5.33b)$$

Then the offset is bounded as

$$\left|e(\infty)\right| \le \left|e_1(\infty)\right| + \left|e_2(\infty)\right| = 0.277, \quad (5.34)$$

which is within 12 % of the maximum shaft torque. Similarly, the offset bounds of the 2nd–6th gears are 0.171, 0.111, 0.0724, 0.0477, and 0.0315, respectively, which are all within 12 % of the maximum shaft torque of the corresponding gear positions.

It should be noted that although the error of 12 % is relatively large for some dynamic control applications (such as shift control), it is a conservative upper bound of the estimation offset, and the real offset could be much smaller, because in the above derivation

(a) The multiple use of inequalities enlarges the calculated result;
(b) Setting disturbances as step signals is also conservative.

It is also worth noting that the shaft torque error could be reduced through improving the estimation accuracy of the engine torque [26]. For example, if the estimation error of the engine torque is assumed to be bounded within ±5 % of the true value, the bounds of $|e(\infty)|$ become 0.131, 0.0823, 0.0543, 0.0356, 0.0235, and 0.0155, respectively, which are less than 5.6 % of the maximum shaft torque of the corresponding gear positions.

5.4 Simulation Results

5.4.1 Powertrain Simulation Model

In this section, the proposed observer of axle shaft torque is evaluated on a powertrain simulation model. The model is established by the commercial simulation software AMESim, which supports the Simulink environment by S-Function. The constructed model can capture the important transient dynamics of the driveline, such as the drive shaft oscillation and the tire slip.

Engine

Because there is no torque converter included in AMT vehicles, the engine model used is a little more precise than that of the last chapters. The engine model based on AMESim submodels gives the output torque, fuel consumption and emissions, etc., according to the accelerator pedal position acted by the driver, engine speed and water temperature.

An initial torque is read in a lookup table, i.e., a map relative to the engine speed and the load requested by the control unit. Then the torque is corrected by considering the friction losses, which is given as a map of the friction mean effective pressure (FMEP) relative to the engine speed and the water temperature. The maps of the engine torque and the friction losses are shown in Fig. 5.4. The lag time from load request to torque generation is also considered, and treated as a first-order lag.

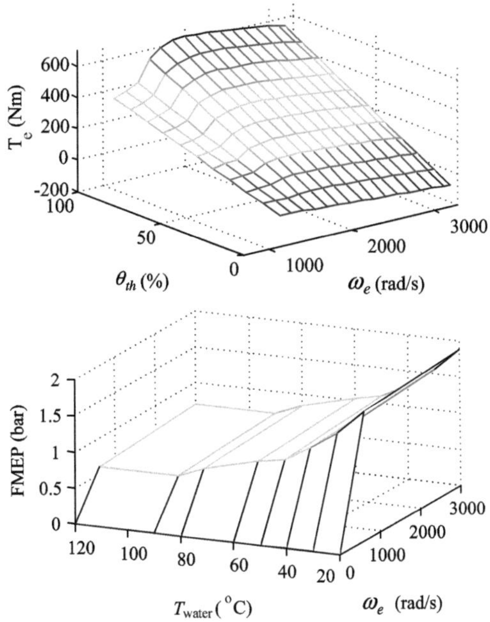

Fig. 5.4 Engine torque map and friction map

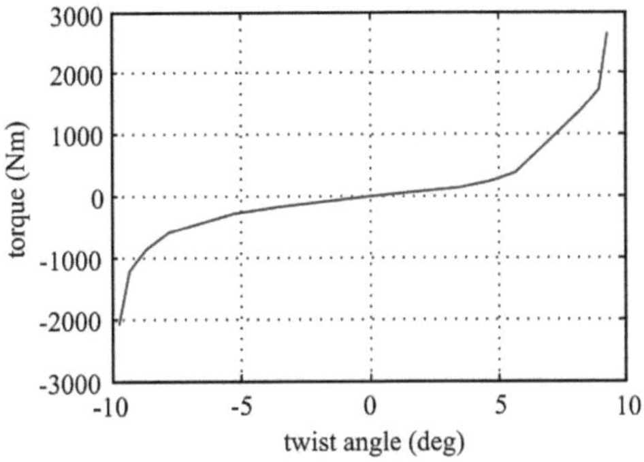

Fig. 5.5 Clutch spring characteristics

Clutch and Transmission

Here an AMT with a 6-speed transmission is used, the speed ratios of which are shown in Table 5.2. The dry clutch is modeled in consideration of the internal damping. The spring characteristics of the clutch are shown in Fig. 5.5.

The parameters used in the powertrain simulation model are listed in Table 5.2. The parameters represent a typical medium-duty truck equipped with a 6.2 l diesel engine.

Table 5.2 Nominal values of simulation model parameters

Engine		
I_e	Inertia of crank and fly wheel	$0.68\ \mathrm{kg\,m^2}$
T_{water}	Water temperature	80 °C
Clutch		
I_c	Inertia of clutch plate	$0.005\ \mathrm{kg\,m^2}$
C_{ct}	Damping of clutch twist motion	0.2 Nm/(rad/s)
$T_{c\max}$	Maximum Coulomb friction torque	900 Nm
Transmission		
I_t	Inertia of transmission input	$0.008\ \mathrm{kg\,m^2}$
C_t	Damping of transmission input	0.05 Nm/(rad/s)
R_1–R_6	Gear ratio	7.57, 5.00, 3.38, 2.25, 1.50, 1.00
Differential gear		
R_{df}	Gear ratio	5.0
Drive axle shaft		
I_s	Equivalent inertia from transmission output shaft to axle shaft	$5.8\ \mathrm{kg\,m^2}$
C_s	Damping of axle shaft input	0.8 Nm/(rad/s)
K_s	Axle shaft stiffness	900 Nm/deg
C_{st}	Damping of axle shaft torsion	200 Nm/(rad/s)
Tire		
I_w	Inertia of one tire	$5\ \mathrm{kg\,m^2}$
R_w	Tire radius	0.51 m
T_w	Resistant moment of tires	300 Nm
dS_x	Longitudinal slip threshold of tire	0.1
$F_{x\max}$	Maximum longitudinal force of tire	18000 N
Vehicle		
m	Vehicle mass	6000 kg
θ_g	Road grade	0°
ρ	Air density	$1.2\ \mathrm{kg / m^3}$
A_A	Front area of vehicle	$6\ \mathrm{m^2}$
C_D	Aerodynamic drag coefficient	0.7

Discrete Speed Sensor Model

Besides the complete powertrain simulation model of the vehicle drivetrain, a discrete speed sensor model is also constructed. The precision of speed measurement greatly influences the observer accuracy. In production vehicles with ABS (Anti-lock Brake System), magnetic pickup sensors are available for the measurement of the clutch output speed and wheel speed. It is more accurate if the measurement noise brought about by this kind of sensors is included in the simulation model.

There are generally two methods to detect the shaft speed by pick-up sensors, one measuring the angle passed in a certain time, and the other measuring the time needed to pass a certain angle. The first method can be used for low speed control systems, such as optimal gear position determination systems. For the highly transient applications, such as ABS and gear shift quality control, the second method is necessary.

When measuring the time interval corresponding to a certain number of teeth, the shaft speed can be calculated from

$$\omega = \frac{2\pi n_{\text{in}}}{\Delta t n}, \tag{5.35}$$

where n is the total number of teeth, n_{in} is the number of teeth corresponding to the time measurement, Δt is the counted time interval.

Measurement delay results from the time required for a new tooth to pass the pickup. Moreover, the irregularities of the teeth position and the randomness of the trigger, which convert the analog signal into the square-wave signal, may introduce random sensor noises. The general method to simulate realistic sensor characteristics is to construct a discrete sensor model [17, 20], where the randomness can be taken into account by adding a random angle to the regular tooth angle.

The speed sensors of this work are assumed to have 48 teeth, and the time interval corresponding to 3 teeth is recorded to calculate the rotational speed. A relative tolerance of teeth location of 0.169 % [17] and a trigger randomness of 1.5 % are considered.

Figure 5.6 shows a simulation example of the output of the speed sensor model and the "true" value of speed.

5.4.2 Simulation Results

Figure 5.7 gives the simulation results of the 1st gear drive with the parameter setting of Table 5.2, based on which the simplified model of observer design is derived. From the speeds of ω_c and ω_w, it can be seen that intensive shaft oscillations are provoked by the variation of the engine torque. The engine torque T_e is shown, as well as the estimated engine torque $\hat{T}_e = f_1(\omega_e, \theta_{th})I_1$ used in the observer. From Fig. 5.7, the maximum estimation error (during 5–6 s) of the engine torque is about 75 Nm, i.e., 12 % of the maximum engine torque. It is large enough to cover the error bound of on-board engine torque estimations [4, 11], which means that the simulation is able to check the performance of the designed observer when the engine torque is not accurately known. On the other hand, the designed observer tracks the shaft torque rapidly and the maximum error is about 2000 Nm, which is 8.5 % of the largest shaft torque of the 1st gear. Note that large estimation error of drive

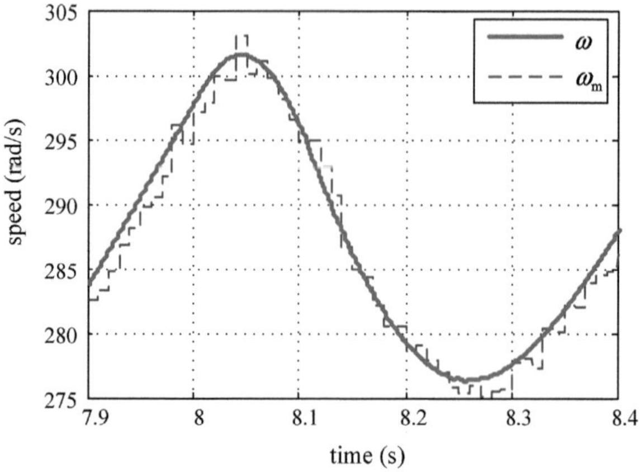

Fig. 5.6 Speed sensor output

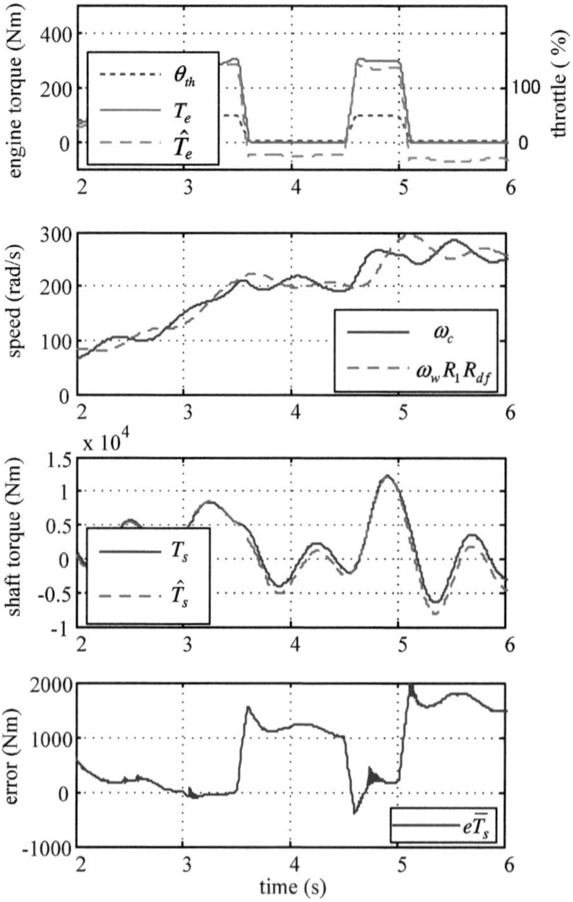

Fig. 5.7 Simulation results (1st gear, $m = 6000$ kg, $\theta_g = 0°$)

Table 5.3 Mean value $E(|e|)$, percentage over corresponding maximum torque $P(|e|)$ and standard deviation $SD(|e|)$ of estimation error

Measure	Fig. 5.7 (1st gear)	Fig. 5.10 (3rd gear)	Fig. 5.11 (6th gear)
$E(\|e\|)$ (Nm)	760.3	637.5	153.8
$P(\|e\|)$	3.24 %	6.08 %	4.96 %
$SD(\|e\|)$ (Nm)	638.7	220.8	79.2

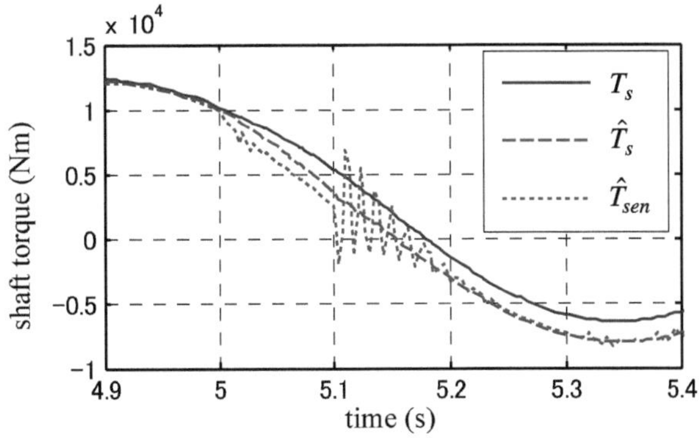

Fig. 5.8 Effectiveness of the proposed observer

shaft torque also appears in 5–6 s, which shows that the estimation error depends largely on the estimation accuracy of the engine torque.

Moreover the mean value and deviation of estimation error are given in Table 5.3 in order to show the overall performance of the observer. It is shown that the mean error is less than about 6 % of the maximum shaft torque of the corresponding gear, which is much less than 12 %, the theoretic result of (5.34) in Sect. 5.3.4.

An alternative method to estimate the drive shaft torque seems to be feasible, i.e., the method of subtracting the inertia torque from the estimated engine torque, which is described by the following equation:

$$\hat{T}_{sen} = (\hat{T}_e - I_1\dot{\omega}_c)R_1 R_{df}. \tag{5.36}$$

The result is plotted in Fig. 5.8 as a dotted line, where the dashed line denotes the result of the proposed observer and the solid line plots the "true" values. Because of the high frequency twist of the clutch springs, serious oscillations are shown in $\hat{T}_{sen}$, which verifies the potential benefits of the proposed observer.

Then the proposed observer is tested under the driving conditions that deviate from the nominal setting. Similarly as Sect. 5.3.4, the vehicle mass and road grade are changed. The results are shown in Fig. 5.9. It can be seen that the variation of the driving condition does not seriously affect the estimation error, and the maximum error is about 2100 Nm, which is still less than 10 % of the maximum shaft torque.

The estimation results of the 3rd gear and the 6th gear are plotted in Fig. 5.10 and Fig. 5.11, respectively. The estimation error converges rapidly, and the error offset (about 1000 Nm in Fig. 5.10, and 200 Nm in Fig. 5.11) is within the anticipated

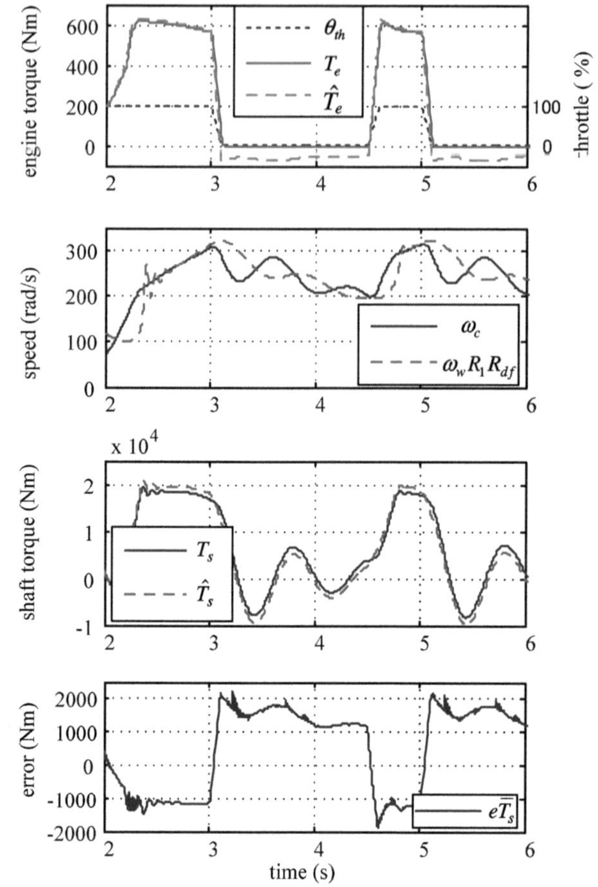

Fig. 5.9 Simulation results (1st gear, $m = 8000$ kg, $\theta_g = 5°$)

levels. Similar to the results of the 1st gear, relatively large steady state error appears because of the large estimation error (somewhat conservative) of the engine torque.

At the same time, there is some large overshoot in the results, and the higher the gear position, the more serious the overshoot. This may motivate to design a switching observer for different gear positions.

Finally, in order to get an in-vehicle assessment of the proposed observer, it is discretized at the sampling frequency 100 Hz [10]. Furthermore, the discrete models of speed sensors [17, 25] are used to give the clutch speed ω_c and the wheel speed ω_w. The speed sensors of this work are assumed to have 48 teeth, and the time interval corresponding to 3 teeth is recorded to calculate the rotational speed. A relative tolerance of teeth location of 0.169 % [17] and a trigger (to convert the analog signal into the square-wave signal) randomness of 1.5 % are considered. Note that because of the strong influence of the sampling time of 10 ms, in the discrete implementation, the observer gain has to be reduced in order to restrain oscillations resulting

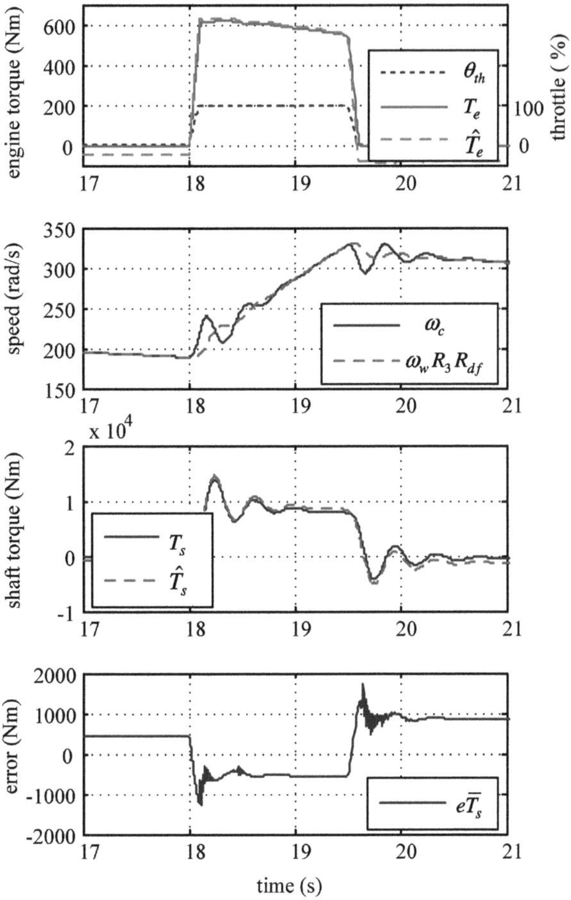

Fig. 5.10 Simulation results (3rd gear, $m = 6000$ kg, $\theta_g = 0°$)

from sampling, and the tuning result is

$$\boldsymbol{L} = \begin{pmatrix} -0.0343 & 0.0041 \end{pmatrix}. \tag{5.37}$$

It is also worth noting that in the discrete implementation, the simulation time steps of the observer model (Simulink) and the vehicle model (AMESim) are different, and the vehicle model is simulated under much shorter time step (less than 1 ms).

The estimation results of the discrete implementation are given in Fig. 5.12. The vehicle is driven in the 1st gear and the driving conditions are the same as in Fig. 5.7. Although the discretization and sensor noise (generated by the constructed speed sensor model) result in some noises, the estimation error is still within the anticipated level.

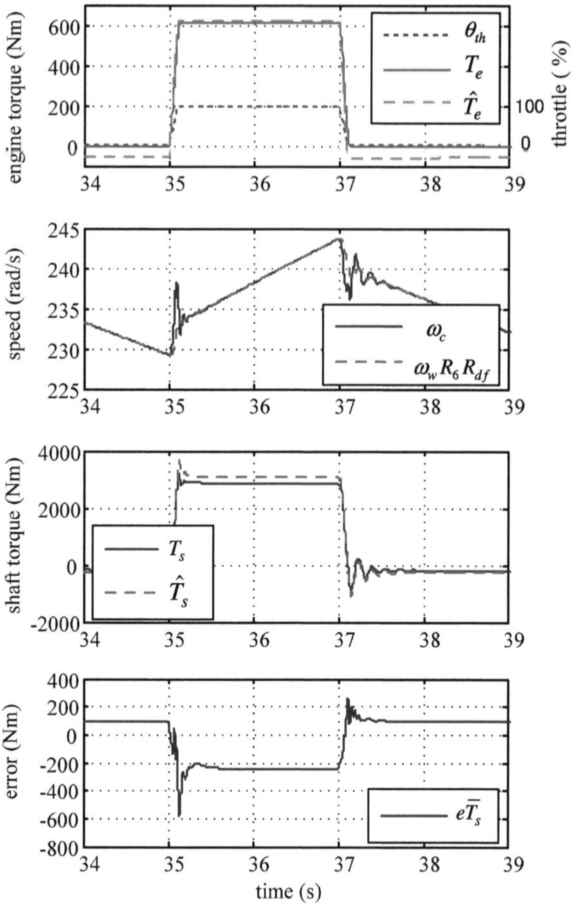

Fig. 5.11 Simulation results (6th gear, $m = 6000$ kg, $\theta_g = 0°$)

5.5 Notes and References

Simulation results show that the proposed observer is robust to driving condition variations, and the observer with constant gain provides satisfying estimation error offset for all gear positions.

However, there is an overshoot that appears in the results of high gear driving. Because the vehicle with step-ratio transmission is a class of switching system, we can solve this problem by designing a switching observer. Concerning switched systems [14], the question of how to design a switching observer with exponential error convergence has been studied by many applications; please refer to [2, 5] for detailed information.

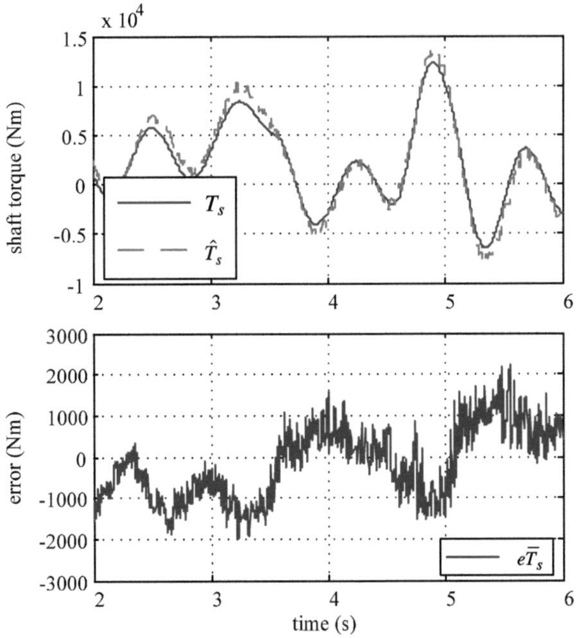

Fig. 5.12 Discrete implementation (1st gear, $m = 6000$ kg, $\theta_g = 0°$)

References

1. Baumann J, Torkzadeh D, Ramstein A, Kiencke U, Schlegl T (2006) Model-based predictive anti-jerk control. Control Eng Pract 14(3):259–266
2. Bejarano FJ, Pisano A (2011) Switched observers for switched linear systems with unknown inputs. IEEE Trans Autom Control 56(3):681–686
3. Berriri M, Chevrel P, Lefebvre D (2008) Active damping of automotive powertrain oscillations by a partial torque compensator. Control Eng Pract 16(7):874–883
4. Brahma I, Sharp M, Frazier T (2008) Estimation of engine torque from a first law based regression model. SAE technical paper 2008-01-1014
5. Chen W, Saif M (2004) Observer design for linear switched control systems. In: Proceeding of the 2004 American control conference, Boston, Massachusetts, June 30–July 2, 2004, pp 5796–5801
6. Dolcini P, Wit CC, Béchart H (2008) Lurch avoidance strategy and its implementation in amt vehicles. Mechatronics 18(5–6):289–300
7. Fredriksson J, Egardt B (2000) Nonlinear control applied to gearshifting in automated manual transmissions. In: Proceedings of the 39th IEEE conference on decision and control, Sydney, Australia, vol 1, pp 444–449
8. Gao B-Z, Chen H, Zhao H-Y, Sanada K (2010) A reduced-order nonlinear clutch pressure observer for automatic transmission. IEEE Trans Control Syst Technol 18(2):446–453
9. Gao B-Z, Chen H, Ma Y, Sanada K (2011) Design of nonlinear shaft torque observer for trucks with automated manual transmission. Mechatronics 21(6):1034–1042
10. Hahn JO, Lee KI (2002) Nonlinear robust control of torque converter clutch slip system for passenger vehicles using advanced torque estimation algorithms. Veh Syst Dyn 37(3):175–192
11. Katsumata M, Kuroda Y, Ohata A (2007) Development of an engine torque estimation model: Integration of physical and statistical combustion model. SAE technical paper 2007-01-1302

12. Krstić M, Kanellakopoulos I, Kokotović P (1995) Nonlinear and adaptive control design. Wiley, New York
13. Lefebvre D, Chevrel P, Richard S (2003) An H-infinity-based control design methodology dedicated to the active control of vehicle longitudinal oscillations. IEEE Trans Control Syst Technol 11(6):948–956
14. Liberzon D (2003) Switching in systems and control. Birkhäuser, Boston
15. Lingman P, Schmidtbauer B (2002) Road slope and vehicle mass estimation using Kalman filtering. Veh Syst Dyn Suppl 37:12–23
16. Lucente G (2007) Modelling of an automated manual transmission system. Mechatronics 17(2–3):73–91
17. Masmoudi RA, Hedrick K (1992) Estimation of vehicle shaft torque using nonlinear observers. ASME J Dyn Syst Meas Control 114:394–400
18. Merry RJE, Molengraft MJG, Steinbuch M (2010) Velocity and acceleration estimation for optical incremental encoders. Mechatronics 20(1):20–26
19. Misawa EA, Hedrick JK (1989) Nonlinear observers—a state-of-the-art survey. ASME J Dyn Syst Meas Control 111:344–352
20. Moskwa JJ, Pan CH (1995) Engine load torque estimation using nonlinear observers. In: Proceedings of the 34th IEEE conference on decision and control, New Orleans, LA, pp 3397–3402
21. Ogata K (2001) Modern control engineering, 4th edn. Prentice Hall, New York
22. Pettersson M (1997) Driveline modeling and control. PhD Thesis, Linköping University, Sweden
23. Pettersson M, Nielsen L (2000) Gear shifting by engine control. IEEE Trans Control Syst Technol 8(3):495–507
24. Pettersson M, Nielsen L (2003) Diesel engine speed control with handling of driveline resonances. Control Eng Pract 11(3):319–328
25. Schwarz R, Nelles O, Scheerer P, Isermann R (1997) Increasing signal accuracy of automotive wheel-speed sensors by on-line learning. In: Proceedings of American control conference, Albuquerque, NM, pp 1131–1135
26. Stotsky AA (2006) Method for estimating engine friction torque. United States Patent No 7,054,738
27. Umbach F, Acker H, Kluge JV, Langheinrich W (2002) Contactless measurement of torque. Mechatronics 12(8):1023–1033
28. Vahidi A, Stefanopoulou A, Peng H (2005) Recursive least squares with forgetting for online estimation of vehicle mass and road grade: theory and experiments. Veh Syst Dyn 43(1):31–55
29. Watechagit S, Srinivasan K (2003) On-line estimation of operating variables for stepped automatic transmissions. In: IEEE conference on control applications (CCA 2003), Istanbul, Turkey, vol 1, pp 279–284
30. Webersinke L, Augenstein L, Kiencke U (2008) Adaptive linear quadratic control for high dynamical and comfortable behavior of a heavy truck. SAE technical paper 2008-01-0534
31. Yi K, Hedrick K, Lee SC (1999) Estimation of tire-road friction using observer based identifiers. Veh Syst Dyn 31(4):233–261

Chapter 6
Clutch Disengagement Timing Control of AMT Gear Shift

6.1 Introduction

Automated Manual Transmissions (AMTs), as shown in Fig. 6.1, are generally constituted by a dry clutch and a multi-speed gearbox, both equipped with electro-mechanical or electro-hydraulic actuators, which are driven by a Transmission Control Unit (TCU). Compared with other topologies of automatic transmissions, AMTs have the advantages of lower weight and higher efficiency [2, 13], and they are widely adopted to offer easy drive and fuel efficiency for trucks. AMTs are also suitable for parallel hybrid electric vehicles [12]. However, one limitation of AMTs is the reduction of driving comfort, caused by the lack of traction during gear shift actuation. The problem is even more serious in the case of heavy duty vehicles, where the driveline torsion is relatively large. Therefore, aiming to improve the shift quality, it is necessary to take into account the reduction of shift time and shift shock [5] in a proper gear shift management.

At the beginning of the gear shift process of AMTs, the torque transmitted by the transmission is decreased and then cut off by active engine control (motor as well in the case of hybrid electric vehicles (HEVs) [11, 12]) and clutch disengagement, then the neutral gear is engaged. Next follows the speed synchronization of the transmission shafts and the engagement of the new gear. Finally, the clutch is engaged, and the engine torque level is recovered as demanded by the driver. Note that some of the above operations may be omitted in some new shift techniques, such as AMTs without synchronizer [3].

The aforementioned actions are usually lumped into 3 phases [16] to reduce shift time. As shown in Fig. 6.2, a typical power-on upshift process, the first phase is the so-called torque control phase, wherein the driveline torque is reduced to zero, and the neutral gear is engaged. Next comes the speed synchronization phase, where the speed difference is synchronized and the new gear is engaged. Finally, during the last phase, the torque level is recovered as demanded by the driver.

The shift shock may be caused during two actuations. First, as it is well known, at the end of the shift, if the clutch is engaged too quickly or the engine torque is recovered too rapidly, driveline resonances may be produced [2, 8]. Besides the above

H. Chen, B. Gao, *Nonlinear Estimation and Control of Automotive Drivetrains*,
DOI 10.1007/978-3-642-41572-2_6,

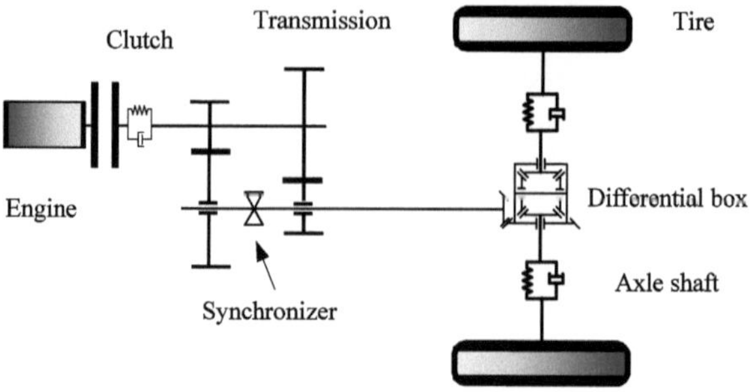

Fig. 6.1 Driveline scheme of AMT

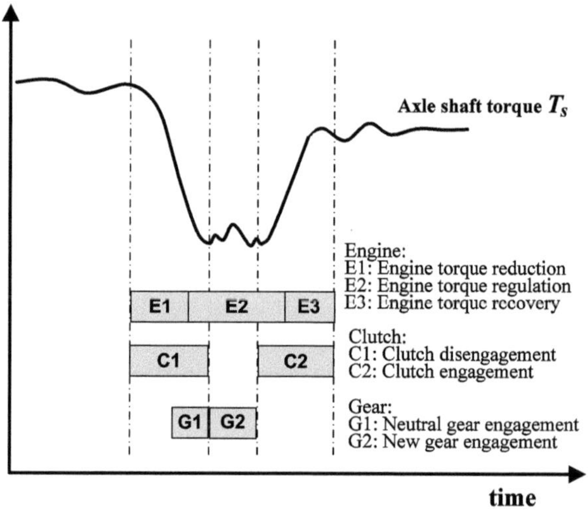

Fig. 6.2 Time sequence of gear upshift process

operations, the clutch disengagement may bring about severe driveline oscillation as well. If the timing of the clutch disengagement is not well controlled, in other words, there is large elastic torsion in the driveline when the traction is interrupted, the potential energy accumulated in the driveline will lead to unwanted driveline and vehicle oscillations.

In order to restrain the driveline oscillation caused by traction drop, it is suggested in [3] that the actuation of traction interruption (clutch disengagement) should be carried out at the moment when the transmission torque (here the "transmission torque" refers to the torque delivered to the clutch) is controlled to zero. In [15–17], based on the general fact that the drive shaft is the main component of the driveline, it is pointed out that the drive shaft torque (here the "drive shaft torque" refers to the torque delivered to the drive axle shaft) can be used instead of the transmission torque. And it is assumed that if the drive shaft torsion is small, the transmission torque is also small. The drive shaft torque is estimated for a full-state feedback controller of active engine control, which aims to damp the driveline resonance as soon as possible.

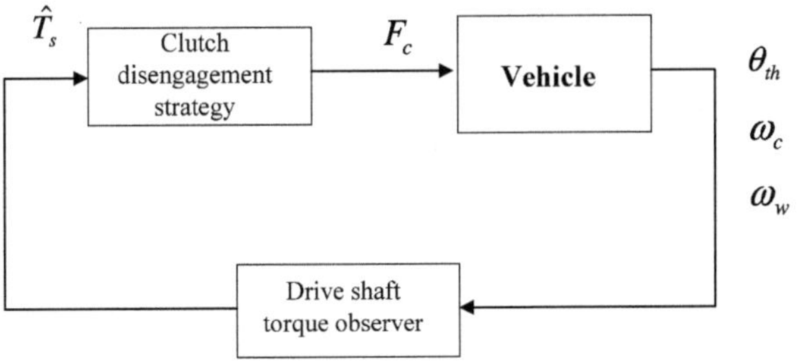

Fig. 6.3 Block diagram of a clutch disengagement system ($\hat{T}_s$, estimated drive shaft torque; F_c, clutch engagement force; θ_{th}, engine throttle; ω_c, ω_w, transmission input speed and wheel speed)

In this chapter, for a further step, the knowledge of drive shaft torque is used to constitute a closed-loop clutch disengagement strategy. Because we prefer short shift time (the requirement is especially strict for heavy-duty vehicles to shift on a slope), it is desired that the engine torque decreases rapidly at the beginning of the shift process. For such a highly transient process, the clutch disengagement strategy is designed so that the clutch is fully disengaged when the drive shaft torque reaches zero for the first time (see Fig. 6.4 for reference). It is reasonable to believe that such a strategy is optimal for the reduction of total shift time while assuring small shift shock because if the clutch is disengaged at an earlier time, the drive shaft torsion may cause severe driveline oscillations, and on the other hand, if the clutch is disengaged until the driveline fluctuations are fully damped out, the total shift time may be prolonged.[1]

6.2 Observer-Based Clutch Disengagement Timing Control

During the first phase, the engine torque is withdrawn, followed by the decrease of the axle shaft (or half shaft) torque T_s. It may take several hundreds of milliseconds for T_s to drop to 0 Nm. During this period, the clutch disengagement and neutral-gear engagement can be carried out simultaneously to reduce the total shift time.

The observer of axle shaft torque designed in the last chapter (Chap. 5) is used for the suggested clutch disengagement strategy, and the block diagram of the proposed system is described in Fig. 6.3.

The block of the "clutch disengagement strategy" controls the clutch engagement force so that the clutch is fully disengaged at the moment when the estimated drive shaft torque $\hat{T}_s$ reaches zero. The vehicle of interest is a medium-duty truck with a 6.2 l diesel engine. The sensors used for rotational speed measurement are Hall-effect pick-up sensors, which are widely used for anti-lock brake systems and automatic transmissions. Because the precision of the proposed drive shaft torque observer relies on the engine torque estimation, a revised observer with switched gains is given as well, which can be activated in the case of large estimation error of the engine torque [4].

[1]This chapter uses the content of [4], with permission from Taylor & Francis.

6.3 Clutch Disengagement Strategy

The clutch engagement force F_c is regarded as the control input. Note that the clutch engagement force is indeed not the initial control variable. In production AMTs, the diaphragm spring of the dry clutch is actuated by electro-mechanical or electro-hydraulic actuators. The characteristics of the hydraulic or electric actuators, however, are specific to a given transmission and implementation method. If the clutch engagement force is assumed to be the control variable, the control strategy could be applicable to various kinds of AMTs, where the used actuator is controlled to deliver the desired force.

When the clutch is slipping, the torque T_c delivered through the clutch is determined by

$$T_c = F_c \mu_d R_c \,\mathrm{sign}(\Delta\omega), \tag{6.1}$$

where μ_d is the dynamic friction coefficient, R_c is the effective radius.

If the clutch is sticking (locked up), the delivered torque is

$$T_c = \frac{\hat{T}_s}{i_i i_{df}} + I_{ict}\dot{\omega}_c, \tag{6.2}$$

where I_{ict} is the inertia from the clutch to the drive shaft, and the subscript i denotes the ith gear position, $i = 1, 2, \ldots, 6$. The delivered torque is no longer determined by the clutch engagement force F_c. However, the maximally transmittable torque for non-slip condition is limited by F_c, i.e.,

$$T_{c\max} = F_c \mu_s R_c \,\mathrm{sign}(T_c), \tag{6.3}$$

where μ_s is the static friction coefficient.

It is desired here that the clutch is disengaged as soon as the drive shaft torque reaches zero, and before that the clutch should be locked up without slipping. Therefore, we design F_c having the following form:

$$F_c = \kappa_c \frac{\hat{T}_s}{i_i i_{df} \mu_d R_c}, \tag{6.4}$$

where κ_c is a coefficient larger than 1. If the value of κ_c is small, clutch slip may be caused before the drive shaft torque reaches zero. On the other hand, if κ_c is too large, the time for clutch actuation will be too short. The tuned value is $\kappa_c = 1.3$. It is clear that by such a clutch disengagement strategy, the clutch will be disengaged when the estimated drive shaft torque $\hat{T}_s$ is approaches zero, and before that the clutch is locked up.

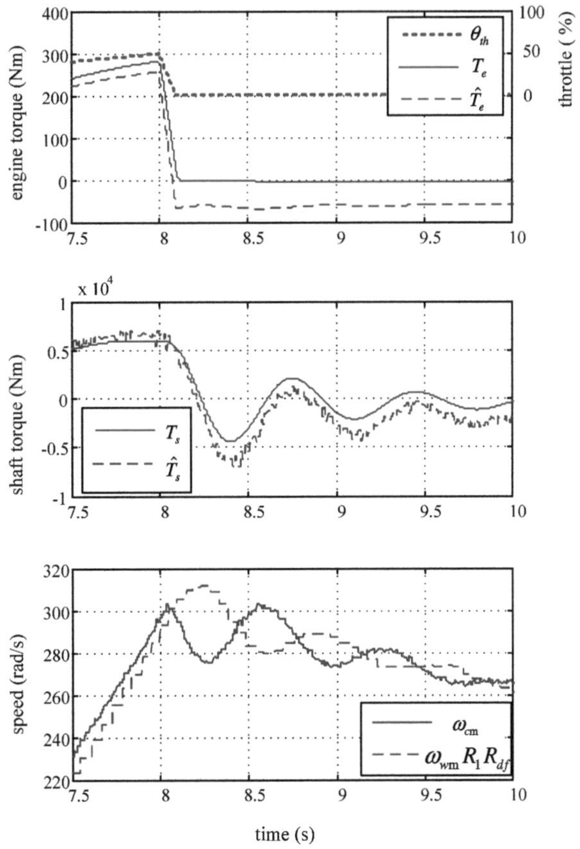

Fig. 6.4 Simulation results (no clutch operation; $m = 6000$ kg, $\theta_g = 0°$, $I_e = 0.68$ kg m^2, $K_s = 900$ Nm/deg)

6.4 Simulation Results

6.4.1 Simulation Results with Constant Observer Gain

The proposed clutch control strategy is tested on the AMESim powertrain simulation model which has been introduced in Chap. 5. In order to get an in-vehicle assessment of the proposed clutch disengagement control system, the designed observer is discretized by a sampling rate of 100 Hz [6] with zero-order holder discretization.

Figure 6.4 gives the simulation results of the engine torque reduction at the beginning of the 1st-to-2nd gear upshift process; however, without clutch operation. The driving condition is the same as the nominal setting, based on which the simplified model and its parameters for the observer design are derived. At 8.0 s, the shift process is started and the engine throttle θ_{th} begins to decrease, followed by the decrease of the engine torque T_e. The estimated engine torque used at the observer, $\hat{T}_e$, is given as well. Note that because of the engine brake effect, there are some negative values in T_e and $\hat{T}_e$ when the throttle angle is small and the engine speed

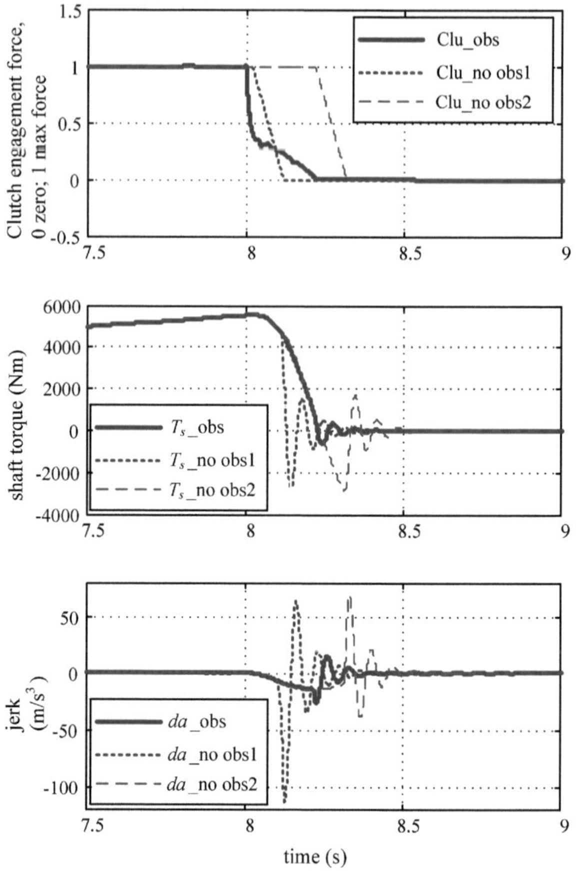

Fig. 6.5 Simulation results (with clutch operation, clutch disengaged at 8.22 s (using observer); 8.12 and 8.32 s (without observer); driving conditions and parameters are the same as those of Fig. 6.4)

is large. A sudden decrease of the engine torque results in severe drive shaft oscillation, which causes uncomfortable driveline shuffle and gear-gap shunt (clonk). The oscillation can be clearly seen from the drive shaft torque T_s, the measured clutch output speed ω_{cm} and the measured wheel speed ω_{wm}. The observer can track drive shaft torque well when there is no large engine torque estimation error.

Then the estimated drive shaft torque is used for the clutch disengagement timing control. The results are plotted as the solid lines in Fig. 6.5, where the clutch is fully disengaged at 8.22 s when the estimated drive shaft torque reaches zero for the first time. For comparison, the results when the clutch is disengaged at 8.12 and 8.32 s are given as well, which are 0.1 s before and after the optimal timing, respectively. The vehicle jerk *da* [5], namely the change rate of the longitudinal acceleration which is used to evaluate the shift shock, is plotted at the bottom of Fig. 6.5. It is clear that smooth clutch disengagement can be assured if an observer is used. Meanwhile, the disengagement based on the observer has the shortest possible shift time because the clutch is disengaged when the shaft torque reaches zero for the first time.

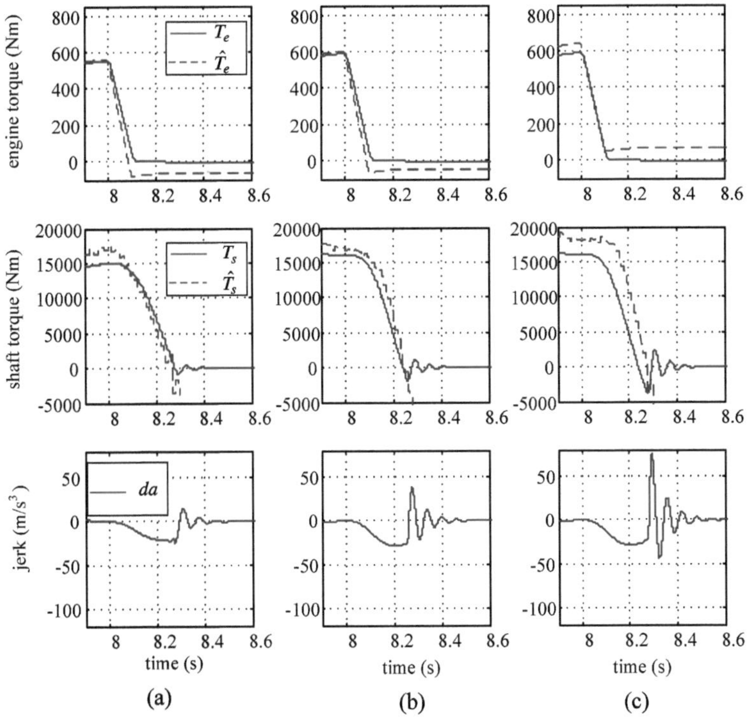

Fig. 6.6 Simulation results under different driving conditions and parameters: (**a**) the same as in Fig. 6.4 except that $m = 8000$ kg, $\theta_g = 5°$; (**b**) the same as in (**a**) except that $I_e = 0.58$ kg m^2, $K_s = 990$ Nm/deg; (**c**) the same as in (**b**) except that the estimation error of T_e is changed

In order to examine the robustness of the proposed control strategy, the driving conditions and parameters are changed step by step, and the results are shown in Fig. 6.6. Figure 6.6(a) is the simulation of the fully loaded vehicle driving on a slope. Although the vehicle mass and the resistance are greatly varied, there is no obviously change of the control performance compared with the solid line of Fig. 6.5. This is because, as shown in the last chapter, the observer gain $\boldsymbol{L} = [l_1, l_2]$ has a relatively small value of l_2, which corresponds to the output side (wheel side) of the drive shaft. Figure 6.6(b) shows the results when the stiffness of the drive shaft K_s and the engine inertia I_e are changed, and the shift shock becomes somewhat larger. It should be noted that the increase of K_s and the decrease of I_e enhance the estimation values of the drive shaft torque simultaneously. Finally, from Fig. 6.6(c), it can be seen that when the estimation error of the engine torque T_e is changed (the steady state error changes from -40 to 80 Nm), the shift shock is greatly affected.

6.4.2 Simulation Results with Switched Observer Gains

Figure 6.6 shows that the observer error may be seriously enlarged by a large engine torque estimation error. Because the engine simulation model in this study is based

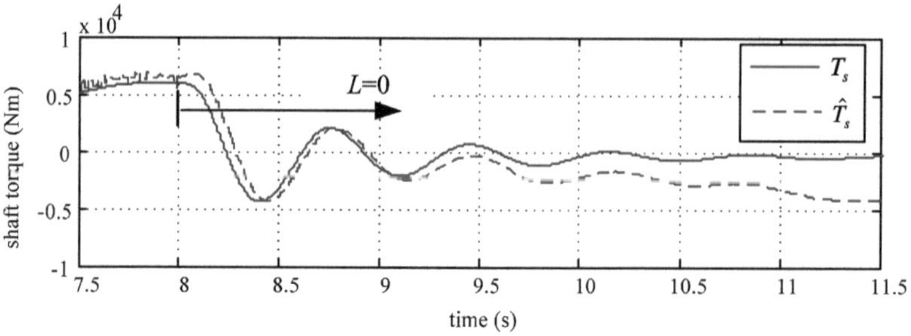

Fig. 6.7 Simulation results with switched observer gain (no clutch operation; driving conditions and parameters are the same as those of Fig. 6.4)

on the torque and friction maps, only a relatively large steady state error is represented. In real engines, however, it is the transient engine torque that is difficult to estimate precisely.

At the beginning of the gear shift of AMT, the engine works in a highly transient state, and if a large estimation error of the transient engine torque is introduced, the shift performance may be furthermore worse than that of Fig. 6.6(c). Therefore, for a easy and low cost implementation, it is preferred that the proposed drive shaft observer can providc an accurate enough estimation even when there exist large estimation errors of the engine torque.

Figure 6.7 is a drive shaft torque estimation using switched observer gains. The simulation condition and the estimation error of the engine torque T_e is the same as those of Fig. 6.4, and the speed sensors have been compensated for teeth partition defects [7]. After 8 s, the observer gain $\boldsymbol{L}$ is changed to be zero, in other words, the shaft torque is estimated using the measured speeds only, without considering a correction term. As expected, the estimation values drift out because of the accumulated sensor error. Fortunately, in a short time, such as 1 s, which is sufficient for clutch disengagement, the method can provide an accurate enough estimation.

Then, the observer with switched gains is used for clutch disengagement control with results shown in Fig. 6.8. The driving condition is set to be the same as those of Fig. 6.6(c). It can be seen that at 8.0 s, when the shift is started, the observer with a normal gain can provide a good enough initial value for the following estimation with the gain of zero. The estimated shaft torque reaches zero at 8.33 s. We can see that the clutch is also fully disengaged at about 8.33 s, and the vehicle jerk is about 50 m/s^3, which is better than in Fig. 6.6(c), and we think it is an acceptable level for a fully loaded truck shifting on a slope. Because the neutral gear is also engaged at 8.33 s, the estimated shaft torque does not track true values anymore after that.

6.5 Notes and References

As shown above, the estimation of drive shaft torque is influenced by the precision of the engine torque values. Since it is not profitable to use torque sensors in production

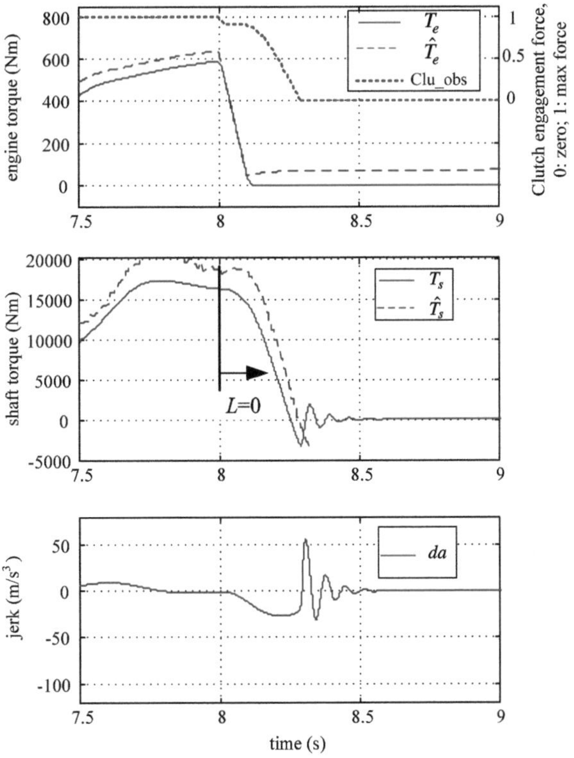

Fig. 6.8 Simulation results with switched gain (driving conditions and parameters are the same as in Fig. 6.6(c), i.e., $m = 8000$ kg, $\theta_g = 5°$, $I_e = 0.58\ \mathrm{kg\,m^2}$, $K_s = 990$ Nm/deg)

engines due to cost and integration complexity, the question of how to accurately estimate the transient engine torque becomes an important issue. If precise engine torque information is available, an observer with a switched gain, which needs a high-precision sensor and sensor calibration [7, 14], will no longer be necessary.

The engine torque can be estimated by mean-value engine models, which represent the engine with look-up tables [9, 10]. It takes much shorter CPU time to run the mean value models. However, if high precision of the engine torque estimation is required, many experiments are needed to calibrate these models; their inputs of contain variables such as engine speed, manifold absolute pressure, injection timing (spark advance for gasoline engines), fuel mass and water temperature.

Thermodynamics laws have also been used to estimate the engine torque [1, 9]. It is reported in [9] that using Wiebe function can give good estimation in both slow and fast engine operation. However, it is also pointed out that when the combustion is fast, the performance of this method may be limited.

By using complex physical models (with a characteristic timescale to the order of every crankshaft degree), the engine torque can be estimated precisely through indicated pressure estimation [20, 21]. The engine torque estimation by these methods, however, is computationally demanding.

Some other estimation methods using the oscillation measurement and Fourier decomposition of flywheel speed [18, 19] have also been proposed. The interested readers are encouraged to refer to these publications.

References

1. Brahma I, Sharp M, Frazier T (2008) Estimation of engine torque from a first law based regression model. SAE technical paper 2008-01-1014
2. Dolcini P, Wit CC, Béchart H (2008) Lurch avoidance strategy and its implementation in amt vehicles. Mechatronics 18(5–6):289–300
3. Fredriksson J, Egardt B (2000) Nonlinear control applied to gearshifting in automated manual transmissions. In: Proceedings of the 39th IEEE conference on decision and control, Sydney, Australia, vol 1, pp 444–449
4. Gao B-Z, Lei Y-L, Ge A-L, Chen H, Sanada K (2011) Observer-based clutch disengagement control during gear shift process of automated manual transmission. Veh Syst Dyn 49(5):685–701
5. Ge A (1993) Theory and design of automatic transmissions. China Machine Press, Beijing. In Chinese
6. Hahn JO, Lee KI (2002) Nonlinear robust control of torque converter clutch slip system for passenger vehicles using advanced torque estimation algorithms. Veh Syst Dyn 37(3):175–192
7. Hellstrom M (2005) Engine speed based estimation of the indicated engine torque. Master Thesis, Linkoping University, Sweden
8. Horn J, Bamberger J, Michau P, Pindl S (2003) Flatness-based clutch control for automated manual transmissions. Control Eng Pract 11(12):1353–1359
9. Katsumata M, Kuroda Y, Ohata A (2007) Development of an engine torque estimation model: Integration of physical and statistical combustion model. SAE technical paper 2007-01-1302
10. Lack AC (2003) Engine torque estimation. United States Patent, No US6584391B2
11. Lawrie RE, Reed RG, Rausen DJ (2000) Automated manual transmission shift sequence controller. United States Patent, No 6019698
12. Lin CC, Peng H, Grizzle JW, Liu J, Busdiecker M (2003) Control system development of an advanced-technology medium-duty hybrid electric truck. SAE technical paper 2003-01-3369
13. Lucente G (2007) Modelling of an automated manual transmission system. Mechatronics 17(2–3):73–91
14. Nishida K, Kaneko T, Takahashi Y, Aoki K (2011) Estimation of indicated mean effective pressure using crankshaft angular velocity variation. SAE technical paper 2011-32-0510
15. Pettersson M (1997) Driveline modeling and control. PhD Thesis, Linköping University, Sweden
16. Pettersson M, Nielsen L (2000) Gear shifting by engine control. IEEE Trans Control Syst Technol 8(3):495–507
17. Pettersson M, Nielsen L (2003) Diesel engine speed control with handling of driveline resonances. Control Eng Pract 11(3):319–328
18. Rizzoni G (1989) Estimate of indicated torque from crankshaft speed fluctuations: a model for the dynamics of the IC engine. IEEE Trans Veh Technol 38(3):168–179
19. Stotsky AA (2009) Automotive engines: control, estimation, statistical detection. Springer, Berlin
20. Zweiri YH, Seneviratne LD (2006) Diesel engine indicated and load torque estimation using a non-linear observer. Proc Inst Mech Eng, Part D, J Automob EngMech 220(6):775–785
21. Zweiri YH, Seneviratne LD (2007) Diesel engine indicated torque estimation based on artificial neural networks. In: Proceedings of the IEEE/ACS international conference on computer systems, vol 1, pp 791–798

Chapter 7
Clutch Engagement Control of AMT Gear Shift

7.1 Introduction

As shown in the last chapter (Chap. 6), for a power-on gear shift sequence of an AMT, the following actions are included: reducing engine torque, disengaging clutch, engaging neutral gear, engaging new gear, engaging clutch, and restoring engine torque. Shift shock may be caused by the operations of clutch disengagement (together with the engine torque reduction) and clutch engagement (together with the engine torque restoration). In Chap. 6, the clutch disengagement control was addressed, and in this chapter, the engagement control will be discussed in detail.

Both upshift and downshift will be considered. The process of downshift is quite different from that of upshift because the engine speed has to be reduced to reach the synchronization speed for gear upshift, while it has to be increased for gear downshift. However, the engine speed cannot be controlled equally fast in both directions [13]. In other words, in the case of downshift, the synchronization speed may already be reached when the clutch is to be engaged, while for the upshift the engine speed cannot be decelerated enough in a short time. Hence, gear upshift and gear downshift have to be addressed in different control schemes.[1]

7.2 Power-On Upshift of AMT

In the gear upshift by engine control [13, 31], the speed synchronization of the new gear is realized by active engine control and the gear is changed without operation of the dry clutch. It is necessary for the engine to decelerate as fast as possible because during the synchronization phase the traction of the wheel is totally cut off.

On the other hand, if the clutch operation is involved [19], it is easier to obtain a short synchronization phase because when the clutch is disengaged the inertia

[1]This chapter uses the content of [16], with permission from Inderscience Enterprises Ltd.

H. Chen, B. Gao, *Nonlinear Estimation and Control of Automotive Drivetrains*,

DOI 10.1007/978-3-642-41572-2_7,

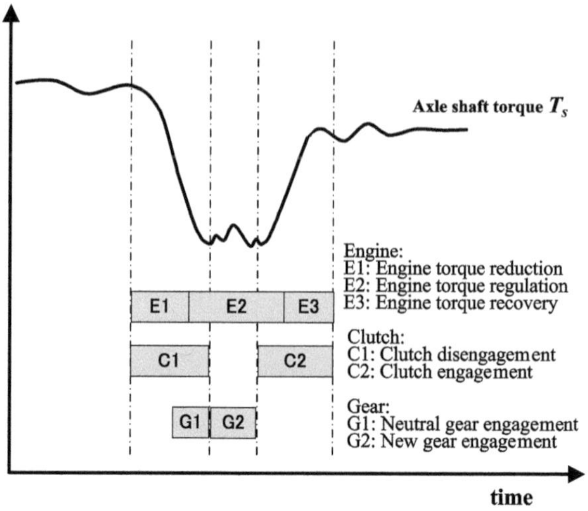

Fig. 7.1 Time sequence of the gear upshift process

moment to be synchronized is very small. Moreover, after the gear synchronization phase, when the clutch is engaged, the friction torque can be used to compensate for the traction interruption, which helps to reduce torque-interruption time. With such a control scheme, a gear upshift process with the shortest possible torque-interruption time is depicted in Fig. 7.1, which has been shown in the last chapter as Fig. 6.2.

During the first phase, the engine torque is withdrawn, and the clutch disengagement and neutral-gear engagement are carried out simultaneously to reduce the total shift time. It has been demonstrated in the last chapter (Chap. 6) that if the clutch is fully disengaged or the neutral gear is engaged just at the moment when T_s decreases to zero, the torque-reduction phase could be finished quickly while no large driveline oscillation is stimulated [15, 31].

In the second phase, the new gear is synchronized, and then engaged, by the synchronizer. Because the clutch has been disengaged, the inertia moment to be synchronized (from the clutch plate to the input shaft of the synchronizer) is very small, such as 0.05 kg m^2 for heavy-duty trucks and 0.005 kg m^2 for micro passenger cars. Hence, the gear synchronization can be finished in a short time, such as less than 0.1 s.

In the last phase, the clutch is engaged and the engine torque is recovered. It is clear that if the clutch is engaged abruptly, or the engine torque is restored too rapidly, the driveline resonances will be produced. On the contrary, if the clutch is engaged too slowly, the torque interruption time will be enlarged. Thus, the control objectives in this phase can be summarized as: (i) minimizing clutch engagement time and friction losses; (ii) keeping clutch friction torque to track the request of the driver; (iii) ensuring smooth acceleration of the vehicle. The first and the second requirements enforce the clutch to engage quickly and recover the traction back as soon as possible. The third request is added to restrain the shift shock, which is evaluated through longitudinal jerk (change rate of acceleration) [11]. During this clutch slipping phase, the cooperation of the engine torque-down control [20, 23] is important, and it can significantly reduce the shift time and shift shock.

It is reasonable to believe that, if done perfectly, the control strategy in Fig. 7.1 provides a shift time as short as possible. However, the different and sometimes conflictive control objectives make successful clutch control a challenge. Because gear shifting involves wide ranges of speed and torque, traditional controller development [12, 26], which is normally based on event-driven (rule-based) control or feedforward control, needs much calibration in order to obtain satisfying multi-objective control performance.

Aiming to reduce the calibration requirements, the model-based control [10] is introduced in the field of automotive control, and among the various model-based controller designs, Model Predictive Control (MPC) (also referred to as moving horizon control or receding horizon control) attracts much attention [4, 6, 7, 16]. MPC became a potential feedback strategy because of its ability to handle multi-variable systems, to take time-domain constraints into account explicitly and to deal with multiple objectives in an optimal sense [1, 5, 29]. Although for a long time MPC has been widely used in process industry where slow dynamics is dominant, thanks to rapid development of computing, MPC is also adopted by fast dynamical systems in recent years, such as in aerospace and defence [8].

In this section, therefore, after the dynamics and the control problems of gear upshift of AMT are investigated and summarized in detail, MPC, together with the observer techniques of the last chapter, are adopted to address these challenging problems in a *torque-based powertrain control scheme*. The vehicle of interest is still the medium-duty truck with a 6.2 l diesel engine and a 6-speed manual transmission.

7.2.1 Dynamics and Control Strategy

Torque Reduction Phase

Once a power-on upshift is initiated, as shown in Fig. 7.1, the torque transmitted to the axle shaft T_s is reduced by the active engine torque control, followed by the clutch disengagement and the neutral-gear engagement. If the timing of clutch disengagement or neutral-gear engagement is not well controlled, the potential energy accumulated in the driveline may lead to unwanted driveline oscillation.

In order to restrain the resonance caused by traction drop, it is suggested that the clutch disengagement should be carried out when the driveline torque is controlled to zero. Based on the general fact that the axle shaft is the main component of the driveline, it is pointed out in [30–32] that if the clutch is fully disengaged or the neutral gear is engaged when the axle shaft torque T_s reaches zero, there will be no severe oscillations.

The control of this torque control phase has been investigated in the last chapter, please refer to Chap. 6 for details.

Gear Synchronization Phase

In the gear synchronization phase, the new gear is synchronized and engaged by the synchronizer. During this phase, the traction is totally cut-off, and it is required to finish this phase as soon as possible. The push force that can be applied on the synchronizer is limited by the capacity of the synchronizer. Generally, the synchronization can be accomplished within 0.1 s without violating the limitation of synchronizer capacity.

Torque Recovery Phase

Because the deceleration rate of the engine speed is limited, after the new gear is engaged, the engine speed is usually larger than the clutch output speed. Then during the torque recovery phase, the clutch slips until the speed difference is synchronized, and at the same time, the engine torque is recovered.

There are many different approaches that have been proposed for clutch slip control, such as fuzzy control [34], μ synthesis [33], map-based calibration [27], sliding mode control [38], supervisory control [22, 25] and backstepping [14]. Because the clutch engagement is expected to satisfy the conflicting requirements of minimizing clutch wear and minimizing shift shock, an optimization-based algorithm becomes a potential solution for this problem. For example, Hybrid Model Predictive Control (HMPC) [4] and Linear Quadratic based optimal control [11, 18] have been used to control the clutch engagement during start-up scenario.

Actually, when the clutch is engaged during gear shifting, the friction torque (the inertia torque needed to pull the engine speed down) can be used as a compensation for the traction loss. Therefore, an important control objective could be added besides the requirements of small friction losses and small shift shock, namely, reconstructing the transmission output torque as soon as possible. This objective, as a part of the *torque-based powertrain control scheme*, is critical to reducing the total time of torque interruption, which helps to significantly improve drivability.

Control-Oriented Modeling In order to simplify the control law, the components of the driveline, including that of the axle shaft, are all neglected, and the driveline is simplified as a two-mass system as shown in Fig. 7.2. The motion of the driveline can be described by the following equations:

$$\dot{\omega}_e = \frac{1}{I_e}T_e - \frac{1}{I_e}T_c, \tag{7.1a}$$

$$\dot{\omega}_c = \frac{1}{I_{v,i}}T_c - \frac{1}{I_{v,i}}T_{v0}, \tag{7.1b}$$

where ω_e is the engine speed, ω_c is the output speed of the clutch, I_e denotes the inertia moment of the engine crank shaft, $I_{v,i}$ denotes the equivalent inertia moment

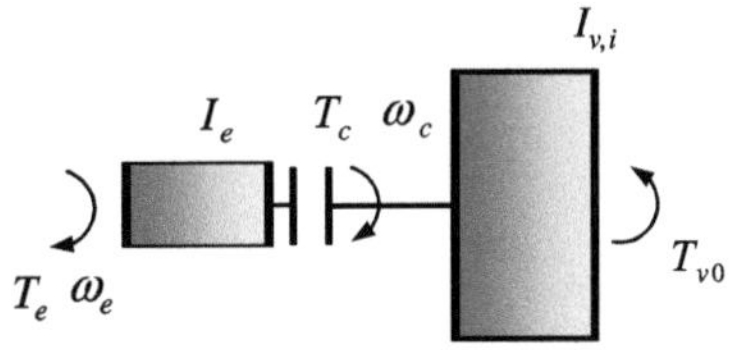

Fig. 7.2 Simplified driveline model of the torque recovery phase

from the clutch output shaft to the vehicle, at the ith gear position, T_e is the engine torque, T_c is the clutch friction torque, and T_{v0} is the converted driving resistance.

Control Strategy The control objectives of the torque recovery phase can be described in detail as

(a) The clutch slip speed is expected to decrease to zero as soon as possible, i.e., minimizing clutch engagement time;
(b) The transmission output torque is recovered back according to the demand of the driver as soon as possible, i.e., keeping the transmission output torque track a reference trajectory;
(c) Vehicle jerk is to be kept small, i.e., ensuring smooth vehicle acceleration.

The engine torque T_e is still regulated by feedforward control, and before the clutch is synchronized, T_e is recovered back according to the demand of the driver in open loop as an increasing ramp. On the other hand, the clutch torque T_c is controlled to deal with the conflicting requirements straightforwardly. Note that the requirement of minimizing friction losses is not directly included because during the gear shifting operation, the clutch engagement time is not so long as that of the start-up maneuver, and furthermore, the friction work energy could be obviously reduced by the cooperation of the engine torque-down control. *The control of the torque recovery phase is regarded as a multi-objective optimization problem, which will be solved in the framework of MPC.*

To represent the control objectives quantitatively, we choose the clutch slip speed $\Delta\omega = \omega_e - \omega_c$ as the system state, and the system dynamics is rewritten in the following state-space form:

$$\begin{aligned} \dot{x} &= A_c x + B_{cu} u + B_{cd} d, \\ y &= C x, \end{aligned} \tag{7.2}$$

with

$$A_c = 0, \tag{7.3a}$$

$$B_{cu} = -\frac{1}{I_e} - \frac{1}{I_{v,i}}, \tag{7.3b}$$

$$B_{cd} = \begin{pmatrix} \frac{1}{I_e} & \frac{1}{I_{v,i}} \end{pmatrix}, \tag{7.3c}$$

$$C = 1. \tag{7.3d}$$

The control input is

$$u = T_c, \tag{7.4}$$

and the measured (estimated) disturbance is

$$d = (T_e\ T_v)^T. \tag{7.5}$$

With the above state-space form, the control requirements could be conveniently represented by a suitably-chosen objective function, which will be seen later.

As shown in Appendix D, the model (7.2) is discretized in time with sampling period T_s, and the discrete time model is given as

$$x(k+1) = Ax(k) + B_u u(k) + B_d d(k), \tag{7.6a}$$

$$y(k) = Cx(k), \tag{7.6b}$$

where

$$A = e^{A_c T_s},$$

$$B_u = \int_0^{T_s} e^{A_c \tau} d\tau \cdot B_{cu},$$

$$B_d = \int_0^{T_s} e^{A_c \tau} d\tau \cdot B_{cd}.$$

In order to introduce the integral action to reduce offset, we rewrite (7.6a), (7.6b) in the incremental form (see Appendix D)

$$\Delta x(k+1) = A\Delta x(k) + B_u \Delta u(k) + B_d \Delta d(k), \tag{7.7a}$$

$$y(k) = C\Delta x(k) + y(k-1), \tag{7.7b}$$

where

$$\Delta x(k) = x(k) - x(k-1),$$

$$\Delta u(k) = u(k) - u(k-1),$$

$$\Delta d(k) = d(k) - d(k-1).$$

The current values of the disturbances T_e and T_v can be estimated, but their future information is not predictable, hence the values of the disturbances in the control horizon are considered constant, i.e.,

$$\Delta d(k+i) = 0 \quad \text{for } i \geq 1. \tag{7.8}$$

The requirement of control objective (a) *concerning the engagement time can be quantitatively represented by adding the penalty item of* $\|\Delta\omega - r_{\Delta\omega}\|^2$, *wherein* $r_{\Delta\omega}$ *may be chosen as 0 rad/s. As to the requirement* (b), *we can add another penalty on* $\|T_c - r_{T_c}\|^2$, *where* r_{T_c} *is determined from the acceleration pedal. Finally, the request* (c) *could be met through introducing* $\|\Delta T_c\|^2$ *into the objective function, where* ΔT_c *is the increment of* T_c. *Because the transmission output torque is determined by* T_c, *it is reasonable to believe that the vehicle jerk could be reduced if* ΔT_c *is restrained.*

Based on the above analysis, the objective function is chosen as

$$J = \sum_{i=1}^{N_p} \left\| \gamma_{\Delta\omega,i} \left(\Delta\omega(k+i|k) - r_{\Delta\omega}(k+i) \right) \right\|^2 + \sum_{i=0}^{N_c-1} \left\| \gamma_{T_c,i} \left(T_c(k+i|k) - r_{T_c}(k+i) \right) \right\|^2 + \sum_{i=0}^{N_c-1} \left\| \gamma_{\Delta T_c,i} \Delta T_c(k+i|k) \right\|^2. \tag{7.9}$$

Along with the constructed model (7.7a), (7.7b), the objective function is rearranged in the vector form as

$$J = \left\| \Gamma_y \left(Y(k+1|k) - R(k+1) \right) \right\|^2 + \left\| \Gamma_u \left(U(k|k) - R_u(k) \right) \right\|^2 + \left\| \Gamma_{\Delta u} \Delta U(k) \right\|^2, \tag{7.10}$$

with

$$Y(k+1|k) = \begin{bmatrix} y(k+1|k) \\ y(k+2|k) \\ \vdots \\ y(k+N_p|k) \end{bmatrix}_{N_p \times 1}, \quad R(k+1) = \begin{bmatrix} r_{\Delta\omega}(k+1) \\ r_{\Delta\omega}(k+2) \\ \vdots \\ r_{\Delta\omega}(k+N_p) \end{bmatrix}_{N_p \times 1}, \tag{7.11a}$$

$$U(k|k) = \begin{bmatrix} u(k|k) \\ u(k+1|k) \\ \vdots \\ u(k+N_c-1|k) \end{bmatrix}_{N_c \times 1}, \quad R_u(k+1) = \begin{bmatrix} r_{T_c}(k+1) \\ r_{T_c}(k+2) \\ \vdots \\ r_{T_c}(k+N_c) \end{bmatrix}_{N_c \times 1}, \tag{7.11b}$$

$$\Delta U(k) = \begin{bmatrix} \Delta u(k|k) \\ \Delta u(k+1|k) \\ \vdots \\ \Delta u(k+N_c-1|k) \end{bmatrix}_{N_c \times 1}, \tag{7.11c}$$

Parameter N_p is the prediction horizon and N_c is the control horizon, which satisfies $N_c \le N_p$.

Matrices Γ_y and Γ_u are the weighting factors, which are shown as

$$\Gamma_y = \begin{bmatrix} \gamma_{\Delta\omega,1} & 0 & \dots & 0 \\ 0 & \gamma_{\Delta\omega,2} & \dots & 0 \\ \vdots & \vdots & \ddots & \vdots \\ 0 & 0 & \dots & \gamma_{\Delta\omega,N_p} \end{bmatrix}_{N_p \times N_p}, \tag{7.12a}$$

$$\Gamma_u = \begin{bmatrix} \gamma_{T_c,1} & 0 & \dots & 0 \\ 0 & \gamma_{T_c,2} & \dots & 0 \\ \vdots & \vdots & \ddots & \vdots \\ 0 & 0 & \dots & \gamma_{T_c,N_c} \end{bmatrix}_{N_c \times N_c}, \tag{7.12b}$$

$$\Gamma_{\Delta u} = \begin{bmatrix} \gamma_{\Delta T_c,1} & 0 & \dots & 0 \\ 0 & \gamma_{\Delta T_c,2} & \dots & 0 \\ \vdots & \vdots & \ddots & \vdots \\ 0 & 0 & \dots & \gamma_{\Delta T_c,N_c} \end{bmatrix}_{N_c \times N_c}. \tag{7.12c}$$

Finally, the optimization problem of the clutch engagement control during gear shifting is described as follows:

$$\min_{\Delta U(k)} J\big(x(k), \Delta U(k), N_p, N_c\big) \tag{7.13a}$$

subject to (7.7a), (7.7b) and

$$\Delta u_{\min}(k+i|k) \le \Delta u(k+i|k) \le \Delta u_{\max}(k+i|k),$$

$$i = 0, 1, \dots, N_c - 1. \tag{7.13b}$$

The input constraint (7.13b) is included because in practice the change rate of clutch torque, $\dot{T}_c$, is restricted.

It is clear that the weighting matrices Γ_y, Γ_u and Γ_{Δ_u} will influence the dynamic behavior of gear shifting. Matrix Γ_y forces the clutch to be engaged as soon as possible, Γ_u keeps the transmission output torque tracking the driver's request, and Γ_{Δ_u} means the penalty on the shift shock. Therefore, relatively higher Γ_y results in a fast gear shift, and lesser Γ_y leads to a slow one, but with smoother dynamic performance.

From now on, the model predictive controller will be derived. According to the basics of MPC (see Appendix D), by iterating (7.7a), (7.7b), we can infer the sequences of outputs to be predicted, and present them in the form of

$$Y(k+1|k) = \mathcal{S}_x \Delta x(k) + \mathcal{I} y(k) + \mathcal{S}_d \Delta d(k) + \mathcal{S}_u \Delta U(k), \tag{7.14}$$

where

$$\mathcal{S}_x = \begin{bmatrix} CA \\ CA^2 + CA \\ \vdots \\ \sum_{i=1}^{N_p} CA^i \end{bmatrix}_{N_p \times 1}, \quad \mathcal{I} = \begin{bmatrix} 1 \\ 1 \\ \vdots \\ 1 \end{bmatrix}_{N_p \times 1}, \quad \mathcal{S}_d = \begin{bmatrix} CB_d \\ CAB_d + CB_d \\ \vdots \\ \sum_{i=1}^{N_p} CA^{i-1} B_d \end{bmatrix}_{N_p \times 2},$$

$$\mathcal{S}_u = \begin{bmatrix} CB_u & \mathbf{0} & \mathbf{0} & \cdots & \mathbf{0} \\ \sum_{i=1}^{2} CA^{i-1} B_u & CB_u & \mathbf{0} & \cdots & \mathbf{0} \\ \vdots & \vdots & \vdots & \ddots & \vdots \\ \sum_{i=1}^{N_c} CA^{i-1} B_u & \sum_{i=1}^{N_c-1} CA^{i-1} B_u & \cdots & \ddots & CB_u \\ \vdots & \vdots & \vdots & \ddots & \vdots \\ \sum_{i=1}^{N_p} CA^{i-1} B_u & \sum_{i=1}^{N_p-1} CA^{i-1} B_u & \cdots & \cdots & \sum_{i=1}^{N_p-N_c+1} CA^{i-1} B_u \end{bmatrix}_{N_p \times N_c}.$$

Moreover, the control vector $U(k|k)$ can be represented in the incremental from

$$U(k|k) = \mathcal{I}_{N_c} u(k-1|k-1) + \mathcal{L} \Delta U(k), \tag{7.16}$$

where

$$\mathcal{I}_{N_c} = \begin{bmatrix} 1 \\ 1 \\ \vdots \\ 1 \end{bmatrix}_{N_c \times 1}, \quad \mathcal{L} = \begin{bmatrix} 1 & \mathbf{0} & \mathbf{0} & \cdots & \mathbf{0} \\ 1 & 1 & \mathbf{0} & \cdots & \mathbf{0} \\ \vdots & \vdots & \ddots & \ddots & \vdots \\ 1 & 1 & \cdots & \cdots & 1 \end{bmatrix}_{N_c \times N_c}. \tag{7.17}$$

Then, if we do not consider the inequality constraints (7.13b), we can solve the optimality problem (7.13a), (7.13b), and get the optimal solution of $\Delta U^*(k) \in \mathbb{R}^{N_c \times 1}$ at time k, by calculating the gradient of the objective function over the independent variable $\Delta U(k)$ and setting it to zero. The result reads

$$\begin{aligned} \Delta U^*(k) = & \left(\mathcal{S}_u^T \Gamma_y^T \Gamma_y \mathcal{S}_u + \Gamma_{\Delta u}^T \Gamma_{\Delta u} + \mathcal{L}^T \Gamma_u^T \Gamma_u \mathcal{L} \right)^{-1} \\ & \times \left[\mathcal{S}_u^T \Gamma_y^T \Gamma_y E_p(k+1|k) + \Gamma_u^T \Gamma_u \mathcal{L} \big(R_u - U(k-1) \big) \right], \end{aligned} \tag{7.18}$$

with $E_p(k+1|k)$ being calculated by

$$E_p(k+1|k) = R(k+1) - \mathcal{S}_x \Delta x(k) - \mathcal{I} y(k) - \mathcal{S}_d \Delta d(k). \tag{7.19}$$

If the input constraint is considered, the optimization problem (7.13a), (7.13b) subject to inequality constraints (7.13b) can be formulated as a quadratic programming (QP) problem

$$\min_{\Delta U(k)} \; \Delta U(k)^T H \Delta U(k) - G(k+1|k)^T \Delta U(k) \tag{7.20a}$$

$$\text{s.t.} \quad C_u \Delta U(k) \geq b(k+1|k), \tag{7.20b}$$

where

$$H = \mathcal{S}_u^T \Gamma_y^T \Gamma_y \mathcal{S}_u + \Gamma_{\Delta u}^T \Gamma_{\Delta u} + \mathcal{L}^T \Gamma_u^T \Gamma_u \mathcal{L},$$

$$G(k+1|k) = 2\big(\mathcal{S}_u^T \Gamma_y^T \Gamma_y E_p(k+1|k) + \Gamma_u^T \Gamma_u \mathcal{L}\big(R_u - \mathcal{I}_{N_c} u(k-1|k-1)\big)\big),$$

$$C_u = \begin{bmatrix} -I_{m\times m} & I_{m\times m} \end{bmatrix}^T,$$

$$b(k+1|k) = \begin{bmatrix} -\Delta u_{\max}(k+1|k) \\ \vdots \\ -\Delta u_{\max}(k+1|k) \\ \Delta u_{\min}(k+1|k) \\ \vdots \\ \Delta u_{\min}(k+1|k) \end{bmatrix}_{2N_c\times 1}.$$

It is clear that $H \geq 0$, hence the optimal solution of the optimization problem exists, which is denoted as $\Delta U^*(k)$. By considering above input constraints and solving the quadratic programming (QP) problem (7.20a), (7.20b), we can get the control sequence $\Delta U^*(k)$. Only the first element of $\Delta U^*(k)$ is used to determine the control signal $u(k)$

$$u(k|k) = \Delta u^*(k|k) + u(k-1|k-1). \tag{7.21}$$

Then, the real control command of clutch torque $T_{c,\mathrm{req}}$ is set as $u(k)$

$$T_{c,\mathrm{req}}(k) = u(k), \tag{7.22}$$

and is applied to the plant. This procedure is repeated at each sampling interval.

After the desired clutch torque $T_{c,\mathrm{req}}$ is determined, the clutch engagement force F_c is calculated by

$$F_c = \frac{T_{c,\mathrm{req}}}{\mu_d R_c}, \tag{7.23}$$

where μ_d is the dynamic friction coefficient. Note that in practice, the torque transmissibility of a dry clutch could be much more complex, please refer to [35] for details.

The control algorithm of the torque recovery phase is finally summarized in the following steps:

Step 1: At time k, determine the desired clutch torque $T_{c,\mathrm{req}}(k)$ by the model predictive controller, namely, by the control law (7.18) or by solving the QP problem (7.20a), (7.20b), and then by (7.21) and (7.22);

Step 2: Implement the clutch engagement force $F_c(k)$ calculated from (7.23);

Step 3: If the clutch has been engaged to a certain level (the clutch will be locked up soon), recover the engine torque T_e by pre-determined feedforward control.

Step 4: If the clutch is locked up (the speed difference reaches zero), finish the gear shift control by ramping up the clutch force and continue to recover the engine torque T_e back to the driver's demand; if the clutch is slipping, go to Step 1 and repeat the optimal calculation at next sampling time $k+1$.

7.2.2 Simulation Results

In this section, the proposed control scheme, including the shaft torque observer and the model predictive controller, is programmed using MATLAB/Simulink and combined with the complete powertrain simulation model used in the previous two chapters through co-simulation.

The simulation results shown in Fig. 7.3 is a power-on 1st-to-2nd gear upshift process, where the periods of 4–4.3 s, 4.3–4.42 s, and 4.42–4.8 s correspond to the torque reduction phase, the gear synchronization phase and the torque recovery phase, respectively, which can be seen clearly from the signal of the axle shaft torque T_s.

From 4 s, the shift process is started by the reduction of engine torque T_e, followed by the decrease of the axle shaft torque T_s. During the torque reduction phase (4–4.3 s), T_s is estimated by the observer designed in Chap. 5, and the clutch engagement force is reduced according to the estimated T_s. The observer gain used here is $\boldsymbol{L} = [-2000, 35]$. At 4.3 s, when the axle shaft torque T_s approaches zero, the clutch is totally disengaged and the synchronizer of the 1st gear is disengaged. The vehicle jerk da is given to evaluate the shift shock, which is less than 15 rad/s^3.

After the transmission is disconnected, the gear synchronization phase begins. It costs the synchronizer 0.12 s, i.e., from 4.3 to 4.42 s, to synchronize and engage the 2nd gear.

Once the 2nd gear is engaged, it enters the torque recovery phase, and the model predictive controller is used to control the clutch. The parameters of the model predictive controller are chosen as follows: the prediction horizon and the control horizon are $N_p = 10$ and $N_c = 2$, respectively; the weighting factors are $\gamma_{\Delta\omega,i} = 25$, $\gamma_{T_c,i} = 15$, $\gamma_{\Delta T_c,i} = 1$; the reference values are $\Delta\omega_{\text{ref}} = 0$ rad/s and $T_{c,\text{ref}} = 200$ Nm; the input constraints are $\Delta u_{\min} = -1200$ Nm/s and $\Delta u_{\max} = 1200$ Nm/s. It can be seen that the clutch torque tracks the desired value $T_{c,\text{ref}}$ rapidly and smoothly, and as a result, severe driveline oscillation [37] is successfully avoided after the clutch is synchronized. Moreover, the tracking control of the clutch friction torque contributes to the fast re-instatement of the vehicle traction.

At last, some important evaluation metrics [37] of shift quality are shown in Table 7.1.

The total shift time, including the three phases, is 0.8 s, and thanks to the fast torque reduction (by the torque observer) and fast torque re-instatement (by MPC), the torque interruption time (defined as the time when the traction torque is less than a half of the full torque) is 0.37 s, which is very short for present AMT vehicles.

It should be noted that in the simulations there is no clutch and gear selection delay (assuming that the clutch wear condition is known precisely, and there is no

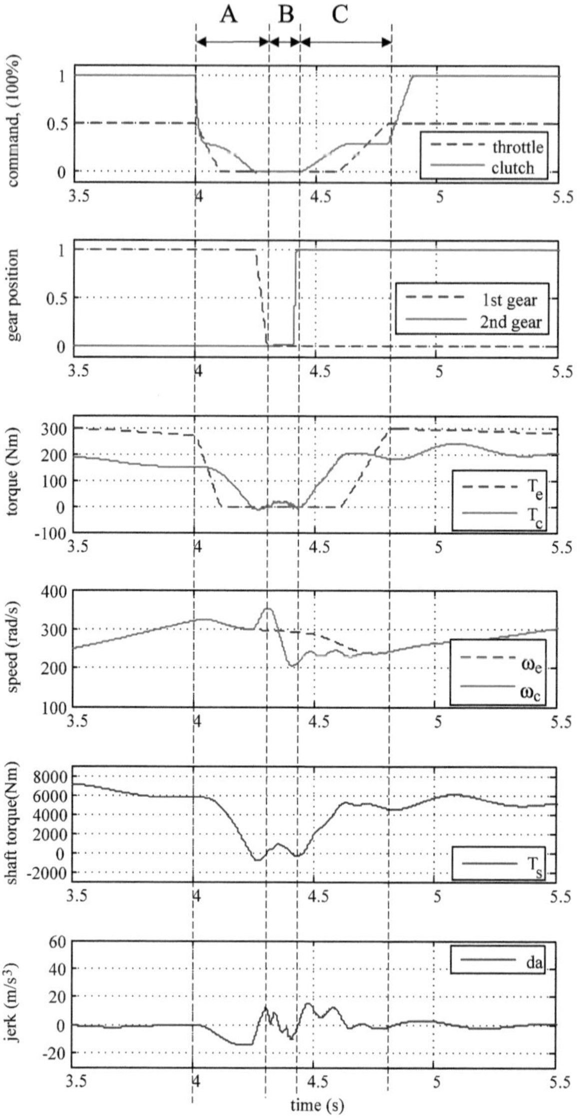

Fig. 7.3 Simulation results of 1st-to-2nd upshift: (**A**) torque reduction phase with axle shaft torque observer; (**B**) gear synchronization phase; (**C**) torque recovery phase with MPC

Table 7.1 Main evaluation metrics of shift quality of 1st-to-2nd upshift

Total shift time	0.8 s
Torque-interruption time	0.37 s
Peak jerk	15 m/s^3
Friction loss	855 J

free clearance between the flywheel and the clutch plate when operating the clutch; moreover, the gear shift actuators are mounted in parallel, and the gear selection operation is not included), and the shift time may by enlarged in practice. Even when

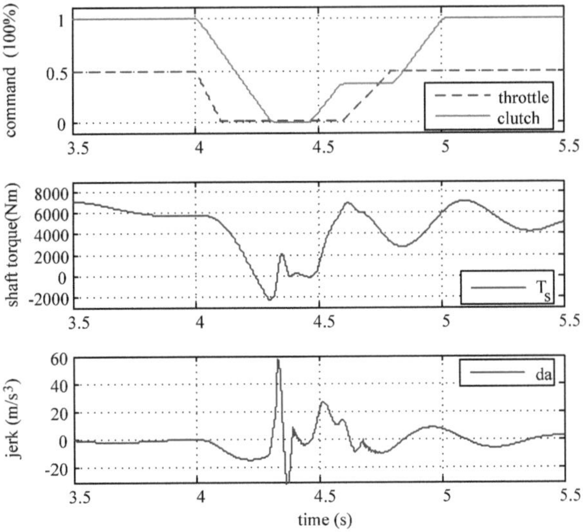

Fig. 7.4 Shift shock of 1st-to-2nd upshift with feedforward control

the clutch and gear selection delay is considered and the shift time is increased by 0.2 s, the shift time is still short enough for the gear shift of trucks with AMT.

On the other hand, the maximum value of the longitudinal jerk during the gear shifting is 15 m/s^3. Generally speaking, the acceptable jerk is 10–25.5 m/s^3 for resonances with frequency of $f \leq 3$ Hz, and 10–37.2 m/s^3 for resonances of frequency $f > 3$ Hz [17]. Hence it can be seen that the jerk level of the shift is good enough. The friction energy of the 1st-to-2nd upshift is 855 J, which is small enough for the shift process of a mid-size truck.

For comparison, Fig. 7.4 gives the results of the 1st-to-2nd upshift without the torque observer and model predictive controller. It can be seen that, if the feedforward control law is not perfectly calibrated, severe shift shock and oscillations of the shaft torque T_s may be introduced.

Power-on 3rd-to-4th gear upshift is also simulated, and the results are shown in Fig. 7.5. The peak jerk is 15 m/s^3, and the torque interruption time is 0.25 s, which satisfies the drivability requirements very well.

7.3 Power-On Downshift of AMT

In the case of downshift, the engine speed must be increased (it is decreased for upshift) to meet the synchronization speed of the new gear. Hence the control scheme of the downshift is much different from that of the upshift. Moreover, in order to minimize the shift time, it is preferable to increase the engine speed as quickly as possible. Therefore, when the clutch begins to engage, even if the synchronization speed is met, the no-lurch condition [16] of the clutch is not satisfied (because the input speed is increasing while the output speed is slightly decreasing). The question

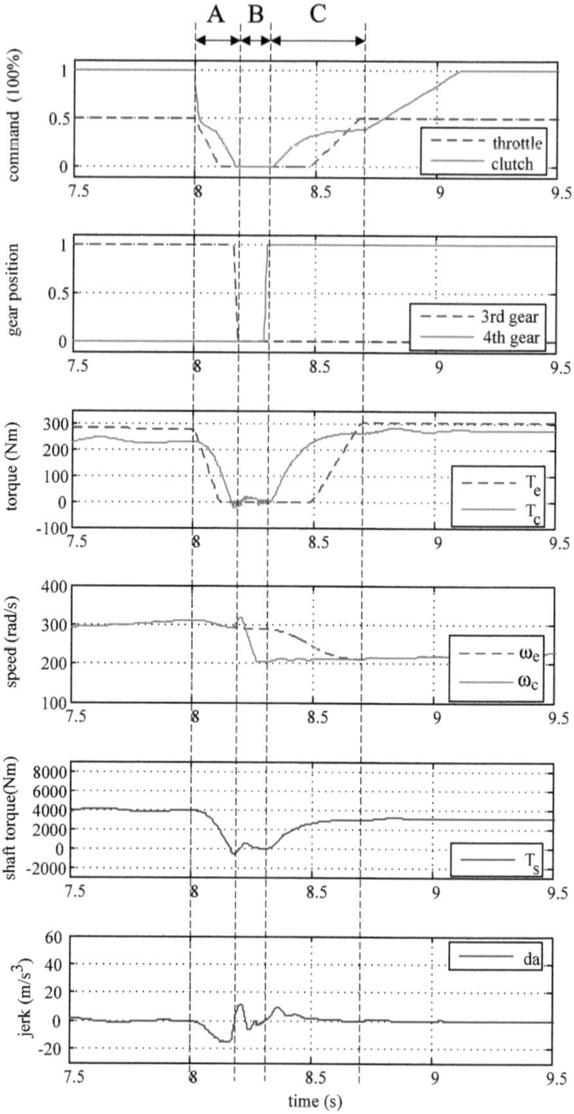

Fig. 7.5 Simulation results of 3rd-to-4th upshift: (**A**) torque reduction phase with axle shaft torque observer; (**B**) gear synchronization phase; (**C**) torque recovery phase with MPC

of how to engage the clutch quickly and smoothly becomes a challenging control task.

In this section, the dynamics and control of AMT downshift will be investigated and summarized in detail. With the proposed control scheme, the engine speed increases quickly to meet the synchronization speed of the new gear, and MPC used in the above will be extended to control the clutch engagement under such a highly transient condition. Because one of the control objectives of MPC is to make the clutch friction torque track the driver's desired value, a gear downshift process with the shortest possible torque-interruption time could be obtained.

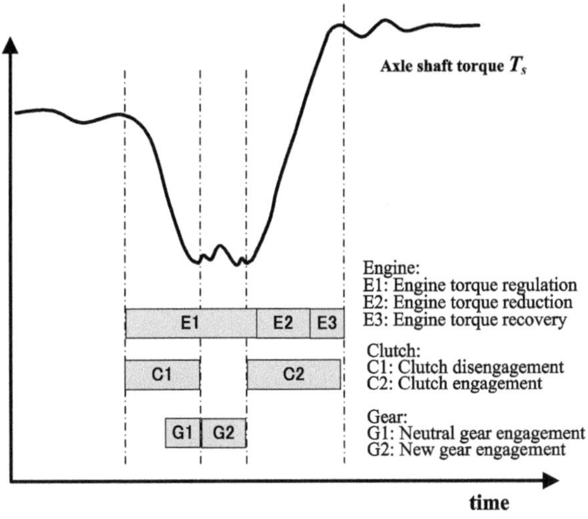

Fig. 7.6 Time sequence of power-on downshift process

7.3.1 Dynamic Process of Power-On Downshift

A power-on downshift sequence is shown in Fig. 7.6. The first phase begins with the clutch disengagement, i.e., slipping-opening of the clutch (C1), and finishes when the neutral gear is engaged (G1). *When the clutch is disengaged, the engine torque is regulated* (E1) *and the engine torque does not necessarily need to be reduced at the beginning of gear shifting because the engine speed has to be improved to reach the synchronization speed (note that in the case of the upshift, the engine torque has to be reduced when the clutch is disengaged).* The neutral gear is engaged at the same time when the clutch is totally disengaged.

In the second phase, the new gear is synchronized, and then engaged, by the synchronizer (G2).

At the beginning of the last phase, the torque recovery phase, it is assumed that the engine speed has reached a level not less than the clutch output speed. Then the clutch slips until it is engaged (C2), and at the same time the engine torque is temporarily reduced to reduce friction loss and shift time, i.e., the co-called "torque-down control" (E2) is applied. Finally, when the clutch is to be locked-up, the authority of the engine torque control is transmitted according to the demand of the driver (E3).

Finally, the whole shift process of AMT downshift is summarized in the following steps:

Step 1: Disengage the clutch, and regulate the engine speed at the same time, so that intensive shift shock could be prevented and the engine speed could be increased to reach the new gear synchronization speed;

Step 2: Engage the neutral gear at the end of clutch disengagement;

Step 3: Engage the new gear and continue to regulate the engine speed to be not less than the clutch output speed;

Step 4: Engage the clutch, and at the same time reduce the engine torque temporarily by pre-determined feedforward control;

Step 5: Recover the engine torque back according to the driver's demand before the clutch is synchronized.

7.3.2 Control Problem Description

During the first phase, the *clutch disengagement phase*, when the clutch state is transferred from slipping to opening, the output torque of the transmission is determined by the friction torque of the clutch. Hence the clutch could be disengaged through feedforward control to provide smooth torque reduction, and consequently, intensive torque fluctuation of the driveline could be avoided. At the same time, the engine torque is regulated to increase the engine speed to reach the target value (not less than the synchronization speed of the new gear). Because this control objective is not very strict, PID control or feedforward control could be used. Here, the focus is put on the third phase (clutch engagement control), and the engine is controlled through feedforward control.

In the second phase, the *speed synchronization phase*, the new gear is engaged, and the engine is still controlled in open loop to increase the engine speed to the target level, namely, not less than the synchronization speed of the new gear.

With the above control scheme of the first two phases, when the third phase, the *torque recovery phase*, begins, the engine speed is increasing quickly. The clutch has to be engaged under such a highly transient state. It is clear that if the clutch is engaged abruptly, or the engine torque is restored too rapidly, the driveline resonances will be produced. On the contrary, if the clutch is engaged too slowly, the torque interruption time will be enlarged. Thus the control objectives in this phase can be summarized as follows:

(i) Minimizing clutch engagement time and friction losses;
(ii) Keeping clutch friction torque to track the driver's request;
(iii) Ensuring smooth acceleration of the vehicle.

The first and the second requirements enforce the clutch to engage quickly and recover the traction back as soon as possible. The third request is added to restrain the shift shock, which is evaluated through longitudinal jerk (change rate of acceleration) [11].

As mentioned in the last section, during this clutch slipping phase, the cooperation of the engine torque-down control [20, 23] is important, and it can reduce the shift time and shift shock significantly. In order to simplify the to-be-designed controller, the engine is still controlled in open loop, and the same as in the last section, MPC will also be adopted to address the multi-objective optimal problem of the clutch engagement control of the AMT downshift process.

7.3.3 Controller Design of Torque Recovery Phase

When the clutch slips, the motion of the driveline can be described by the same equations as in the above upshift process:

$$\dot{\omega}_e = \frac{1}{I_e} T_e - \frac{1}{I_e} T_c, \tag{7.24a}$$

$$\dot{\omega}_c = \frac{1}{I_{v,i}} T_c - \frac{1}{I_{v,i}} T_{v0}. \tag{7.24b}$$

Then based on the simplified model and the MPC design method of the above section, the clutch control strategy of the torque recovery phase of the downshift could be obtained, which is omitted here.

7.3.4 Simulation Results

Power-on downshift always happens under the following two driving conditions:

(A) The driver pushes the accelerator pedal suddenly to get large driving torque;
(B) The vehicle is driven under large throttle angle, but the driving resistance becomes large (such as entering a road with slope), and the present gear cannot provide enough driving torque.

Simulation Results of Maneuver A

Figure 7.7 gives the simulation results of the first maneuver. The transmission is shifted from the 2nd gear to the 1st gear, wherein the fully loaded vehicle is driving on a slope of 5 degrees. The periods of 13.2–13.4 s, 13.4–13.5 s, and 13.5–14 s are respectively the torque reduction phase, the gear synchronization phase, and the torque recovery phase.

Before 13.2 s, the vehicle is driven in the 2nd gear, and the accelerator pedal angle is small. From 13.2 s, the driver wants to accelerate fast and presses the accelerator pedal. Then from 13.2 s, the shift process begins, and the clutch is disengaged using open-loop control. At the same time, the engine torque is also regulated in open-loop to increase the engine speed. At 13.4 s, the clutch is fully disengaged, and the neutral gear is also engaged.

Next, from 13.4 to 13.5 s follows the gear synchronization phase, and the 1st gear is engaged. After that, the engine speed is increased to 230 rad/s, which is greater than the clutch output speed 200 rad/s.

Finally, one has the torque recovery phase, and the designed model predictive controller is used to control the clutch. The parameters of the model predictive controller are chosen as follows: the prediction horizon and the control horizon are

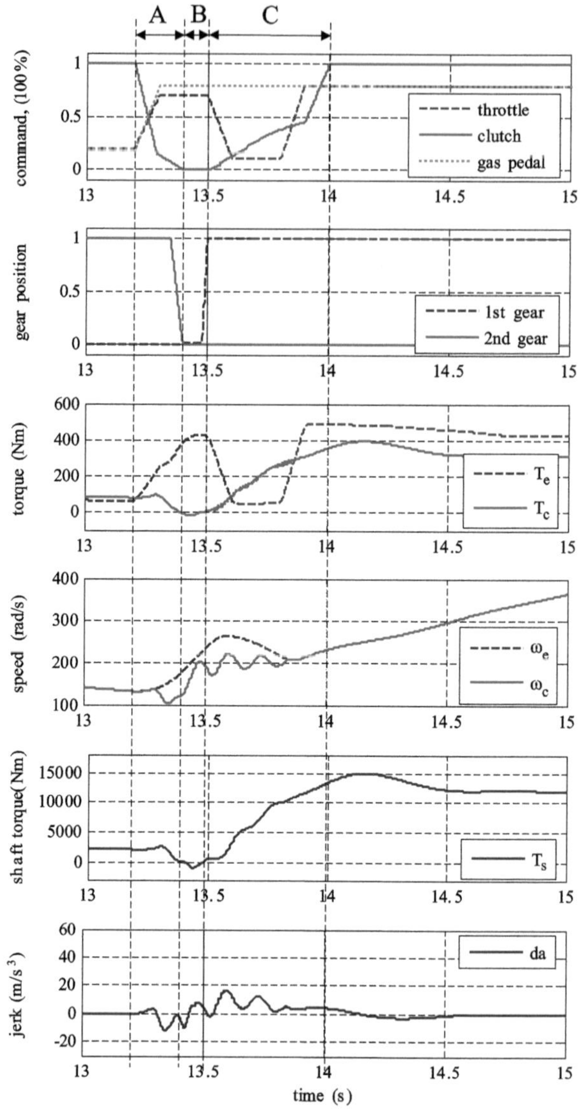

Fig. 7.7 Simulation results of power-on 2nd-to-1st downshift (pressing of gas pedal)

$N_p = 10$ and $N_c = 2$; the weighting factors are $\gamma_{\Delta\omega,i} = 10$, $\gamma_{T_c,i} = 7$, $\gamma_{\Delta T_c,i} = 1$; the reference values are $\Delta\omega_{\text{ref}} = 0$ rad/s and $T_{c,\text{ref}} = 400$ Nm; the input constraints are $\Delta u_{\min} = -1000$ Nm/s and $\Delta u_{\max} = 1000$ Nm/s. It can be seen that the clutch torque tracks the desired value $T_{c,\text{ref}}$ rapidly and smoothly, and as a result, severe driveline oscillation [37] is successfully avoided after the clutch is synchronized. Moreover, the tracking control of clutch friction torque contributes to the fast re-instatement of the vehicle traction.

Some important evaluation metrics [37] of shift quality are shown in Table 7.2.

Table 7.2 Main evaluation metrics of shift quality of 2nd-to-1st downshift

Total shift time	0.8 s
Torque-interruption time	0.35 s
Peak jerk	17.5 m/s^3
Friction loss	1984 J

The total shift time, including the three phases, is 0.8 s, and thanks to fast torque re-instatement (by MPC), the torque interruption time (defined as the time when the traction torque is less than a half of the full torque) is 0.35 s, which is very short for present AMT vehicles.

The maximum value of the longitudinal jerk during the gear shifting is 17.5 m/s^3, and the friction energy of the 1st-to-2nd upshift is 1984 J, which are acceptable for a gear shift of mid-size trucks.

Simulation Results of Maneuver B

Figure 7.8 gives the simulation results of the second maneuver, wherein the vehicle drives from flat to grade road. At first, the vehicle is driving in the 4th gear on a flat road, and from 13 s, it enters a slope with an angle of 5 degrees. From 15 s, it is judged by the transmission control unit (TCU) that the 4th gear cannot provide enough driving torque anymore, and it begins to shift to the 3rd gear.

The torque reduction phase, the gear synchronization phase and the torque recovery phase correspond respectively to the periods of 15–15.2 s, 15.2–15.3 s and 15.3–15.7 s.

At last, some important evaluation metrics [37] of shift quality are shown in Table 7.3.

7.4 Notes and References

The dynamics and control of the gear shift of AMT vehicles are described and addressed under the torque-based powertrain control scheme. One of the findings is to use the clutch friction torque as compensation for the traction interruption, which is realized through the critical enabling technology, i.e., model predictive control (MPC), which contributes to a shorter torque interruption time.

In the future work, multi-model MPC or Hybrid MPC [2, 3, 24] can be used to address the clutch engagement control problem because the engagement process from slipping to locked-up is essentially a process of state switching [4].

Another important issue of MPC in automotive drivetrain is solving the MPC optimization problem online at each sampling time. In order to speed up the computation of MPC, hardware architectures which are capable of parallel computation are under active investigation [9]. Field programmable gate array (FPGA) [21] provides a compromise between the special-purpose application-specific integrated circuit hardware and general-purpose processors, and implementing MPC on a FPGA

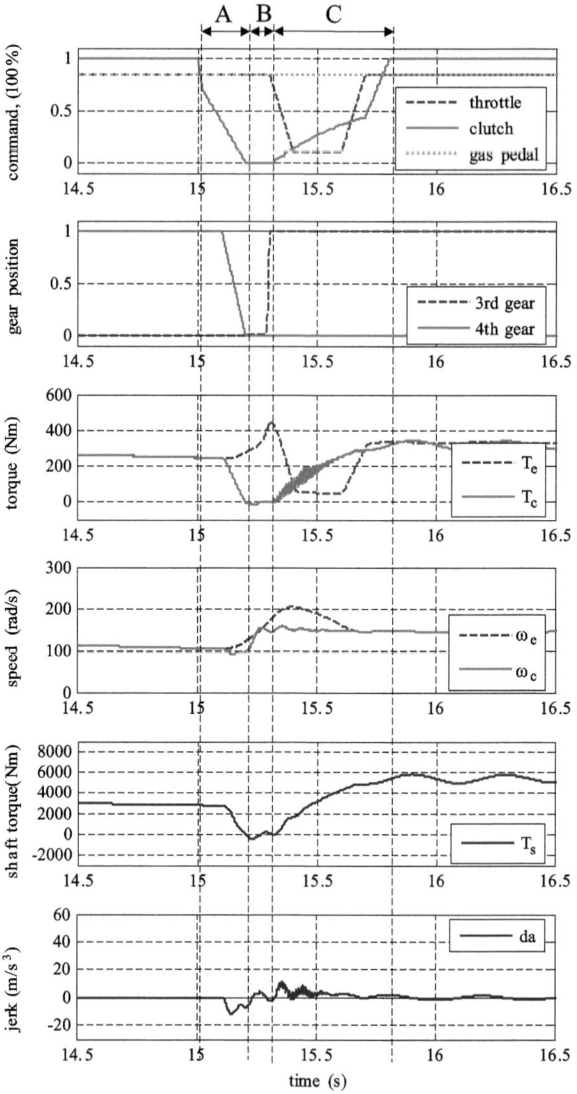

Fig. 7.8 Simulation results of power-on 4th-to-3rd downshift (driving into a slope)

Table 7.3 Main evaluation metrics of shift quality of 4th-to-3rd downshift

Total shift time	0.8 s
Torque-interruption time	0.3 s
Peak jerk	11 m/s^3
Friction loss	1582 J

sounds promising. Some early attempts in this direction are reported in [28, 36], wherein the authors explore the implementation of the MPC technology into an FPGA chip.

References

1. Allgöwer F, Badgwell TA, Qin JS, Rawlings JB, Wright SJ (1999) Nonlinear predictive control and moving horizon estimation—an introductory overview. In: Frank PM (ed) Advances in control, highlights of ECC'99. Springer, Berlin, pp 391–449
2. Balluchi A, Benvenuti L, Ferrari A, Sangiovanni-Vincentelli AL (2006) Hybrid systems in automotive electronics design. Int J Control 79(5):375–394
3. Bemporad A, Morari M (1999) Control of systems integrating logic, dynamics, and constraints. Automatica 35:407–427
4. Bemporad A, Borrelli F, Glielmo L, Vasca F (2001) Hybrid control of dry clutch engagement. In: Proceedings of the European control conference, Porto, Portugal
5. Bemporad A, Morari M, Dua V, Pistikopoulos EN (2002) The explicit linear quadratic regulator for constrained systems. Automatica 38(1):3–20
6. Bengtsson J, Strandh P, Johansson R (2006) Multi-output control of a heavy duty HCCI engine using variable valve actuation and model predictive control. SAE technical paper 2006-01-0873
7. Cairano SD, Yanakiev D, Bemporad A, Kolmanovsky IV, Hrovat D (2008) An MPC design flow for automotive control and applications to idle speed regulation. In: Proceedings of the 47th IEEE conference on decision and control, pp 5692–5697
8. Chen H, Scherer CW (2006) Moving horizon H_∞ control with performance adaptation for constrained linear systems. Automatica 42(6):1033–1040
9. Chen H, Xu F, Xi Y (2012) Field programmable gate array/system on a programmable chip-based implementation of model predictive controller. IET Control Theory Appl 6(8):1055–1063
10. Cho D (1987) Nonlinear control methods for automotive powertrain systems. PhD Thesis, MIT
11. Dolcini P, Wit CC, Béchart H (2008) Lurch avoidance strategy and its implementation in amt vehicles. Mechatronics 18(5–6):289–300
12. Dourra H, Mourtada A (2008) Adaptive nth order lookup table used in transmission double swap shift control. SAE technical paper 2008-01-0538
13. Fredriksson J, Egardt B (2003) Active engine control for gearshifting in automated manual transmissions. Int J Veh Des 32(3/4):216–230
14. Gao B-Z, Chen H, Sanada K, Hu Y-F (2011) Design of clutch slip controller for automatic transmission using backstepping. IEEE/ASME Trans Mechatron 16(3):498–508
15. Gao B-Z, Lei Y-L, Ge A-L, Chen H, Sanada K (2011) Observer-based clutch disengagement control during gear shift process of automated manual transmission. Veh Syst Dyn 49(5):685–701
16. Gao B-Z, Lu X-H, Chen H, Lu X-T, Li J (2013) Dynamics and control of gear upshift in automated manual transmissions. Int J Veh Des 63(1):61–83
17. Ge A (1993) Theory and design of automatic transmissions. China Machine Press, Beijing. In Chinese
18. Glielmo L, Vasca F (2000) Optimal control of dry clutch engagement. SAE technical paper 2000-01-0837
19. Glielmo L, Iannelli L, Vacca V, Vasca F (2006) Gearshift control for automated manual transmissions. IEEE/ASME Trans Mechatron 11(1):17–26
20. Goetz M, Levesley MC, Crolla DA (2005) Dynamics and control of gearshifts on twin-clutch transmissions. Proc Inst Mech Eng, Part D, J Automob EngMech 219(8):951–963
21. Guo HY, Chen H, Xu F, Wang F, Lu GY (2013) Implementation of ekf for vehicle velocities estimation on fpga. IEEE Trans Ind Electron 60(9):3823–3839
22. Hahn JO, Lee KI (2002) Nonlinear robust control of torque converter clutch slip system for passenger vehicles using advanced torque estimation algorithms. Veh Syst Dyn 37(3):175–192
23. Haj-Fraj A, Pfeiffer F (2002) A model based approach for the optimisation of gearshifting in automatic transmissions. Int J Veh Des 28(1–3):171–188

24. Heijden ACVD, Serrarens AFA, Camlibel MK, Nijmeijer H (2007) Hybrid optimal control of dry clutch engagement. Int J Control 80(11):1717–1728
25. Kim DH, Yang KJ, Hong KS, Hahn JO, Lee KI (2003) Smooth shift control of automatic transmissions using a robust adaptive scheme with intelligent supervision. Int J Veh Des 32(3/4):250–272
26. Kulkarni M, Shim T, Zhang Y (2007) Shift dynamics and control of dual-clutch transmissions. Mech Mach Theory 42(2):168–182
27. Lei YL, Gao BZ, Tian H, Ge AL, Yan S (2005) Throttle control strategies in the process of integrated powertrain control. Chin J Mech Eng 18(3):429–433 (English Edition)
28. Ling KV, Yue SP, Maciejowski JM (2006) A FPGA implementation of model predictive control. In: Proceedings of American control conference, Minnesota, USA, pp 1930–1935
29. Mayne DQ, Rawlings JB, Rao CV, Scokaert POM (2000) Constrained model predictive control: stability and optimality. Automatica 36(6):789–814
30. Pettersson M (1997) Driveline modeling and control. PhD Thesis, Linköping University, Sweden
31. Pettersson M, Nielsen L (2000) Gear shifting by engine control. IEEE Trans Control Syst Technol 8(3):495–507
32. Pettersson M, Nielsen L (2003) Diesel engine speed control with handling of driveline resonances. Control Eng Pract 11(3):319–328
33. Sanada K, Kitagawa A (1998) A study of two-degree-of-freedom control of rotating speed in an automatic transmission, considering modeling errors of a hydraulic system. Control Eng Pract 6:1125–1132
34. Tanaka H, Wada H (1995) Fuzzy control of engagement for automated manual transmission. Veh Syst Dyn 24(4/5):365–376
35. Vasca F, Iannelli L, Senatore A, Reale G (2011) Torque transmissibility assessment for automotive dry-clutch engagement. IEEE/ASME Trans Mechatron 16(3):564–573
36. Vouzis PD, Bleris LG, Arnold MG, Kothare MV (2009) A system on-a-chip implementation for embedded real-time model predictive control. IEEE Trans Control Syst Technol 17(5):1006–1016
37. Wheals JC, Crewe C, Ramsbottom M, Rook S, Westby M (2002) Automated manual transmissions—a European survey and proposed quality shift metrics. SAE technical paper 2002-01-0929
38. Yokoyama M (2008) Sliding mode control for automatic transmission systems. J Jpn Fluid Power Syst Soc 39(1):34–38. In Japanese

Chapter 8
Data-Driven Start-Up Control of AMT Vehicle

8.1 Introduction

As shown in Chaps. 4 and 7, clutch engagement is an important and difficult control issue. The performance requirements for the clutch engagement during the start-up process include: minimizing clutch lockup time, minimizing friction losses during the slipping phase and ensuring smooth acceleration of the vehicle, i.e., enhancing the driving comfort. However, these requirements are sometimes conflicting, for example, the drivability enhancement results in longer clutch lockup time. Although there are many different control strategies proposed for the control of dry clutch engagement in the literatures, the control methods are model-based, which rely heavily on the explicit process modeling. The characteristics of AMT clutch during start-up process are complex; moreover, the system characteristics change along with the variation of driving conditions and long-term aging. For example, the damping coefficients of rotational shafts change greatly according to environmental temperature. Long-term aging and variation of driving conditions still bring about significant modeling errors. Due to these characteristics of the start-up process of AMT vehicles, an explicit model of the system is hard to construct. Even though one can be obtained, it is not easy to deal with the high order of the system. Moreover, due to the physical constraints of the driveline system mechanism, the maximum friction clutch torque provided from the clutch is restricted, and the range of engine speed is limited.

With the development of computer technology, a lot of data can be obtained in modern industries, thus, the data-driven methods present not only a new avenue but also new challenges both in theories and applications [5, 6, 10, 12]. In [2, 4], data-driven predictive control algorithm is proposed as an example of the efficient data-driven control methods. It elegantly combines data-driven subspace identification and predictive control. For its inherent characteristics, a data-driven predictive controller computes the control action directly based on the input–output data only and does not require any explicit model of the system.

In this chapter, for the control problem of the start-up process of AMT vehicles, a data-driven predictive controller is designed. Moreover, the time-domain hard con-

H. Chen, B. Gao, *Nonlinear Estimation and Control of Automotive Drivetrains*,

DOI 10.1007/978-3-642-41572-2_8,

straints of the input and the output are taken into account. The simulation results show that AMT clutch with the data-driven predictive controller works very well, and this process meets the control requirements, i.e., fast clutch lockup time, small friction losses, and preservation of driving comfort.[1]

8.2 Control Requirements

The start-up of AMT vehicles, using a friction clutch, is a very important process for drivability and fuel consumption. The core of the starting control is the control of the clutch, and during the start-up process of AMT vehicles from stop, the driveline performance described above heavily depends on the engagement of the dry clutch. The considered control problem of the clutch engagement during the start-up process of AMT vehicles is that the clutch speed ω_c has to track the engine speed ω_e by the effect of the friction clutch torque T_c. Moreover, the clutch engagement is expected to satisfy the different and sometimes conflicting objectives:

- Fast clutch lockup time;
- Small friction losses during the slipping phase;
- Preservation of driver comfort, i.e., smooth acceleration of the vehicle.

In addition, as for the physical constraints of the driveline system mechanism, time-domain output constraints are represented by the restricted engine speed ω_e. The friction clutch torque T_c, considered as a control input, is bounded because of the actuator saturation. Moreover, because the frequency response of the actuator is limited, the control move cannot change very quickly.

The criterion for minimizing the difference between a given reference and a predicted output is usually chosen in a quadratic form, and a control move $u(\cdot)$ can also be included to penalize control efforts. In order to achieve the start-up control, we choose the friction clutch torque T_c as the control input u, the clutch slip speed $\Delta\omega = \omega_e - \omega_c$ as the output y and the engine speed as the constrained output y_b. Then, the optimization problem with input and constrained output constraints of clutch control during start-up process is described as follows:

Problem 8.1

$$\min_{\Delta u_f(k)} J\big(y(k), \Delta u_f(k), N_p, N_c\big)$$

subject to

$$u_{\min}(k+m) \le u(k+m) \le u_{\max}(k+m), \quad m = 0, 1, \ldots, N_c - 1, \tag{8.1a}$$

$$\Delta u_{\min}(k+m) \le \Delta u(k+m) \le \Delta u_{\max}(k+m), \quad m = 0, 1, \ldots, N_c - 1, \tag{8.1b}$$

$$y^b_{\min}(k+q) \le y_b(k+q) \le y^b_{\max}(k+q), \quad q = 1, 2, \ldots, N_p, \tag{8.1c}$$

[1]This chapter uses the content of [7], with permission from IEEE.

where

$$J = \left\| \Gamma_y \left(\hat{y}_f(k+1) - R_e(k+1) \right) \right\|^2 + \left\| \Gamma_u \Delta u_f(k) \right\|^2, \tag{8.2}$$

the predictive control output sequence $\hat{y}_f(k+1)$, *predictive constrained output sequence* $\hat{y}_f^b(k+1)$ *and the future input sequence* $u_f(k)$ *are defined as follows:*

$$\hat{y}_f(k+1|k) \triangleq \begin{bmatrix} \hat{y}(k+1|k) \\ \hat{y}(k+2|k) \\ \vdots \\ \hat{y}(k+N_p|k) \end{bmatrix}, \qquad \hat{y}_f^b(k+1|k) \triangleq \begin{bmatrix} \hat{y}_b(k+1|k) \\ \hat{y}_b(k+2|k) \\ \vdots \\ \hat{y}_b(k+N_p|k) \end{bmatrix},$$

$$u_f(k) \triangleq \begin{bmatrix} u(k|k) \\ u(k+1|k) \\ \vdots \\ u(k+N_c-1|k) \end{bmatrix}.$$

At time k, the control input sequence $\Delta u_f(k)$ *to be optimized can be defined as*

$$\Delta u_f(k) \triangleq \begin{bmatrix} \Delta u(k|k) \\ \Delta u(k+1|k) \\ \vdots \\ \Delta u(k+N_c-1|k) \end{bmatrix}.$$

The reference sequence of the output is as follows:

$$R_e(k+1) = \begin{bmatrix} r(k+1) & r(k+2) & \ldots & r(k+N_p) \end{bmatrix}^T,$$

$$\Gamma_y = \begin{bmatrix} \gamma_{y,1} & 0 & \ldots & 0 \\ 0 & \gamma_{y,2} & \ldots & 0 \\ \vdots & \vdots & \ddots & \vdots \\ 0 & 0 & \ldots & \gamma_{y,N_p} \end{bmatrix}, \qquad \Gamma_u = \begin{bmatrix} \gamma_{u,1} & 0 & \ldots & 0 \\ 0 & \gamma_{u,2} & \ldots & 0 \\ \vdots & \vdots & \ddots & \vdots \\ 0 & 0 & \ldots & \gamma_{u,N_c} \end{bmatrix}.$$

N_p *and* N_c *are the prediction horizon and the control horizon, respectively, satisfying* $N_p \geq N_c$.

From the analysis of physical meanings, it is clear that

- $J_1 = \|\Gamma_y(\hat{y}_f(k+1) - R_e(k+1))\|^2$ *forces the slip speed to converge to zero, i.e., to minimize clutch lockup time and minimize the friction losses;*
- $J_2 = \|\Gamma_u \Delta u_f(k)\|^2$ *controls the change rate of the control action, and ensures smooth acceleration of the vehicle because* T_c *determines the acceleration of the vehicle, and hence vehicle jerk can be reflected by* $\dot{T}_c$, *which corresponds to* $\Delta u_f(k)$;
- *The constraints of control input* $u(k)$ *and* $\Delta u(k)$ *reflect the ability of actuator, and the constraints of* $\Delta u(k)$ *has the effect of keeping the clutch engagement smooth and the jerk small; the constraints of system output* $y_b(k)$ *avoids stalling the engine.*

Obviously, it is contradictive to minimize J_1 and J_2 simultaneously, so in order to trade off the two objectives, the weighting factors Γ_y and Γ_u are given. They are chosen to ensure small facing wear and good dynamic performance. Matrix Γ_y forces the launch process to be finished, while Γ_u means the penalty on the shift shock.

Therefore, based on input–output data obtained from the driveline simulation model, which was constructed by AMESim and has been used in the last three chapters, the data-driven predictive control method is adopted to control the start-up process of AMT vehicles. Moreover, it takes time-domain constraints into account explicitly and deals with multiple objectives in a somehow optimal sense.

8.3 Data-Driven Start-Up Predictive Controller of AMT Vehicle

In order to deal with the control problem arising from the start-up process of AMT vehicles and to meet the control requirements mentioned above, a data-driven predictive controller will be designed based on the input–output data in this section. The subspace predictor is obtained directly from the input–output data (the clutch friction torque T_c and the clutch slip speed $\Delta\omega$) without an explicit physical model of the system, which predicts the future dynamic behavior of the system. Moreover, the predictive output equation is derived based on predictive control methodology. Considering the system constraints, the optimal control sequence is determined by solving the optimization problem online. The optimal output is applied to the driveline system as the feedback control signal. According to the basic principles of predictive control, this process is repeated at each sampling time. The details of the design process of the data-driven predictive controller will be given next.

8.3.1 Subspace Linear Predictor

A data-driven predictive control algorithm combines the results from subspace identification methods within the field of predictive control, an illustration of the data-driven predictive control method is shown in Fig. 8.1. The novelty of the data-driven predictive control algorithm over other control methods is that it does not use the traditional, explicit parametric description of the system such as transfer function or state-space model in the development of the controller. Instead, it uses the subspace linear predictor to predict the future output values of the system. The derivation of the subspace linear predictor via input–output data is presented in Appendix F.

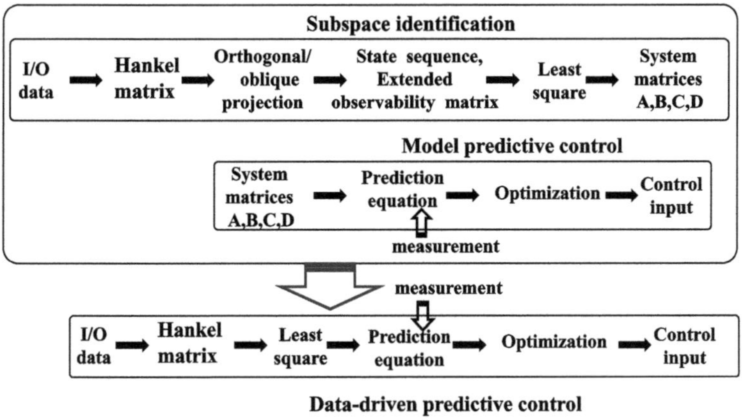

Fig. 8.1 Illustration of the data-driven predictive control method

Fig. 8.2 Input and output data for identification

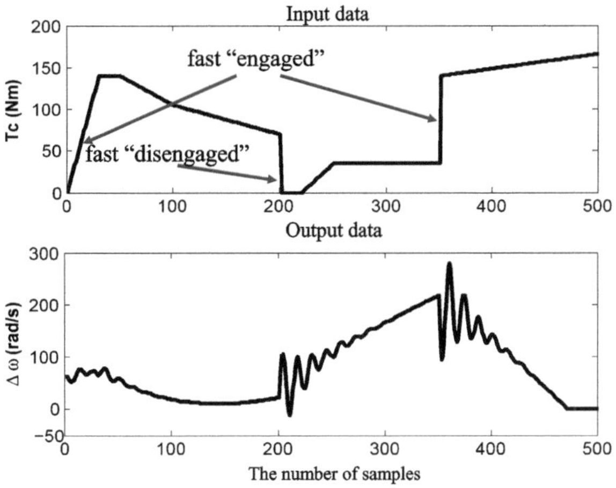

8.3.2 Data-Driven Start-Up Predictor

Design of Input–Output Identification Data

In order to achieve the data-driven start-up predictive controller, we will design the identification data which can excite dynamics relevant to the control goal, such as the vibration of clutch, drive shaft and tyre. The absolute exciting signals for the AMT clutch are designed during the vehicle start-up and applied to the complete AMESim powertrain model. The identification data are obtained in conditions of straight flat road ($\alpha = 0°$), fixed throttle opening ($\theta_{th} = 60°$), invariable gear ratio ($i_{t1} = 7.57$) and lightly loaded vehicle ($m = 6000$ kg). The open-loop data of the input T_c and output $\Delta\omega$ for identification are shown in Fig. 8.2.

Data-Driven Predictor

The open-loop system measurements of the input, the output and the constrained output $u(k)$, $y(k)$ and $y_b(k)$ for $k \in \{0, 1, 2, \ldots, 2i + j - 2\}$ are collected through the simulation results of Fig. 8.2. The data block Hankel matrices U_p, U_f, Y_p and Y_f for $u(k)$ and $y(k)$ are constructed as follows:

$$U_p = \begin{bmatrix} u_0 & u_1 & \ldots & u_{j-1} \\ u_1 & u_2 & \ldots & u_j \\ \vdots & \vdots & \ddots & \vdots \\ u_{i-1} & u_i & \ldots & u_{i+j-2} \end{bmatrix}, \quad U_f = \begin{bmatrix} u_i & u_{i+1} & \ldots & u_{i+j-1} \\ u_{i+1} & u_{i+2} & \ldots & u_{i+j} \\ \vdots & \vdots & \ddots & \vdots \\ u_{2i-1} & u_{2i} & \ldots & u_{2i+j-2} \end{bmatrix}, \tag{8.3}$$

$$Y_p = \begin{bmatrix} y_0 & y_1 & \ldots & y_{j-1} \\ y_1 & y_2 & \ldots & y_j \\ \vdots & \vdots & \ddots & \vdots \\ y_{i-1} & y_i & \ldots & y_{i+j-2} \end{bmatrix}, \quad Y_f = \begin{bmatrix} y_i & y_{i+1} & \ldots & y_{i+j-1} \\ y_{i+1} & y_{i+2} & \ldots & y_{i+j} \\ \vdots & \vdots & \ddots & \vdots \\ y_{2i-1} & y_{2i} & \ldots & y_{2i+j-2} \end{bmatrix}, \tag{8.4}$$

where p and f denote the past and future block observations. The matrices above have i-block rows and j-block columns. The constrained output Hankel matrices Y_p^b and Y_f^b for $y_b(k)$ can be formed by the same way.

According to the deviation of subspace linear predictor presented in Appendix F, we will recursively develop the subspace input–output matrix equations in the field of subspace identification as follows:

$$\hat{Y}_f = L_w W_p + L_u U_f, \tag{8.5}$$

$$\hat{Y}_f^b = L_w^b W_p^b + L_u^b U_f, \tag{8.6}$$

In terms of Eqs. (F.29) and (F.30), we can obtain the subspace predictor coefficients L_w and L_u. The terms $\hat{Y}_f^b$, L_w^b, W_p^b and L_u^b of constrained output $y_b(k)$ can be obtained in the way of (F.14) to (F.30). The data-driven predictor (8.5) and (8.6) is applied to predict the output of the system by the paste input and output data as well as the future input data.

Validation Data

In order to validate the effectiveness of the predictor (8.5), that is, whether it can reflect the system dynamics, a group of signal data are used to test the identified subspace matrices, and the data are plotted in Fig. 8.3. From Fig. 8.3 it is clear that the identified predictive outputs $\Delta\omega^*$ (the dotted line) matches the true outputs $\Delta\omega$ (the solid line) of the model very well. It shows that the predictor can accurately predict the future output values of the system.

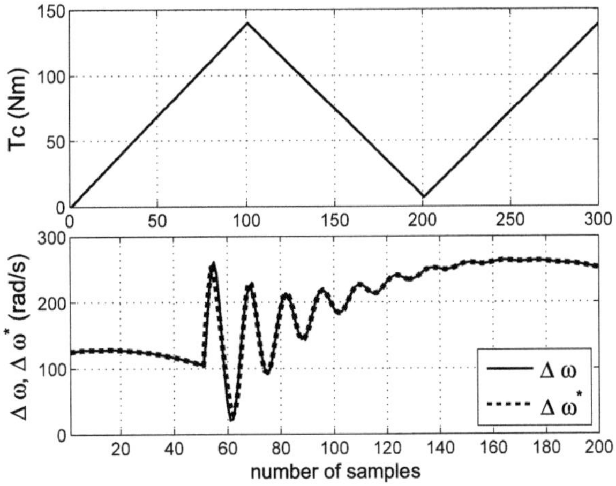

Fig. 8.3 Validation data

8.3.3 Predictive Output Equation

Aiming to deal with the optimization problem described in Sect. 8.2, this section will derive the predictive output equation based on the data-driven method and the predictive control method. To guarantee regulation with zero steady-state error for the reference input, the subspace matrix incremental input–output expressions for the system are

$$\Delta\hat{Y}_f = L_w \Delta W_p + L_u \Delta U_f,$$

$$\Delta\hat{Y}_f^b = L_w^b \Delta W_p^b + L_u^b \Delta U_f,$$

and

$$\Delta\hat{y}_f(k) = L_w(1:N_p,:)\begin{bmatrix}\Delta y_p\\ \Delta u_p\end{bmatrix} + L_u(1:N_p,1:N_c)\Delta u_f,$$

$$\Delta\hat{y}_f^b(k) = L_w^b(1:N_p,:)\begin{bmatrix}\Delta y_p^b\\ \Delta u_p\end{bmatrix} + L_u^b(1:N_p,1:N_c)\Delta u_f(k),$$

where

$$\Delta W_p = \begin{bmatrix}\Delta Y_p\\ \Delta U_p\end{bmatrix}, \tag{8.7}$$

$$\Delta y_p = \begin{bmatrix}\Delta y(k-i+1) & \Delta y(k-i+2) & \dots & \Delta y(k)\end{bmatrix}^T, \tag{8.8}$$

$$\Delta u_p = \begin{bmatrix}\Delta u(k-i) & \Delta u(k-i+1) & \dots & \Delta u(k-1)\end{bmatrix}^T, \tag{8.9}$$

ΔW_p^b and Δy_p^b can be obtained in the same way as Eqs. (8.7) to (8.9).

Then, the vector of the optimal prediction of the future outputs can be expressed as follows [4]:

$$\hat{y}_f(k+1) = \mathbf{y}(k) + L_w^{\Delta}(1:N_p,:)\begin{bmatrix}\Delta y_p \\ \Delta u_p\end{bmatrix} + S_{N_p,N_c}\Delta u_f(k)$$
$$= F + S_{N_p,N_c}\Delta u_f(k), \tag{8.10}$$
$$\hat{y}_f^b(k+1) = \mathbf{y}_b(k) + L_w^{b\Delta}(1:N_p,:)\begin{bmatrix}\Delta y_p^b \\ \Delta u_p\end{bmatrix} + S_{N_p,N_c}^b\Delta u_f(k)$$
$$= F_b + S_{N_p,N_c}^b\Delta u_f(k),$$

where S_{N_p,N_c} is the $N_p m \times N_c l$ dynamic matrix containing the step response coefficients/Markov parameters and formed from L_u,

$$S_{N_p,N_c} = L_u(1:N_p,1:N_c)\begin{bmatrix} 1 & 0 & \dots & 0 \\ 1 & 1 & \dots & 0 \\ \vdots & \vdots & \ddots & \vdots \\ 1 & 1 & \dots & 1 \end{bmatrix}, \tag{8.11}$$

$$\mathbf{y}(k) = \begin{bmatrix} y(k) & y(k) & \dots & y(k) \end{bmatrix}^T, \tag{8.12}$$

L_w^{Δ} is constructed from L_w, and F is the free response for the case of measured disturbances,

$$L_w^{\Delta}(k,:) = \sum_{i=1}^{k} L_w(k,:) \quad 1 \le k \le N_p, \tag{8.13}$$

$$F = \mathbf{y}(k) + L_w^{\Delta}(1:N_p,:)\begin{bmatrix}\Delta y_p \\ \Delta u_p\end{bmatrix}. \tag{8.14}$$

In the same way, $\mathbf{y}_b(k)$, $L_w^{b\Delta}(1:N_p h,:)$, F_b and S_{N_p,N_c}^b about $\hat{y}_f^b$ can be obtained as Eqs. (8.11) to (8.14).

8.3.4 Data-Driven Predictive Controller Without Constraints

It is assumed that the constraints of the system are first neglected. By differentiating (8.2) with respect to $\Delta u_f(k)$ and equating it to zero, the feedback plus feedforward control law becomes

$$\Delta u_f(k) = \left(S_{N_p,N_c}^T \Gamma_y^T \Gamma_y S_{N_p,N_c} + \Gamma_u^T \Gamma_u\right)^{-1} S_{N_p,N_c}^T \left(R_e(k+1) - F\right).$$

Only the first element of the $\Delta u_f(k)$ is implemented, and the calculation is repeated at each time instant. Hence, at time instant k, the control law is given as follows:

$$\Delta u_k = K_{\Delta}\left(R_e(k+1) - F\right), \tag{8.15}$$

where

$$K_{\Delta} = \begin{bmatrix} 1 & 0 & \dots & 0 \end{bmatrix}\left(S_{N_p,N_c}^T \Gamma_y^T \Gamma_y S_{N_p,N_c} + \Gamma_u^T \Gamma_u\right)^{-1} S_{N_p,N_c}^T.$$

8.3.5 Data-Driven Predictive Controller with Constraints

Now, it is considered that the constraints of the system take the form given in Eqs. (8.1a) to (8.1c). When Eq. (8.10) is substituted into (8.2), the following equation is obtained:

$$\begin{aligned}J &= \big(R_e(k+1) - F - S_{N_p,N_c}\Delta u_f(k)\big)^T \Gamma_y^T \Gamma_y \big(R_e(k+1) \\ &\quad - F - S_{N_p,N_c}\Delta u_f(k)\big) + \Delta u_f(k)^T \Gamma_u^T \Gamma_u \Delta u_f(k) \\ &= \Delta u_f(k)^T S_{N_p,N_c}^T \Gamma_y^T \Gamma_y S_{N_p,N_c} \Delta u_f(k) \\ &\quad + \Delta u_f(k)^T \Gamma_u^T \Gamma_u \Delta u_f(k) + E(k+1)^T \Gamma_y^T \Gamma_y E(k+1) \\ &\quad - 2E(k+1)^T \Gamma_y^T \Gamma_y S_{N_p,N_c} \Delta u_f(k),\end{aligned} \tag{8.16}$$

where $E(k+1) = R_e(k+1) - F$. By arranging Eq. (8.16) and combining the constraints, the optimization problem (8.1a)–(8.1c) can be formulated as a quadratic programming (QP) problem [8, 9]:

$$\begin{aligned}&\min_{\Delta u_f(k)} \frac{1}{2}\Delta u_f(k)^T H \Delta u_f(k) + G(k+1|k)^T \Delta u_f(k), \\ &\text{s.t.} \quad C_u \Delta u_f(k) \geq b(k+1|k),\end{aligned} \tag{8.17}$$

where

$$H = 2\big(S_{N_p,N_c}^T \Gamma_y^T \Gamma_y S_{N_p,N_c} + \Gamma_u^T \Gamma_u\big), \tag{8.18a}$$

$$G(k+1|k) = -2S_{N_p,N_c}^T \Gamma_y^T \Gamma_y E(k+1), \tag{8.18b}$$

$$C_u = \begin{bmatrix} -I & I & -L^T & L^T & (-S_{N_p,N_c}^b)^T & (S_{N_p,N_c}^b)^T \end{bmatrix}^T, \tag{8.19}$$

$$b(k+1|k) = \begin{bmatrix} -\Delta u_{\max}(k) \\ \vdots \\ -\Delta u_{\max}(k+m-1) \\ -\Delta u_{\min}(k) \\ \vdots \\ -\Delta u_{\min}(k+m-1) \\ u(k-1) - u_{\max}(k) \\ \vdots \\ u(k-1) - u_{\max}(k+m-1) \\ u_{\min}(k) - u(k-1) \\ \vdots \\ u_{\min}(k+m-1) - u(k-1) \\ F_b - Y_{\max}^b(k+1) \\ -F_b + Y_{\min}^b(k+1) \end{bmatrix}. \tag{8.20}$$

Moreover, L, $Y^b_{\max}$ and $Y^b_{\min}$ in (8.19) and (8.20) are given by

$$L = \begin{bmatrix} 1 & 0 & \dots & 0 \\ 1 & 1 & \dots & 0 \\ \vdots & \vdots & \ddots & \vdots \\ 1 & 1 & \dots & 1 \end{bmatrix},$$

$$Y^b_{\min}(k+1) = \begin{bmatrix} y^b_{\min}(k+1) \\ y^b_{\min}(k+2) \\ \vdots \\ y^b_{\min}(k+p) \end{bmatrix}, \qquad Y^b_{\max}(k+1) = \begin{bmatrix} y^b_{\max}(k+1) \\ y^b_{\max}(k+2) \\ \vdots \\ y^b_{\max}(k+p) \end{bmatrix}.$$

It is clear from (8.18a), (8.18b) that $H \geq 0$, hence the optimal solution of the optimization problem exists, which is denoted as $\Delta u_f(k)$. Moreover, it is strictly convex if $H > 0$. By considering input and output variable constraints (8.1a)–(8.1c) and solving the quadratic programming (QP) problem (8.17), the control sequence can be obtained. Only the first element of $\Delta u_f(k)$ is used to determine the control signal $u_f(k)$, and then, it is applied to the plant. This procedure is repeated at each sampling interval.

8.4 Simulation Results

The designed data-driven controller is programmed using MATLAB/Simulink and tested on the complete AMESim powertrain model used in the past three chapters. Before the actual simulation results are presented, the choices for the simulation settings and tuning parameters will be described first. The parameters of data-driven predictive controller are chosen as follows: the numbers of rows and columns of the data block Hankel matrices in Eq. (8.3) are $i = 50$, $j = 400$; the prediction horizon and the control horizon are $N_p = 50$ and $N_c = 10$; the weighting factors (8.2) are $\gamma_{y,i} = 0.13$, $\gamma_{u,i} = 1$. The main control requirement is to make the clutch speed ω_c track the engine speed ω_e as soon as possible, i.e., to make the clutch slip speed $\Delta\omega$ converge to the demanded reference value $R_e(k+1)$. Taking the rapidity and smoothness of the start-up process into account, $r(k+i) = \beta^i y(k)$, $i = 1, 2, \dots, N_p$ are defined, where $\beta \in [0, 1]$ is an adjustable parameter. The smaller the β, the faster the start-up.

8.4.1 *Controller Test Under Nominal Conditions*

The friction work W_f of clutch losses is one of the most important control objectives during the start-up process of vehicles. The friction work W_f of the clutch under

Table 8.1 The friction work of clutch W_f (in J)

Fig. 8.4	Fig. 8.5	Fig. 8.8	Fig. 8.9	Fig. 8.10
4352	8967	2211	4413	25165

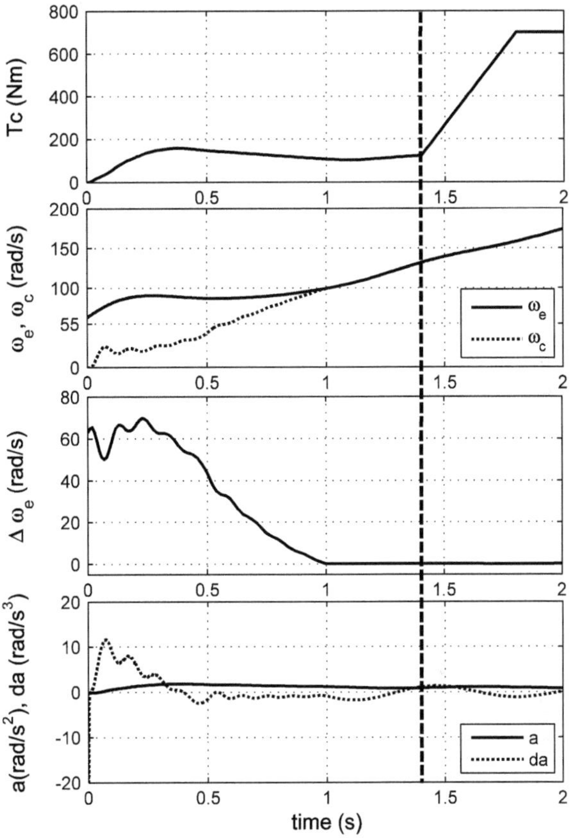

Fig. 8.4 Simulation results of the nominal driving condition with the controller neglecting constraints (fixed throttle opening $\theta_{th} = 60°$; vehicle mass $m = 6000$ kg; road slope $\alpha = 0°$)

different driving conditions is calculated as follows:

$$W_f = \int_{t_0}^{t_f} T_c \Delta\omega \, dt, \tag{8.21}$$

where t_0 and t_f are the start and stop time of clutch engagement, respectively. The clutch friction work W_f of the whole simulation results based on the designed controller are given in Table 8.1.

In this part, the designed data-driven predictive controllers are tested under the nominal conditions. First, the start-up of a vehicle with the designed data-driven predictive controller (8.15) which neglects constraints is simulated, and the results are shown in Fig. 8.4.

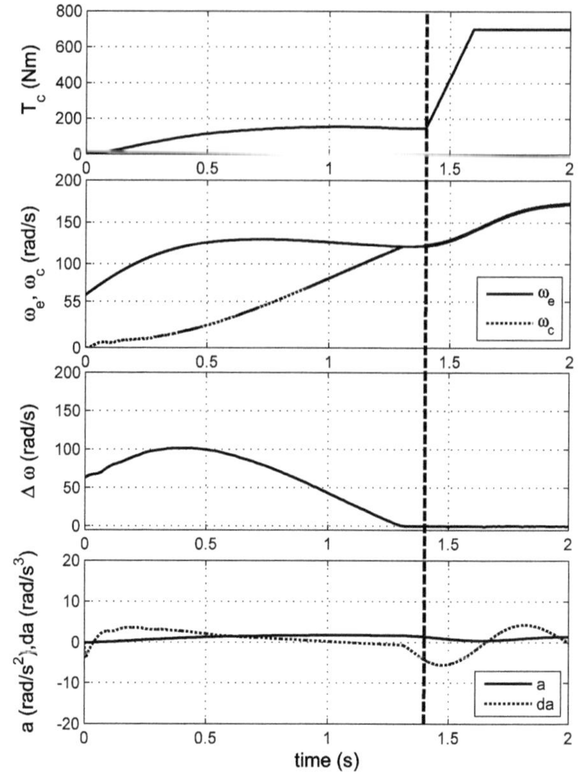

Fig. 8.5 Simulation results of the nominal driving condition with the controller considering constraints (fixed throttle opening $\theta_{th} = 60°$; vehicle mass $m = 6000$ kg; road slope $\alpha = 0°$)

It can be seen that the end-time of the launch process is at about 0.8 s, which shows that the clutch engages rapidly, and the total friction work W_f is about 4352 J shown in Table 8.1, but the maximum jerk da (change rate of the acceleration) is up to about 12 m/s^3. It should be noted that after the clutch is engaged, at about 1.4 s, the clutch torque T_c is increased by a pre-determined pattern to make the clutch lock up reliably, this process is shown in the following figures after the vertical dashed line.

At the same time, due to the vehicle physical constructive characteristics, the engine speed ω_e and the clutch friction torque T_c are restricted. For a typical medium-duty truck, the maximum engine speed ω_e is about 300 rad/s and the minimum is about 55 rad/s which ensures that the engine works normally, and the maximum friction clutch torque is about 700 Nm. The control input constraints are $u_{\min} = 0$ Nm and $u_{\max} = 700$ Nm, and the amplitudes of constrained output $y^b_{\min} = 55$ rad/s and $y^b_{\max} = 300$ rad/s. Moreover, according to the frequency bandwidth of the clutch actuator, the constraints of the control incremental are chosen as $\Delta u_{\min} = -30$ Nm and $\Delta u_{\max} = 30$ Nm. Under the same driving conditions of Fig. 8.4, the start-up of the vehicle having the designed data-driven predictive controller with constraints by solving (8.17) is simulated, and the results are shown in Fig. 8.5.

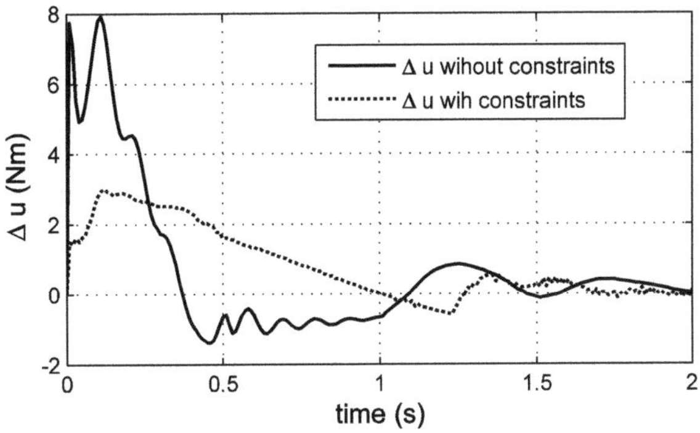

Fig. 8.6 Δu in case of data-driven predictive controller with (without) constrains

Compared to Fig. 8.4, it can be seen that the clutch is engaged smoothly, and the maximum jerk da is reduced from about 12 m/s^3 to about 3 m/s^3, i.e., the comfort requirement of the driver is well satisfied. The reason for this is that the constraints of Δu play a major role in these processes, the specific values of Δu in the case of a data-driven predictive controller with (without) constrains are shown in Fig. 8.6. From Fig. 8.6, it is clear that the maximum Δu of the data-driven constrained predictive controller is about 3 Nm, but in the case of a data-driven predictive controller without constraints, the maximum Δu is 8 Nm.

Moreover, Table 8.1 shows that the total friction work W_f of Fig. 8.5 is 8967 J, which is more than total friction work W_f of Fig. 8.4 because the data-driven constrained predictive controller aims to make the clutch engagement smooth at the expense of some total friction work.

8.4.2 Controller Test Under Changed Conditions

The data-driven predictive controller is gained from the input–output data from the driveline model. It is necessary to test the proposed controller under the driving conditions that deviate from the nominal driving setting, wherein the vehicle mass, the road grade, the engine throttle angle are varied.

Medium Throttle Opening

Compared with the nominal conditions, the engine throttle angle θ_{th} is adjusted as shown in Fig. 8.7, and it tries to simulate the operation of a driver.

Based on the data-driven predictive controller without constraints, the simulation results are obtained and shown in Fig. 8.8.

From Fig. 8.8, it is clear that the engine speed w_e is smaller than the minimum 55 rad/s, which violates the constraints and will make the engine stop, meaning a failed launch. At the same time, the jerk of the driveline during the start-up process

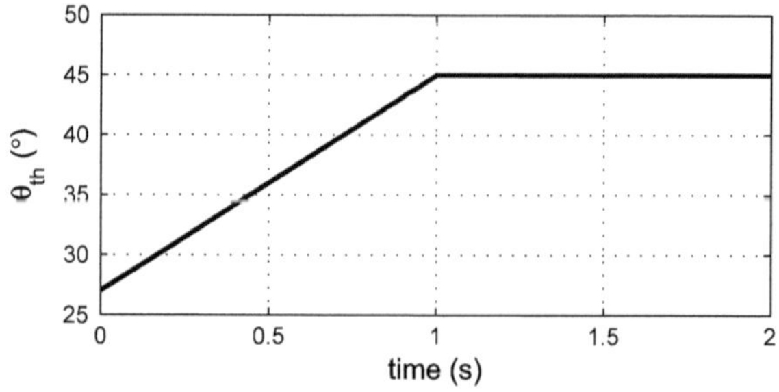

Fig. 8.7 Medium throttle opening

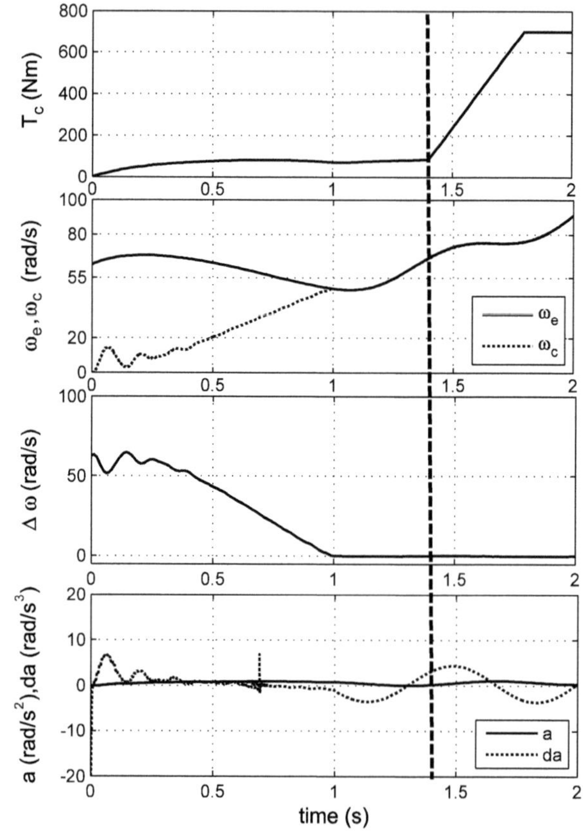

Fig. 8.8 Simulation results of the different driving conditions with the controller neglecting constraints (medium throttle opening $\theta_{th} = 45°$; vehicle mass $m = 6000$ kg; road slope $\alpha = 0°$)

is about 8 m/s^3, the clutch engagement is at about 1 s and the total friction work W_f is 2211 J.

Under the same driving conditions of Fig. 8.8 and considering the same constraints as in Fig. 8.5, the simulation results obtained by using the data-driven constrained predictive controller are shown in Fig. 8.9.

Comparing the results of Fig. 8.9 with those of Fig. 8.8, the most important result is that the engine speed w_e does not touch the lower bound of 55 rad/s which ensures

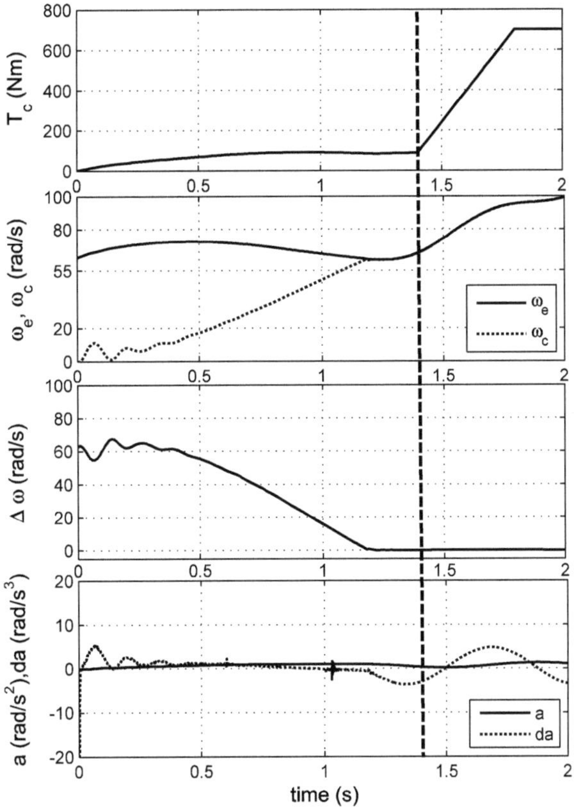

Fig. 8.9 Simulation results of the different driving conditions with the controller considering constraints (medium throttle opening $\theta_{th} = 45°$; vehicle mass $m = 6000$ kg; road slope $\alpha = 0°$)

that the engine works in good operation region, and the vehicle is launched successfully, though the total friction work W_f is bigger than that in Fig. 8.8. Furthermore, it shows that the maximum jerk da in Fig. 8.9 is obviously smaller than the one in Fig. 8.8. The good results of the data-driven predictive controller mentioned above are credited to its capacity of dealing with constraints. On the whole, the data-driven constrained predictive controller is better than the one without constraints, due to the proposed controller's natural ability to handle physical constraints arising in the actual application.

Full Throttle Opening, Fully Loaded Mass and Steep Road Slope

The variation of the vehicle mass and the road slope results in a change of the vehicle dynamics. Hence, a simulation of a fully loaded vehicle launched on a slope is carried out. The results are shown in Fig. 8.10, where the driving condition settings are as follows: the engine throttle opening is adjusted the same as Fig. 8.7, but the opening is full at 1 s, i.e., $\theta_{th} = 90°$; the vehicle mass is increased from 6000 to 8000 kg; the road grade angle is varied from 0° to 5°.

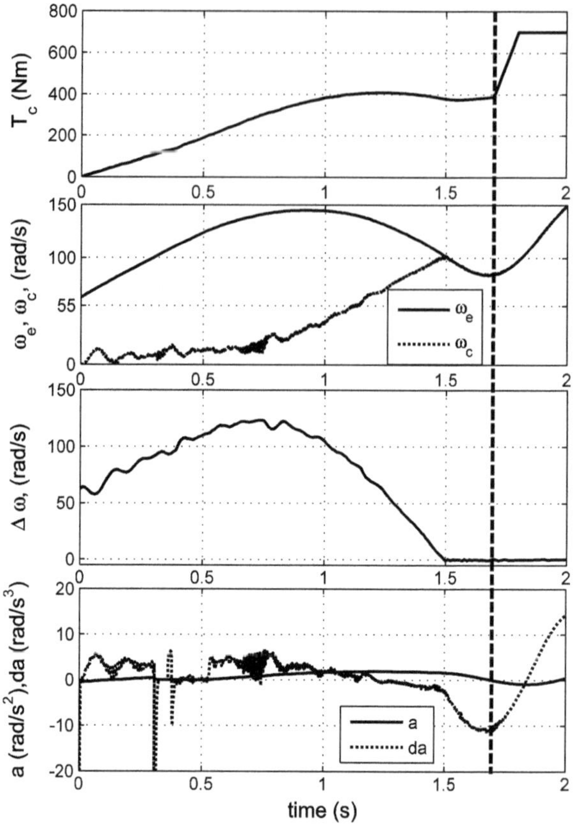

Fig. 8.10 Simulation results of the different driving conditions with the controller considering constraints (full throttle opening $\theta_{th} = 90°$; fully loaded mass $m = 8000$ kg; steep road slope $\alpha = 5°$)

It should be noted that at the beginning of the start-up, because of the large change of the vehicle mass and the road slope, the variation of the clutch speed is not smooth. However, it is shown that the vehicle can still be launched successfully with acceptable driveline shock da and friction losses shown in Table 8.1. This illustrates the potential benefits of the designed data-driven predictive controller under different uncertainties.

At last, from Table 8.1, it is clear that the friction work of Figs. 8.8 and 8.9 are much less than that of Figs. 8.4 and 8.5, respectively. The reason is that in Figs. 8.4 and 8.5, the nominal setting, the engine throttle angle is set to a greater value of 60°, which requires larger clutch torque, and consequently results in much larger friction losses. Therefore, the friction work of Fig. 8.10 is up to 25165 J, which is reasonable for the full-throttle opening.

8.5 Notes and References

Although this chapter has considered some uncertainties, such as throttle opening, vehicle mass and road grade, and tested the robustness of the proposed controller, much deeper investigations are needed, including data-driven controller based on the online input–output data, which is used to deal with the uncertainties and disturbances resulting from the variation of driving conditions and long-term aging. The data-driven predictive control method is being applied to solve many industry issues in recent years [1, 3, 11]. For more detailed information about the data-driven predictive control method, please refer to [2, 4–6, 10, 12].

References

1. Chiera BA, White LB (2005) A subspace predictive controller for End-to-End TCP congestion control. In: AusCTW, pp 42–48
2. Favoreel W, De Moor B (1998) SPC: subspace predictive control. In: Technical report. Katholieke Universiteit, Leuven, pp 49–98
3. Hallouzie R, Verhaegen M (2008) Fault-tolerant subspace predictive control applied to a Boeing 747 model. J Guid Control Dyn 31(4):873–891
4. Kadali K, Huang B, Rossiter A (2003) A data driven subspace approach to predictive controller design. Control Eng Pract 11(3):261–278
5. Lee JM, Lee JH (2005) Approximate dynamic programming-based approaches for input-output data-driven control of nonlinear processes. Automatica 41:1281–1288
6. Lespinats S, Verleysen A, Giron M, Fertil B (2007) DD-HDS: a method for visualization and exploration of high-dimensional data. IEEE Trans Neural Netw 18(5):1265–1279
7. Lu XH, Chen H, Wang P, Gao BZ (2011) Design of a data-driven predictive controller for start-up process of AMT vehicles. IEEE Trans Neural Netw 22(12):2201–2212
8. Maciejowski JM (2002) Predictive control: with constraints. Prentice Hall, New York
9. Morari M, Lee JH (1997) Model predictive control: past, present and future. In: Proc PSE'97-ESCAPE-7 symposium, Trondheim
10. Qin SJ, Lin W, Ljung L (2005) A novel subspace identification approach with enforced causal models. Automatica 41(12):2043–2053
11. Wang X, Huang B, Chen T (2007) Data-driven predictive control for solid oxide fuel cells. J Process Control 17(2):103–114
12. Xu JX, Hou ZS (2009) Notes on data-driven system approaches. Acta Autom Sin 35(6):668–675

Appendix A
Lyapunov Stability

Lyapunov stability theory [1] plays a central role in systems theory and engineering. An equilibrium point is stable if all solutions starting at nearby points stay nearby; otherwise, it is unstable. It is asymptotically stable if all solutions starting at nearby points not only stay nearby, but also tend to the equilibrium point as time approaches infinity. These notions are made precise as follows.

Definition A.1 Consider the autonomous system [1]

$$\dot{x} = f(x), \tag{A.1}$$

where $f \colon D \to \mathbb{R}^n$ is a locally Lipschitz map from a domain D into $\mathbb{R}^n$.

The equilibrium point $x = 0$ is

- Stable if, for each $\epsilon > 0$, there is $\delta = \delta(\epsilon) > 0$ such that

$$\|x(0)\| < \delta \Rightarrow \|x(t)\| < \epsilon, \quad \forall t \geq 0;$$

- Unstable if it is not stable;
- Asymptotically stable if it is stable and δ can be chosen such that

$$\|x(0)\| < \delta \Rightarrow \lim_{t\to\infty} x(t) = 0.$$

Theorem A.1 *Let $x = 0$ be an equilibrium point for system* (A.1) *and $D \subset \mathbb{R}^n$ be a domain containing $x = 0$. Let $V : D \to \mathbb{R}$ be a continuously differentiable function such that*

$$V(0) = 0 \quad and \quad V(x) > 0 \quad in\ D - \{0\}, \tag{A.2a}$$

$$\dot{V}(x) \leq 0 \quad in\ D. \tag{A.2b}$$

Then, $x = 0$ is stable. Moreover, if

$$\dot{V}(x) < 0 \quad in\ D - \{0\}, \tag{A.3}$$

then $x = 0$ is asymptotically stable.

H. Chen, B. Gao, *Nonlinear Estimation and Control of Automotive Drivetrains*,
DOI 10.1007/978-3-642-41572-2,

A function $V(x)$ satisfying (A.2a) is said to be positive definite. A continuously differentiable function $V(x)$ satisfying (A.2a) and (A.2b) is called a Lyapunov function. A function $V(x)$ satisfying

$$V(x) \to \infty \quad \text{as } \|x\| \to \infty \tag{A.4}$$

is said to be radially unbounded.

Theorem A.2 *Let $x = 0$ be an equilibrium point for system* (A.1). *Let $V : \mathbb{R}^n \to \mathbb{R}$ be a continuously differentiable, radially unbounded, positive definite function such that*

$$\dot{V}(x) < 0, \quad \forall x \neq 0. \tag{A.5}$$

Then $x = 0$ is globally asymptotically stable.

In some cases of $\dot{V}(x) \leq 0$, asymptotical stability of $x = 0$ can be proved if we can establish that no system trajectory stays forever at points where $\dot{V}(x) = 0$ except at $x = 0$. This follows from LaSalle's invariance principle. Before stating LaSalle's invariance principle, we give the notation of invariance.

A set M is said to be an invariant set with respect to system (A.1) if

$$x(0) \in M \Rightarrow x(t) \in M, \quad \forall t \in \mathbb{R}. \tag{A.6}$$

That is, if a solution of (A.1) belongs M at some time instant, it belongs to M for all future and past time.

A set M is said to be a positively invariant set with respect to system (A.1) if

$$x(0) \in M \Rightarrow x(t) \in M, \quad \forall t \geq 0. \tag{A.7}$$

That is, any trajectory of (A.1) starting from M stays in M for all future time.

Theorem A.3 *Let D be a compact (closed and bounded) set with the property that every trajectory of system* (A.1) *starting from D remains in D for all future time. Let $V\colon D \to \mathbb{R}$ be a continuously differentiable positive definite function such that*

$$\dot{V}(x) \leq 0 \quad \textit{in } D. \tag{A.8}$$

Let E be the set of all points in D where $\dot{V}(x) = 0$ and M be the largest invariant set in E. Then every trajectory of (A.1) *starting from D approaches M as $t \to \infty$.*

Corollary A.1 *Let $x = 0$ be an equilibrium point for system* (A.1) *and $D \subset \mathbb{R}^n$ be a domain containing $x = 0$. Let $V\colon D \to \mathbb{R}$ be a continuously differentiable positive definite function such that*

$$\dot{V}(x) \leq 0 \quad \textit{in } D. \tag{A.9}$$

Let $S = \{x \in D \mid \dot{V}(x) = 0\}$, and suppose that no solution other than the trivial solution can forever stay in S. Then $x = 0$ is asymptotically stable.

Corollary A.2 *Let $x = 0$ be an equilibrium point for system* (A.1). *Let $V: \mathbb{R}^n \to \mathbb{R}$ be a continuously differentiable, radially unbounded, positive definite function such that*

$$\dot{V}(x) \le 0 \quad \textit{for all } x \in \mathbb{R}^n. \tag{A.10}$$

Let $S = \{x \in \mathbb{R}^n \mid \dot{V}(x) = 0\}$, and suppose that no solution other than the trivial solution can forever stay in S. Then $x = 0$ is globally asymptotically stable.

If $\dot{V}(x)$ is negative definite, $S = \{0\}$. Then, Corollaries A.1 and A.2 coincide with Theorems A.1 and A.2.

Example A.1 Consider the pendulum equation with friction as follows:

$$\dot{x}_1 = x_2, \tag{A.11a}$$

$$\dot{x}_2 = -\frac{g}{l}\sin x_1 - \frac{k}{m}x_2. \tag{A.11b}$$

We respectively apply Corollary A.1 and Theorem A.1 to discuss the stability properties of this system.

Let us first choose the energy as a Lyapunov function candidate, namely

$$V(x) = \frac{g}{l}(1 - \cos x_1) + \frac{1}{2}x_2^2 \tag{A.12}$$

which satisfies $V(0) = 0$ and $V(x) > 0$ in $D - \{0\}$ with $D = \{x \in \mathbb{R}^2 | -\pi < x_1 < \pi\}$. Differentiating $V(x)$ along the trajectories of the system leads to

$$\begin{aligned}\dot{V}(x) &= \frac{g}{l}\dot{x}_1 \sin x_1 + x_2\dot{x}_2 \\ &= \frac{g}{l}x_2 \sin x_1 + x_2\left(-\frac{g}{l}\sin x_1 - \frac{k}{m}x_2\right) \\ &= -\frac{k}{m}x_2^2,\end{aligned}$$

which implies that $\dot{V}(x) \le 0$ in D. Let S be the set in D which contains all states where $\dot{V}(x) = 0$ is maintained, i.e., $S = \{x \in D \mid \dot{V}(x) = 0\}$. For the pendulum system, we infer from $\dot{V}(x) \equiv 0$ that

$$x_2(t) \equiv 0 \quad \Rightarrow \quad \dot{x}_1(t) \equiv 0 \quad \Rightarrow \quad x_1(t) \equiv \textit{constant} \tag{A.13}$$

and

$$x_2(t) \equiv 0 \quad \Rightarrow \quad \dot{x}_2(t) \equiv 0 \quad \Rightarrow \quad \sin x_1 = 0. \tag{A.14}$$

The only point on the segment $-\pi < x_1 < \pi$ rendering $\sin x_1 = 0$ is $x_1 = 0$. Hence, no trajectory of the pendulum system (A.11a), (A.11b) other than the trivial solution can forever stay in S, i.e., $S = \{0\}$. Then, $x = 0$ is asymptotically stable by Corollary A.1.

We can also show the stability property by choosing other Lyapunov function candidates. One possibility is to replace the term $\frac{1}{2}x_2^2$ in (A.12) by the quadratic form $\frac{1}{2}x^T Px$ for some positive definite matrix P. That is, we choose

$$V(x) = \frac{g}{l}(1 - \cos x_1) + \frac{1}{2}x^T Px \tag{A.15}$$

as a Lyapunov function candidate, where

$$P = \begin{pmatrix} p_{11} & p_{12} \\ p_{12} & p_{22} \end{pmatrix}$$

is to be determined. In order to ensure that $\frac{1}{2}x^T Px$ is positive definite, the elements of P have to satisfy

$$p_{11} > 0, \; p_{22} > 0, \quad p_{11}p_{22} - p_{12}^2 > 0. \tag{A.16}$$

Differentiating $V(x)$ along the trajectories of the system leads to

$$\begin{aligned}
\dot{V}(x) &= \frac{g}{l}\dot{x}_1 \sin x_1 + x^T P\dot{x} \\
&= \frac{g}{l}x_2 \sin x_1 + p_{11}x_1x_2 - \frac{g}{l}p_{12}x_1 \sin x_1 - \frac{k}{m}p_{12}x_1x_2 \\
&\quad + p_{12}x_2^2 - \frac{g}{l}p_{22}x_2 \sin x_1 - \frac{k}{m}p_{22}x_2^2 \\
&= \frac{g}{l}(1 - p_{22})x_2 \sin x_1 - \frac{g}{l}p_{12}x_1 \sin x_1 + \left(p_{11} - \frac{k}{m}p_{12}\right)x_1x_2 \\
&\quad + \left(p_{12} - \frac{k}{m}p_{22}\right)x_2^2.
\end{aligned}$$

Now we can choose p_{11}, p_{12} and p_{22} to ensure $\dot{V}(x)$ is negative definite. First of all, we take

$$p_{22} = 1, \qquad p_{11} = \frac{k}{m}p_{12} \tag{A.17}$$

to cancel the cross-product terms $x_2 \sin x_1$ and x_1x_2 and arrive at

$$\dot{V}(x) = -\frac{g}{l}p_{12}x_1 \sin x_1 + \left(p_{12} - \frac{k}{m}p_{22}\right)x_2^2. \tag{A.18}$$

By combining (A.16) and (A.17), we have

$$0 < p_{12} < \frac{k}{m}. \tag{A.19}$$

Let us take $p_{12} = 0.5\frac{k}{m}$. Then, we have

$$\dot{V}(x) = -\frac{gk}{2lm}x_1 \sin x_1 - \frac{k}{2m}x_2^2. \tag{A.20}$$

The term $x_1 \sin x_1 > 0$ for all $-\pi < x_1 < \pi$. Defining $D = \{x \in \mathbb{R}^2 | -\pi < x_1 < \pi\}$, we can conclude that, with the chosen P, $V(x)$ is positive definite and $\dot{V}(x)$ is negative definite on D. Hence, $x = 0$ is asymptotically stable by Theorem A.1.

Results for autonomous systems can be extended to non-autonomous systems [1].

Theorem A.4 *Consider the non-autonomous system*

$$\dot{x} = f(t, x), \tag{A.21}$$

where $f\colon [0, \infty) \times D \to \mathbb{R}^n$ *is piecewise continuous in* t *and locally Lipschitz in* x *on* $[0, \infty) \times D$*, and* $D \subset \mathbb{R}^n$ *is a domain that contains the origin* $x = 0$*.*

Let $x = 0$ *be an equilibrium point for* (A.21) *and* $D \subset \mathbb{R}^n$ *be a domain containing* $x = 0$*. Let* $V : [0, \infty) \times D \to \mathbb{R}$ *be a continuously differentiable function such that*

$$W_1(x) \le V(t, x) \le W_2(x), \tag{A.22a}$$

$$\frac{\partial V}{\partial t} + \frac{\partial V}{\partial x} f(t, x) \le 0 \tag{A.22b}$$

for all $t \ge 0$ *and for all* $x \in D$*, where* $W_1(x)$ *and* $W_2(x)$ *are continuous positive definite functions on* D*. Then,* $x = 0$ *is uniformly stable.*

References

1. Khalil HK (2002) Nonlinear Systems. Prentice Hall, New York

Appendix B
Input-to-State Stability (ISS)

For a linear time-invariant system

$$\dot{x} = Ax + Bw \tag{B.1}$$

with a Hurwitz matrix A, we can write the solution as

$$x(t) = e^{(t-t_0)A}x(t_0) + \int_{t_0}^{t} e^{(t-\tau)A}Bw(\tau)\,d\tau \tag{B.2}$$

and have

$$\|x(t)\| \le \beta(t)\|x(t_0)\| + \gamma\|w\|_\infty, \tag{B.3}$$

where

$$\beta(t) = \|e^{A(t-t_0)}\| \to 0 \quad \text{and} \quad \gamma = \|B\|\int_{t_0}^{\infty}\|e^{A(s-t_0)}\|\,ds < \infty.$$

Moreover, we use the bound $\|e^{(t-t_0)A}\| \le ke^{-\lambda(t-t_0)}$ to estimate the solution by

$$\|x(t)\| \le ke^{-\lambda(t-t_0)}\|x(t_0)\| + \frac{k\|B\|}{\lambda}\|w\|_\infty, \tag{B.4}$$

where λ could be given by $\lambda := \min\{|\mathrm{Re}\{\mathrm{eig}(A)\}|\}$. Since $\lambda > 0$, the above estimate shows that the zero-input response decays exponentially, while the zero-state response is bounded for every bounded input, that is, has a bounded-input–bounded-state property.

For general nonlinear systems, however, it should not be surprising that these properties may not hold [1]. ISS (Input-to-state stability) is a notation of stability of nonlinear systems, which is suggested by Eduardo Sontag [2, 3] and merges two different views of stability, namely state space approach usually associated with the name of Lyapunov and the operator approach of which George Zames is one of the main initiators [4]. The Lyapunov concept addresses the stability property of systems without external inputs, while the operator concept studies the I/O properties

H. Chen, B. Gao, *Nonlinear Estimation and Control of Automotive Drivetrains*,
DOI 10.1007/978-3-642-41572-2,

of systems under different external input signals including $\mathcal{L}_2$ and $\mathcal{L}_\infty$ and provides elegant results for linear systems.

B.1 Comparison Functions

Definition B.1 [1] A continuous function $\alpha : [0, a) \to [0, \infty)$ is said to belong to class $\mathcal{K}$ if it is strictly increasing and $\alpha(0) = 0$. It is said to belong to class $\mathcal{K}_\infty$ if $a = \infty$, and $\alpha(r) \to \infty$ when $r \to \infty$.

Definition B.2 A continuous function $\beta : [0, a) \times [0, \infty) \to [0, \infty)$ is said to belong to class $\mathcal{KL}$ if, for each fixed s, the mapping $\beta(r, s)$ belongs to class $\mathcal{K}$ with respect to r and, for each fixed r, the mapping $\beta(r, s)$ is decreasing with respect to s and $\beta(r, s) \to 0$ as $s \to \infty$.

Using these comparison functions, we can restate the stability definition in Appendix A in a more precise fashion [1].

Definition B.3 The equilibrium point $x = 0$ of (A.21) is

- Uniformly stable if there exists a class $\mathcal{K}$ function $\gamma(\cdot)$ and a positive constant δ, independent of t_0, such that

$$\left|x(t)\right| \le \gamma\left(\left|x(t_0)\right|\right), \quad \forall t \ge t_0 \ge 0,\ \forall x(t_0) \in \{x|\ |x| < \delta\}; \tag{B.5}$$

- Uniformly asymptotically stable if there exists a class $\mathcal{K}L$ function $\beta(\cdot, \cdot)$ and a positive constant δ, independent of t_0, such that

$$\left|x(t)\right| \le \beta\left(\left|x(t_0)\right|, t - t_0\right), \quad \forall t \ge t_0 \ge 0,\ \forall x(t_0) \in \{x|\ |x| < \delta\}; \tag{B.6}$$

- Exponentially stable if (B.6) is satisfied with $\beta(r, s) = kre^{-\alpha s}, k > 0, \alpha > 0$;
- Globally uniformly stable, if (B.5) is satisfied with $\gamma \in \mathcal{K}_\infty$ for any initial state $x(t_0)$;
- Globally uniformly asymptotically stable if (B.6) is satisfied with $\beta \in \mathcal{K}L_\infty$ for any initial state $x(t_0)$;
- Globally exponentially stable if (B.6) is satisfied with $\beta(r, s) = kre^{-\alpha s}, k > 0$, $\alpha > 0$ for any initial state $x(t_0)$.

Some properties of comparison functions are summarized as follows [1, 4]:

- If $V : \mathbb{R}^n \to \mathbb{R}$ is continuous, define two functions

$$\underline{\alpha}(r) := \min_{|x| \ge r} \left|V(x)\right|, \qquad \overline{\alpha}(r) := \max_{|x| \le r} \left|V(x)\right|.$$

  Then, they are of class $\mathcal{K}_\infty$ and

$$\underline{\alpha}\left(|x|\right) \le \left|V(x)\right| \le \overline{\alpha}\left(|x|\right), \quad \forall x \in \mathbb{R}^n;$$

- Suppose $\alpha, \gamma \in \mathcal{K}_\infty$, $\beta \in \mathcal{KL}$, then
  - $\alpha^{-1}(\beta(\cdot,\cdot)) \in \mathcal{KL}$;
  - $\alpha^{-1}(\gamma(\cdot)) \in \mathcal{K}_\infty$;
  - $\sup_{s\in[0,t]} \gamma(|u(s)|) = \gamma(\|u_{[0,t]}\|_\infty) := \gamma(\sup_{s\in[0,t]}(|u(s)|))$ due to the monotonic increase of γ;
  - $\gamma(a+b) \le \gamma(2a) + \gamma(2b)$ for $a, b \in \mathbb{R}_{\ge 0}$;
  - If $\alpha(r) \le \beta(s,t) + \gamma(t)$, then $r \le \alpha^{-1}(\beta(s,t) + \gamma(t)) \le \alpha^{-1}(2\beta(s,t)) + \alpha^{-1}(2\gamma(t))$.

B.2 Input-to-State Stability

Consider the nonlinear system

$$\dot{x} = f(x, w), \tag{B.7}$$

where $x \in \mathbb{R}^n$ and $w \in \mathbb{R}^m$ are the state and the external input, respectively, and f is locally Lipschitz in x and w. One has the following definitions [4].

Definition B.4 The system (B.7) is said to be input-to-state stable (ISS) if there exist a class $\mathcal{KL}$ function β and a class $\mathcal{K}$ function γ such that, for any $x(0)$ and for any input $w(\cdot)$ continuous and bounded on $[0,\infty)$, the solution of (B.7) satisfies

$$x(t) \le \beta\big(|x(0)|, t\big) + \gamma\big(\|w\|_\infty\big), \quad \forall t \ge 0. \tag{B.8}$$

Definition B.5 The system (B.7) is said to be integral-input-to-state stable (iISS) if there exist a class $\mathcal{KL}$ function β and class $\mathcal{K}_\infty$ functions α and γ such that, for any $x(0)$ and for any input $w(\cdot)$ continuous and bounded on $[0,\infty)$, the solution of (B.7) satisfies

$$\alpha\big(|x(t)|\big) \le \beta\big(|x_0|, t\big) + \int_0^t \gamma\big(|w(s)|\big)\, ds, \quad t \ge 0. \tag{B.9}$$

The following results are stated in [4].

Theorem B.1 *For the system* (B.7), *the following properties are equivalent*:

- *It is ISS*;
- *There exists a smooth positive definite function* $V : \mathbb{R}^n \to \mathbb{R}_+$ *such that for all* $x \in \mathbb{R}^n$ *and* $w \in \mathbb{R}^m$,

$$\alpha_1\big(|x|\big) \le V(x) \le \alpha_2\big(|x|\big), \tag{B.10a}$$

$$|x| \ge \gamma\big(|w|\big) \Rightarrow \frac{\partial V}{\partial x} f(x, w) \le -\alpha_3\big(|x|\big), \tag{B.10b}$$

where α_1, α_2, *and* γ *are class* $\mathcal{K}_\infty$ *functions and* α_3 *is a class* $\mathcal{K}$ *function*;

Fig. B.1 Cascade of systems

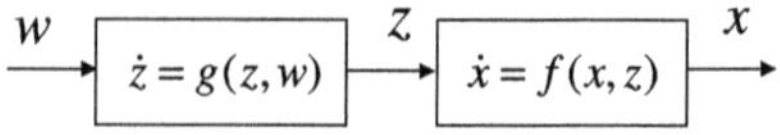

- *There exists a smooth positive definite radially unbounded function V and class $\mathcal{K}_\infty$ functions α and γ such that the following dissipation inequality is satisfied*

$$\frac{\partial V}{\partial x} f(x, w) \leq -\alpha\big(|x|\big) + \gamma\big(|w|\big). \tag{B.11}$$

A continuously differentiable function $V(x)$ satisfying (B.11) is called an ISS-Lyapunov function.

Theorem B.2 *For the system* (B.7), *the following properties are equivalent*:

- *It is iISS*;
- *There exists a smooth positive definite radially unbounded function $V : \mathbb{R}^n \to \mathbb{R}_+$, a class $\mathcal{K}_\infty$ function γ and a positive definite function $\alpha : \mathbb{R}_+ \to \mathbb{R}_+$ such that the following inequality is satisfied*:

$$\frac{\partial V}{\partial x} f(x, w) \leq -\alpha\big(|x|\big) + \gamma\big(|w|\big). \tag{B.12}$$

A continuously differentiable function $V(x)$ satisfying (B.12) is called an iISS-Lyapunov function.

Theorem B.3 *The cascaded system in the form of*

$$\dot{x} = f(x, z),$$
$$\dot{z} = g(z, w),$$

shown in Fig. B.1, *is ISS with input w, if the x-subsystem is ISS with z being viewed as input and the z-subsystem is ISS with input w.*

Theorem B.4 *For the cascaded system in the form of*

$$\dot{x} = f(x, z),$$
$$\dot{z} = g(z),$$

one has the following:

- *It is globally asymptotically stable if the x-subsystem is ISS with z being viewed as input and the z-subsystem is globally asymptotically stable*;
- *It is globally asymptotically stable if the x-subsystem is affine in z and iISS with z being viewed as input, and the z-subsystem is globally asymptotically stable and locally exponentially stable.*

Fig. B.2 Uncertain system

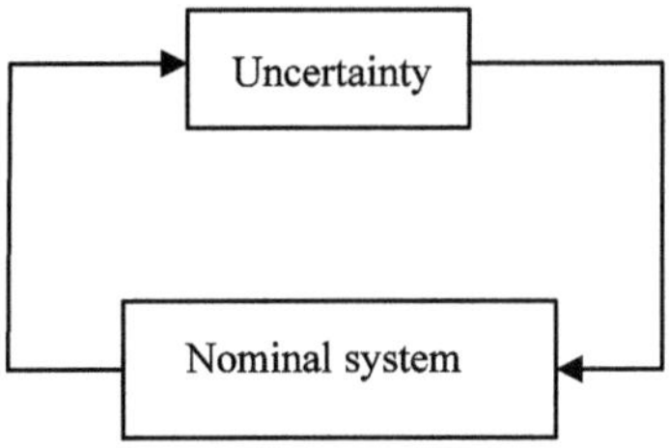

Some of equivalences for ISS are listed as follows:

- (Nonlinear superposition principle) A system is ISS if and only if it is zero-input stable and satisfies the asymptotic gain property. A system satisfies the asymptotic gain property if there is some $\gamma \in \mathcal{K}_\infty$ such that

$$\lim_{t\to\infty} \sup\left|x(t)\right| \le \gamma\left(\|w\|_\infty\right), \quad \forall x(0), w(\cdot), t \ge 0;$$

- A system is ISS if and only if it is robustly stable, where the system is perturbed by the uncertainty as shown in Fig. B.2;
- A system is ISS if and only if it is dissipative with the supply function of $s(x, w) = \gamma(|w|) - \alpha(|x|)$, i.e., the following dissipation inequality

$$V\left(x(t_2)\right) - V\left(x(t_1)\right) \le \int_{t_1}^{t_2} s\left(x(\tau), w(\tau)\right) d\tau$$

holds along all trajectories of the system and for some $\alpha, \gamma \in \mathcal{K}_\infty$;
- A system is ISS if and only if it satisfies the following $\mathcal{L}_2 \to \mathcal{L}_2$ estimate

$$\int_0^t \alpha_1\left(\left|x(\tau)\right|\right) d\tau \le \alpha_0\left(\left|x(0)\right|\right) + \int_0^t \gamma\left(\left|w(\tau)\right|\right) d\tau$$

along all trajectories of the system and for some $\alpha_0, \alpha_1, \gamma \in \mathcal{K}_\infty$.

B.2.1 Useful Lemmas

Lemma B.1 (*Young's Inequality* [p. 75, 1]) *If the constants $p > 1$ and $q > 1$ are such that $(p-1)(q-1) = 1$, then for all $\epsilon > 0$ and all $(x, y) \in \mathbb{R}^2$ we have*

$$xy \le \frac{\epsilon^p}{p}|x|^p + \frac{1}{q\epsilon^q}|y|^q.$$

Choosing $p = q = 2$ and $\epsilon^2 = 2\kappa$, the above inequality becomes

$$xy \le \kappa x^2 + \frac{1}{4\kappa}y^2.$$

Lemma B.2 ([*pp*. 495, 505, 1]) *Let v and ρ be real-valued functions defined on $\mathbb{R}_+$, and let b and c be positive constants. If they satisfy the differential inequality*

$$\dot{v} \le -cv + b\rho(t)^2, \quad v(0) \ge 0 \tag{B.13}$$

then

(a) *The following integral inequality holds*:

$$v(t) \le v(0)e^{-ct} + b\int_0^t e^{-c(t-\tau)}\rho(\tau)^2\, d\tau; \tag{B.14}$$

(b) *If, in addition, $\rho \in \mathcal{L}_2$, then $v \in \mathcal{L}_1$ and*

$$\|v\|_1 \le \frac{1}{c}\left(v(0) + b\|\rho\|_2^2\right); \tag{B.15}$$

(c) *If $\rho \in \mathcal{L}_\infty$, then $v \in \mathcal{L}_\infty$ and*

$$v(t) \le v(0)e^{-ct} + \frac{b}{c}\|\rho\|_\infty^2; \tag{B.16}$$

(d) *If $\rho \in \mathcal{L}_2$, then $v \in \mathcal{L}_\infty$ and*

$$v(t) \le v(0)e^{-ct} + b\|\rho\|_2^2. \tag{B.17}$$

References

1. Khalil HK (2002) Nonlinear Systems. Prentice Hall, New York
2. Sontag ED (1989) Smooth stabilization implies coprime factorization. IEEE Trans Autom Control 34:435–443
3. Sontag ED, Wang Y (1996) New characterizations of input-to-state stability. IEEE Trans Autom Control 41:1283–1294
4. Sontag ED (2008) Input to state stability: basic concepts and results. In: Cachan JM, Groningen FT, Paris BT (eds) Nonlinear and optimal control theory. Lecture notes in mathematics. Springer, Berlin, pp 163–220

Appendix C
Backstepping

In control theory, backstepping [1–3] is a technique for designing stabilizing controllers for a special class of nonlinear dynamical systems. The designer can start the design process at the known-stable system and "back out" new controllers that progressively stabilize each outer subsystem. The process terminates when the final external control is reached. Hence, this process is known as backstepping. Because in each step, Control Lyapunov Function (CLF) is constructed to obtain the virtual control, the control law obtained by backstepping is globally asymptotically stable.

C.1 About CLF

Definition C.1 For a time-invariant nonlinear system [1]

$$\dot{x} = f(x,u), \quad x \in \mathbb{R}^n, \; u \in \mathbb{R}, \; f(0,0) = 0, \tag{C.1}$$

a smooth positive definite and radially unbounded function $V : \mathbb{R}^n \to \mathbb{R}_+$ is called a control Lyapunov function (CLF) if

$$\inf_{u \in \mathbb{R}} \left(\frac{\partial V}{\partial x}(x) f(x,u) \right) < 0, \quad \forall x \neq 0. \tag{C.2}$$

For systems affine in the control,

$$\dot{x} = f(x) + g(x)u, \quad f(0) = 0, \tag{C.3}$$

the CLF inequality becomes

$$\frac{\partial V}{\partial x} f(x) + \frac{\partial V}{\partial x} g(x)\alpha(x) \le -W(x), \tag{C.4}$$

where $\alpha(x)$ is the control law designed for u, and $W : \mathbb{R}^n \to \mathbb{R}$ is positive definite (or positive semi-definite, in this case one needs to apply Theorem A.3 to discuss stability).

H. Chen, B. Gao, *Nonlinear Estimation and Control of Automotive Drivetrains*,
DOI 10.1007/978-3-642-41572-2,

If $V(x)$ is a CLF for (C.3), then a stabilizing control law $\alpha(x)$, smooth for all $x \neq 0$, is given by Sontag's formula [1, 4]

$$\alpha(x) = \begin{cases} -\frac{\frac{\partial V}{\partial x} f + \sqrt{(\frac{\partial V}{\partial x} f)^2 + (\frac{\partial V}{\partial x} g)^4}}{\frac{\partial V}{\partial x} g}, & \frac{\partial V}{\partial x} g \neq 0, \\ 0, & \frac{\partial V}{\partial x} f = 0, \end{cases} \tag{C.5}$$

witch results in

$$W(x) = \sqrt{\left(\frac{\partial V}{\partial x} f\right)^2 + \left(\frac{\partial V}{\partial x} g\right)^4} > 0, \quad \forall x \neq 0.$$

C.2 Backstepping Design

We first use an example to introduce the idea of backstepping.

Example C.1 Consider the system [1]

$$\dot{x}_1 = x_1^2 - x_1^3 + x_2, \tag{C.6a}$$

$$\dot{x}_2 = u. \tag{C.6b}$$

We start with the first equation $\dot{x}_1 = x_1^2 - x_1^3 + x_2$, where x_2 is viewed as virtual control, and proceed to design the feedback control $x_2 = \alpha(x_1)$ to stabilize the origin $x_1 = 0$. With

$$x_2 = -x_1^2 - x_1 \tag{C.7}$$

we cancel the nonlinear term x_1^2 to obtain

$$\dot{x}_1 = -x_1 - x_1^3 \tag{C.8}$$

and infer $V(x_1) = \frac{1}{2}x_1^2$ satisfying

$$\dot{V} = -x_1^2 - x_1^4, \quad \forall x_1 \in \mathbb{R} \tag{C.9}$$

along the solution of (C.8). This implies that $\dot{V}$ is negative definite. Hence, the origin of (C.8) is globally exponentially stable.

Indeed, x_2 is one of system states and $x_2 \neq \alpha(x_1)$. To backstep, we define the error between x_2 and the desired value $\alpha(x_1)$ as follows:

$$z_2 = x_2 - \alpha(x_1) = x_2 + x_1 + x_1^2, \tag{C.10}$$

which serves as a change of variables to transform the system into

$$\dot{x}_1 = -x_1 - x_1^3 + z_2, \tag{C.11a}$$

$$\dot{z}_2 = u + (1 + 2x_1)\left(-x_1 - x_1^3 + z_2\right). \tag{C.11b}$$

Taking

$$V_c(x) = \frac{1}{2}x_1^2 + \frac{1}{2}z_2^2 \tag{C.12}$$

as a composite Lyapunov function, we obtain for the transformed system (C.11a), (C.11b) that satisfies

$$\dot{V}_c = -x_1^2 - x_1^4 + z_2\big(x_1 + (1+2x_1)\big(-x_1 - x_1^3 + z_2\big) + u\big). \tag{C.13}$$

Choosing

$$u = -x_1 - (1+2x_1)\big(-x_1 - x_1^3 + z_2\big) - z_2 \tag{C.14}$$

yields

$$\dot{V}_c = -x_1^2 - x_1^4 - z_2^2, \tag{C.15}$$

which is negative definite. Hence, the origin of the transformed system (C.11a), (C.11b), and hence the original system (C.6a), (C.6b), is globally stable.

Now we consider the system having the following strict feedback form [1, 5]:

$$\dot{x} = f(x) + g(x)\xi_1, \tag{C.16a}$$

$$\dot{\xi}_1 = f_1(x, \xi_1) + g_1(x, \xi_1)\xi_2, \tag{C.16b}$$

$$\dot{\xi}_2 = f_2(x, \xi_1, \xi_2) + g_2(x, \xi_1, \xi_2)\xi_3, \tag{C.16c}$$

$$\vdots \tag{C.16d}$$

$$\dot{\xi}_k = f_k(x, \xi_1, \ldots, \xi_k) + g_k(x, \xi_1, \ldots, \xi_k)u, \tag{C.16e}$$

where $x \in \mathbb{R}^n$ and $\xi_1, \ldots, \xi_k$, which are scalars, are the state variables. Many physical systems can be represented as a strict feedback system, such as the system described in Chap. 4.

The whole design process is based on the following assumption:

Assumption C.1 Consider the system

$$\dot{x} = f(x) + g(x)u, \quad f(0) = 0, \tag{C.17}$$

where $x \in \mathbb{R}^n$ is the state and $u \in \mathbb{R}$ is the scalar control input. There exists a continuously differentiable feedback control law

$$u = \alpha(x), \quad \alpha(0) = 0, \tag{C.18}$$

and a smooth, positive definite, radially unbounded function $V : \mathbb{R}^n \to \mathbb{R}$ such that

$$\frac{\partial V}{\partial x}(x)\big[f(x) + g(x)\alpha(x)\big] \le -W(x) \tag{C.19}$$

with $W:\mathbb{R}^n\to\mathbb{R}$ positive definite (or positive semi-definite, in this case one needs to apply Theorem A.3 to discuss stability).

A special case is when the x-state has dimension 1, i.e., $n=1$. Then, by constructing a Lyapunov function

$$V(x)=\frac{1}{2}x^2 \tag{C.20}$$

for (C.16a), where ξ_1 is regarded as the virtual control, the control law $\alpha(x)$ is determined to satisfy (C.4), i.e.,

$$x\bigl(f(x)+g(x)\alpha(x)\bigr)\le -W(x) \tag{C.21}$$

with $W:\mathbb{R}^n\to\mathbb{R}$ positive definite (or positive semi-definite). Hence, one choice of $\alpha(x)$ is

$$\alpha(x)=\begin{cases}-\frac{W(x)+xf(x)}{xg(x)}, & x\neq 0,\\ 0, & x=0.\end{cases} \tag{C.22}$$

More specially, one can choose $W(x)=k_1x^2$ with $k_1>0$ for simplicity to get

$$\alpha(x)=\begin{cases}-\frac{k_1x+f(x)}{g(x)}, & x\neq 0,\\ 0, & x=0,\end{cases} \tag{C.23}$$

if $g(x)\neq 0$ for all x. Note that the x-subsystem is uncontrollable at the points of $g(x)=0$.

In the following, we assume that Assumption C.1 is satisfied in general.

Since ξ_1 is just a state variable and not the control, we define e_1 as the deviation of ξ_1 from its desired value $\alpha(x)$:

$$e_1=\xi_1-\alpha(x) \tag{C.24}$$

and infer

$$\dot V=\frac{\partial V}{\partial x}\dot x=\frac{\partial V}{\partial x}\bigl(f(x)+g(x)\alpha(x)+g(x)e_1\bigr)\le -W(x)+\frac{\partial V}{\partial x}g(x)e_1. \tag{C.25}$$

Then, the second step begins, and the second Lyapunov function is defined as

$$V_1(x,\xi_1)=V(x)+\frac{1}{2}e_1^2. \tag{C.26}$$

Let the desired value of ξ_2 be $\alpha_1(x,\xi_1)$, and introduce the second error

$$e_2=\xi_2-\alpha_1(x,\xi_1). \tag{C.27}$$

Then we have

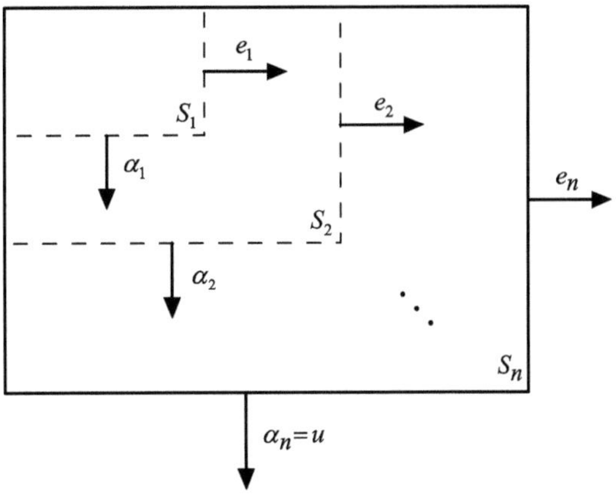

Fig. C.1 Backstepping design procedure [1]

$$
\begin{aligned}
\dot{V}_1 &= \dot{V} + e_1\dot{e}_1 \\
&\leq -W(x) + \frac{\partial V}{\partial x} g(x)e_1 + e_1\left(\dot{\xi}_1 - \frac{\partial \alpha(x)}{\partial x}\dot{x}\right) \\
&\leq -W(x) + e_1\left\{\frac{\partial V}{\partial x} g(x) + f_1(x,\xi_1) + g_1(x,\xi_1)\alpha_1(x,\xi_1)\right. \\
&\quad \left. + g_1(x,\xi_1)e_2 - \frac{\partial \alpha}{\partial x}\big[f(x) + g(x)\xi_1\big]\right\}.
\end{aligned} \tag{C.28}
$$

If $g_1(x,\xi_1) \neq 0$ for all x and ξ_1, the choice of

$$
\alpha_1(x,\xi_1) = \frac{1}{g_1(x,\xi_1)}\left(-c_1e_1 - \frac{\partial V}{\partial x} g(x) - f_1(x,\xi_1) + \frac{\partial \alpha}{\partial x}\big(f(x) + g(x)\xi_1\big)\right) \tag{C.29}
$$

with $c_1 > 0$ leads to

$$
\dot{V}_1 \leq -W_1(x,\xi_1) + \frac{\partial V_1}{\partial \xi_1} g_1(x,\xi_1)e_2, \tag{C.30}
$$

where $W_1(x,e_1) = W(x) + c_1e_1^2$ is positive definite.

Repeat the procedure step by step for the other subsystems (see Fig. C.1) until the final external control is reached. Lyapunov functions are defined as

$$
V_j(x,\xi_1,\ldots,\xi_j) = V(x) + \frac{1}{2}\sum_{i=1}^{l} e_i^2, \quad j = 1,2,\ldots,k \tag{C.31}
$$

with $e_j = \xi_j - \alpha_{j-1}(x,\xi_1,\ldots,\xi_{j-1})$. If the non-singularity conditions

$$
g_j(x,\xi_1,\ldots,\xi_j) \neq 0, \quad \forall x \in \mathbb{R}^n,\ \forall \xi_i \in \mathbb{R}^n,\ i = 1,2,\ldots,j, \tag{C.32}
$$

hold then a final control law can be given by

$$u=\frac{1}{g_k}\left(-c_k e_k-\frac{\partial V_{k-1}}{\partial \xi_{k-1}} g_{k-1}-f_k+\frac{\partial \alpha_{k-1}}{\partial X_{k-1}}\left(F_{k-1}+G_{k-1} \xi_k\right)\right), \tag{C.33}$$

where

$$X_j=\begin{pmatrix} X_{j-1} \\ \xi_j \end{pmatrix}, \quad j=2,3, \ldots, k-1, \tag{C.34a}$$

$$F_j(X_j)=\begin{pmatrix} F_{j-1}(X_{j-1})+G_{j-1}(X_{j-1}) \xi_j \\ f_j(X_{j-1}, \xi_j) \end{pmatrix} \tag{C.34b}$$

$$G_j(X_j)=\begin{pmatrix} 0 \\ g_j(X_{j-1}, \xi_j) \end{pmatrix}, \tag{C.34c}$$

and

$$X_1=\begin{pmatrix} x \\ \xi_1 \end{pmatrix}, \quad F_1(X_1)=\begin{pmatrix} f(x)+g(x) \xi_1 \\ f_1(x, \xi_1) \end{pmatrix}, \quad G_1(X_1)=\begin{pmatrix} 0 \\ g_1(x, \xi_1) \end{pmatrix}. \tag{C.35}$$

Under the above control law, the system is globally asymptotically stable, since

$$\dot{V}_k(x, e_1, \ldots, e_k) \leq -W_k(x, e_1, \ldots, e_k) \tag{C.36}$$

with

$$W_j(x, e_1, \ldots, e_j)=W_{j-1}+c_j e_j^2, \quad j=2,3, \ldots, k, \tag{C.37}$$

which are positive definite.

Applying the backstepping technique, results for some special forms of systems are listed as follows.

(a) **Integrator Backstepping.** Consider the system

$$\dot{x}=f(x)+g(x) \xi, \tag{C.38a}$$

$$\dot{\xi}=u \tag{C.38b}$$

and suppose that the first equation satisfies Assumption C.1 with $\xi \in \mathbb{R}$ as control. If $W(x)$ is positive definite, then

$$V_a(x, \xi)=V(x)+\frac{1}{2}(\xi-\alpha(x))^2 \tag{C.39}$$

is a CLF for the whole system, and one of controls rendering the system asymptotically stable is given by

$$u=-c(\xi-\alpha(x))+\frac{\partial \alpha}{\partial x}(f(x)+g(x) \xi)-\frac{\partial V}{\partial x} g(x), \quad c>0. \tag{C.40}$$

(b) **Linear Block Backstepping.** Consider the cascade system

$$\dot{x} = f(x) + g(x)y, \quad f(0) = 0, \ x \in \mathbb{R}^n, \ y \in \mathbb{R}, \tag{C.41a}$$

$$\dot{\xi} = A\xi + bu, \quad y = h\xi, \ \xi \in \mathbb{R}^q, \ u \in \mathbb{R}, \tag{C.41b}$$

where the linear subsystem is a minimum phase system of relative degree one ($hb \neq 0$). If the x-subsystem satisfies Assumption C.1 with $y \in \mathbb{R}$ as control and $W(x)$ is positive definite, then there exists a feedback control guaranteeing that the equilibrium $x = 0, \xi = 0$ is globally asymptotically stable. One of such controls is

$$u = \frac{1}{hb}\left(-c\big(y - \alpha(x)\big) - hA\xi + \frac{\partial \alpha}{\partial x}\big(f(x) + g(x)y\big) - \frac{\partial V}{\partial x}g(x)\right), \quad c > 0. \tag{C.42}$$

(c) **Nonlinear Block Backstepping.** Consider the cascade system

$$\dot{x} = f_x(x) + g_x(x)y, \quad f(0) = 0, \ x \in \mathbb{R}^n, \ y \in \mathbb{R}, \tag{C.43a}$$

$$\dot{\xi} = f_\xi(x,\xi) + g_\xi(x,\xi)u, \quad y = h(\xi), \ h(0) = 0, \ \xi \in \mathbb{R}^q, \ u \in \mathbb{R}, \tag{C.43b}$$

where the ξ-subsystem has globally defined and relative degree 1 uniformly in x and its zero dynamics is input-to-state stable with respect to x and y as its inputs. If the x-subsystem satisfies Assumption C.1 with $y \in \mathbb{R}$ as control and $W(x)$ is positive definite, then there exists a feedback control guaranteeing that the equilibrium $x = 0, \xi = 0$ is globally asymptotically stable. One of such controls is

$$\begin{aligned} u = {} & \left(\frac{\partial h}{\partial \xi} g_\xi(x,\xi)\right)^{-1} \\ & \times \left(-c\big(y - \alpha(x)\big) - \frac{\partial h}{\partial \xi} f_\xi(x,\xi) + \frac{\partial \alpha}{\partial x}\big(f(x) + g(x)y\big) - \frac{\partial V}{\partial x}g(x)\right), \\ & c > 0. \end{aligned} \tag{C.44}$$

C.3 Adaptive Backstepping

Some systems consist of unknown constant parameters which appear linearly in the system equations. In the presence of such parametric uncertainties, we will be able to achieve both boundedness of the closed-loop states and convergence of the tracking error to zero.

Consider the nonlinear system

$$\dot{x}_1 = x_2 + \theta\varphi(x_1), \tag{C.45a}$$

$$\dot{x}_2 = u. \tag{C.45b}$$

If θ were known, we would apply the backstepping technique to design a stabilizing controller. First, we view x_2 as virtual control and design

$$\alpha_1(x_1, \theta) = -c_1 x_1 - \theta\varphi(x_1) \tag{C.46}$$

to achieve the derivative of the Lyapunov function

$$V_0(x_1) = \frac{1}{2}x_1^2 \tag{C.47}$$

negative definite as follows:

$$\dot{V}_0 = x_1\dot{x}_1 = x_1\big(-c_1x_1 - \theta\varphi(x_1) + \theta\varphi(x_1)\big) = -c_1x_1^2, \tag{C.48}$$

where $c_1 > 0$. Then, we define the difference between x_2 and $\alpha_1(x_1, \theta)$ as

$$z_1 = x_2 - \alpha_1(x_1, \theta) \tag{C.49}$$

to reformulate the system as

$$\dot{x}_1 = z_1 - c_1x_1, \tag{C.50a}$$

$$\dot{z}_1 = \dot{x}_2 - \frac{\partial\alpha_1}{\partial x_1}\dot{x}_1 - \frac{\partial\alpha_1}{\partial\theta}\dot{\theta} = u - \frac{\partial\alpha_1}{\partial x_1}(z_1 - c_1x_1), \tag{C.50b}$$

where $\dot{\theta} = 0$ is used (θ is assumed to be constant). By defining the Lyapunov function

$$V_1(x_1, z_1) = \frac{1}{2}x_1^2 + \frac{1}{2}z_1^2 \tag{C.51}$$

and differentiating it along the above system, we arrive at

$$\begin{aligned}\dot{V}_1(x_1, z_1) &= x_1\dot{x}_1 + z_1\dot{z}_1\\ &= z_1x_1 - c_1x_1^2 + z_1\left(u - \frac{\partial\alpha_1}{\partial x_1}(z_1 - c_1x_1)\right)\\ &= -c_1x_1^2 + z_1\left(u + x_1 - \frac{\partial\alpha_1}{\partial x_1}(z_1 - c_1x_1)\right).\end{aligned} \tag{C.52}$$

If we choose u such that

$$u + x_1 - \frac{\partial\alpha_1}{\partial x_1}(z_1 - c_1x_1) = -c_2z_1 \tag{C.53}$$

with $c_2 > 0$, then

$$\dot{V}_1(x_1, z_1) = -c_1x_1^2 - c_2z_1^2, \tag{C.54}$$

which implies $\dot{V}_1$ is negative definite. Hence, the control law is given as

$$u = c_2\big(x_2 - \alpha_1(x_1, \theta)\big) - x_1 + \frac{\partial\alpha_1}{\partial x_1}\big(x_2 + \theta\varphi(x_1)\big). \tag{C.55}$$

Indeed, θ is unknown, we cannot implement this control law. However, we can apply the idea of backstepping to handle this issue.

We start with x_2 being a virtual control to design an adaptive control law, i.e., we start with

$$\dot{x}_1 = v + \theta\varphi(x_1). \tag{C.56}$$

If θ were known, the control

$$v = -c_1 x_1 - \theta\varphi(x_1) \tag{C.57}$$

would render the derivative of

$$V_0(x_1) = \frac{1}{2}x_1 \tag{C.58}$$

negative definite as $\dot{V}_0 = -c_1 x_1^2$. Since θ is unknown, we apply the certainty-equivalence principle to modify the control law as

$$v = -c_1 x_1 - \hat{\theta}_1\varphi(x_1), \tag{C.59}$$

where $\hat{\theta}_1$ could be an estimate of θ. Then, we obtain

$$\dot{x}_1 = -c_1 x_1 + \tilde{\theta}_1\varphi(x_1) \tag{C.60}$$

with

$$\tilde{\theta}_1 = \theta - \hat{\theta}_1 \tag{C.61}$$

being the parameter estimation error. We extend the Lyapunov function V_0 as

$$V_1(x_1, \tilde{\theta}_1) = \frac{1}{2}x_1 + \frac{1}{2\gamma}\tilde{\theta}_1^2, \tag{C.62}$$

where $\gamma > 0$. Its derivative becomes

$$\begin{aligned} \dot{V}_1 &= x_1\dot{x}_1 + \frac{1}{\gamma}\tilde{\theta}_1\dot{\tilde{\theta}}_1 \\ &= -c_1 x_1^2 + x_1\tilde{\theta}_1\varphi(x_1) + \frac{1}{\gamma}\tilde{\theta}_1\dot{\tilde{\theta}}_1 \\ &= -c_1 x_1^2 + \tilde{\theta}_1\left(x_1\varphi(x_1) + \frac{1}{\gamma}\dot{\tilde{\theta}}_1\right). \end{aligned} \tag{C.63}$$

If we choose

$$x_1\varphi(x_1) + \frac{1}{\gamma}\dot{\tilde{\theta}}_1 = 0, \tag{C.64}$$

then we have the seminegative definite property of V_1 as

$$\dot{V}_1 = -c_1 x_1^2. \tag{C.65}$$

With the assumption of $\dot{\theta} = 0$, the adaptive law is then given by

$$\dot{\hat{\theta}}_1 = \gamma x_1 \varphi(x_1). \tag{C.66}$$

Hence, the adaptive virtual control for x_2 is given as

$$\alpha_1(x_1, \hat{\theta}_1) = -c_1 x_1 - \hat{\theta}_1 \varphi(x_1), \tag{C.67a}$$

$$\dot{\hat{\theta}}_1 = \gamma x_1 \varphi(x_1), \tag{C.67b}$$

and the x_1-equation in (C.45a), (C.45b) becomes

$$\dot{x}_1 = -c_1 x_1 + (\theta - \hat{\theta}_1)\varphi(x_1). \tag{C.68}$$

Now we define the difference between x_2 and $\alpha_1(x_1, \hat{\theta}_1)$ as

$$z_2 = x_2 - \alpha_1(x_1, \hat{\theta}_1), \tag{C.69}$$

then

$$\dot{x}_1 = z_2 + \alpha_1(x_1, \hat{\theta}_1) + \theta\varphi(x_1) = z_2 - c_1 x_1 + \tilde{\theta}_1 \varphi(x_1) \tag{C.70}$$

and

$$\begin{aligned} \dot{z}_2 &= \dot{x}_2 - \frac{\partial \alpha_1}{\partial x_1}\dot{x}_1 - \frac{\partial \alpha_1}{\partial \hat{\theta}_1}\dot{\hat{\theta}}_1 \\ &= u - \frac{\partial \alpha_1}{\partial x_1}\big(z_2 - c_1 x_1 + \tilde{\theta}_1 \varphi(x_1)\big) - \frac{\partial \alpha_1}{\partial \hat{\theta}_1}\gamma x_1 \varphi(x_1). \end{aligned} \tag{C.71}$$

We augment the Lyapunov function V_1 as

$$V_2(x_1, \tilde{\theta}_1, z_2) = \frac{1}{2}x_1^2 + \frac{1}{2\gamma}\tilde{\theta}_1^2 + \frac{1}{2}z_2^2 \tag{C.72}$$

and infer

$$\begin{aligned} \dot{V}_2 &= x_1\dot{x}_1 + \frac{1}{\gamma}\tilde{\theta}_1\dot{\tilde{\theta}}_1 + z_2\dot{z}_2 = x_1 z_2 - c_1 x_1^2 + x_1\tilde{\theta}_1\varphi(x_1) - \tilde{\theta}_1 x_1 \varphi(x_1) \\ &\quad + z_2\left(u - \frac{\partial \alpha_1}{\partial x_1}\big(z_2 - c_1 x_1 + \tilde{\theta}_1\varphi(x_1)\big) - \frac{\partial \alpha_1}{\partial \hat{\theta}_1}\gamma x_1 \varphi(x_1)\right) \\ &= -c_1 x_1^2 + z_2\left(u + x_1 - \frac{\partial \alpha_1}{\partial x_1}\big(z_2 - c_1 x_1 + \tilde{\theta}_1\varphi(x_1)\big) - \frac{\partial \alpha_1}{\partial \hat{\theta}_1}\gamma x_1 \varphi(x_1)\right), \end{aligned}$$

which is rendered negative semidefinite as

$$\dot{V}_2 = -c_1 x_1^2 - c_2 z_2^2 \tag{C.73}$$

if we choose the control law such that

$$u + x_1 - \frac{\partial \alpha_1}{\partial x_1}\big(z_2 - c_1 x_1 + \tilde{\theta}_1 \varphi(x_1)\big) - \frac{\partial \alpha_1}{\partial \hat{\theta}_1}\gamma x_1 \varphi(x_1) = -c_2 z_2. \tag{C.74}$$

This leads to the following control law

$$\begin{aligned} u &= -c_2 z_2 - x_1 + \frac{\partial \alpha_1}{\partial x_1}\big(z_2 - c_1 x_1 + \tilde{\theta}_1 \varphi(x_1)\big) + \frac{\partial \alpha_1}{\partial \hat{\theta}_1}\gamma x_1 \varphi(x_1) \\ &= -c_2 z_2 - x_1 + \frac{\partial \alpha_1}{\partial x_1}\big(x_2 + \theta \varphi(x_1)\big) + \frac{\partial \alpha_1}{\partial \hat{\theta}_1}\gamma x_1 \varphi(x_1) \end{aligned} \tag{C.75}$$

which is still not implementable due to the unknown θ. Hence, we need a new estimate $\hat{\theta}_2$ to build

$$u = -c_2 z_2 - x_1 + \frac{\partial \alpha_1}{\partial x_1}\big(x_2 + \hat{\theta}_2 \varphi(x_1)\big) + \frac{\partial \alpha_1}{\partial \hat{\theta}_1}\gamma x_1 \varphi(x_1). \tag{C.76}$$

With this choice, $\dot{z}_2$ becomes

$$\dot{z}_2 = -x_1 - c_2 z_2 - (\theta - \hat{\theta}_2)\frac{\partial \alpha_1}{\partial x_1}\varphi(x_1). \tag{C.77}$$

Defining an augmented Lyapunov function V_3 as

$$V_3\big(x_1, \tilde{\theta}_1, z_2, (\theta - \hat{\theta}_2)\big) = \frac{1}{2}x_1^2 + \frac{1}{2\gamma}\big(\tilde{\theta}_1^2 + (\theta - \hat{\theta}_2)^2\big) + \frac{1}{2}z_2^2, \tag{C.78}$$

we derive

$$\begin{aligned} \dot{V}_2 &= x_1 \dot{x}_1 + \frac{1}{\gamma}\big(\tilde{\theta}_1 \dot{\tilde{\theta}}_1 + (\theta - \hat{\theta}_2)(\dot{\theta} - \dot{\hat{\theta}}_2)\big) + z_2 \dot{z}_2 \\ &= x_1 z_2 - c_1 x_1^2 + x_1 \tilde{\theta}_1 \varphi(x_1) - \tilde{\theta}_1 x_1 \varphi(x_1) - \frac{1}{\gamma}(\theta - \hat{\theta}_2)\dot{\hat{\theta}}_2 \\ &\quad + z_2\left(-x_1 - c_2 z_2 - (\theta - \hat{\theta}_2)\frac{\partial \alpha_1}{\partial x_1}\varphi(x_1)\right) \\ &= -c_1 x_1^2 - c_2 z_2^2 - (\theta - \hat{\theta}_2)\left(\frac{1}{\gamma}\dot{\hat{\theta}}_2 + z_2 \frac{\partial \alpha_1}{\partial x_1}\varphi(x_1)\right). \end{aligned}$$

By choosing the second update law as

$$\dot{\hat{\theta}}_2 = -\gamma z_2 \frac{\partial \alpha_1}{\partial x_1}\varphi(x_1), \tag{C.79}$$

we arrive at

$$\dot{V}_3 = -c_1 x_1^2 - c_2 z_2^2, \tag{C.80}$$

which is negative semidefinite. Hence, (C.66), (C.67a), (C.76), and (C.79) construct the final adaptive controller for the system (C.45a), (C.45b).

References

1. Krstić M, Kanellakopoulos I, Kokotović P (1995) Nonlinear and adaptive control design. Wiley, New York
2. Kokotovic PV (1992) The joy of feedback: nonlinear and adaptive. In: IEEE control systems
3. Kokotovic Krstic M PV, Kanellakopoulos I (1992) Backstepping to passivity: recursive design of adaptive systems. In: Proc 31st IEEE conf decision contr. IEEE Press, New Orleans, pp 3276–3280
4. Sontag ED (1989) A 'universal' construction of Artstein's theorem on nonlinear stabilization. Syst Control Lett 13:117–123
5. Khalil HK (2002) Nonlinear Systems. Prentice Hall, New York

Appendix D
Model Predictive Control (MPC)

In general, model predictive control is formulated as solving online a finite horizon (open-loop) optimal control problem subject to system dynamics and constraints involving states and controls [1–4]. The methodology of all the controllers belonging to the MPC family is characterized by the following strategy, represented in Fig. D.1 [5].

Based on the current measurement, say at time t, the controller uses a dynamic model (called a prediction model) to predict the future dynamic behavior of the system over a prediction horizon T_p, and determines (over a control horizon $T_c \leq T_p$) the control input $\bar{u}$ such that a pre-specified performance objective is optimized (for example, an integrated square error between the predicted output and the setpoint). Note that the control input between T_c and T_p may be assumed constant and equal to the control at the end of the control horizon in the case of $T_c < T_p$. If there are no disturbances and model–plant mismatch, and/or if the optimization problem can be solved for infinite prediction and control horizons, then, we can apply the input function found at time $t = 0$ to the system for all the time $t \geq 0$. However, this is not possible in general. Due to the existence of disturbances and/or model–plant mismatch, the real system behavior is different from the predicted behavior. Since finding a solution over the infinite horizon to the optimization problem is also impossible in general, we do not have a control input being available forever. Thus, the control input obtained by solving the optimization problem will be implemented only until the next measurement becomes available. We assume that this will be the case every δ time-units, where δ denotes the "sampling time". Updated with the new measurement, at time $t + \delta$, the whole procedure—prediction and optimization—is repeated to find a new control input, with the control and prediction horizons moving forward (for this reason, MPC is also referred to as moving horizon control or receding horizon control). This results in a discrete feedback control with an implicit control law because closed-loop control inputs are calculated by solving online the optimization problem at each sampling time. Hence, MPC is characterized by the following points:

- **Model-based prediction.** In contrast to other feedback controllers that calculate the control action based on the present or past state information, model predictive

H. Chen, B. Gao, *Nonlinear Estimation and Control of Automotive Drivetrains*,
DOI 10.1007/978-3-642-41572-2,

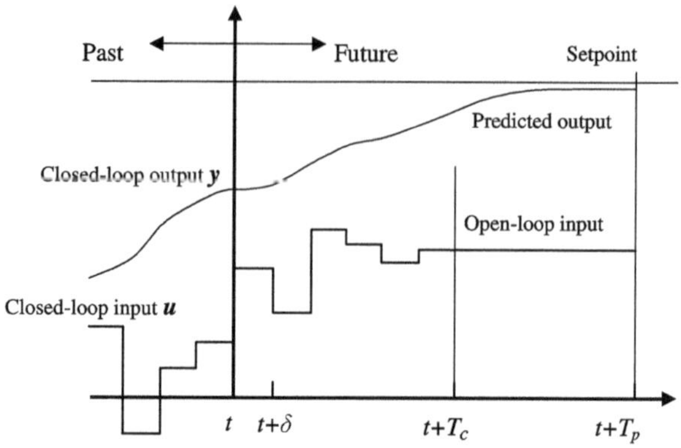

Fig. D.1 Principle of model predictive control

controllers determine the control action based on the predicted future dynamics of the to be controlled system starting from the current state. The model used to complete the prediction can be linear or nonlinear, time-continuous or discrete-time, deterministic or stochastic, etc. We emphasize here the functionality of the model, namely it is able to predict the future dynamics of the to be controlled system, while we do not care for the form of the model. Hence, any kind of models, based on which the system dynamics can be computed, can be used as a prediction model. Some of them are listed as follows:

- Convolution models, including step response and impulse response models;
- First principle models (state space model);
- Fuzzy models;
- Neural network models;
- Data-based models;
- ...

- **Handling of constraints.** In practice, most systems have to satisfy time-domain constraints on inputs and states. For example, an actuator reaches saturation and some states such as temperature and pressure are not allowed to exceed their limitations for the reason of safe operation, or some variables have to be held under certain threshold values to meet environmental regulations. Moreover, when modeling chemical processes from mass, momentum and energy conservation laws, algebraic equations may arise from phase equilibrium calculations and other phenomenological and thermodynamic correlations [6]. These algebraic equations may also be considered as constraints on the dynamics of the process. It is clear that time-domain constraints impose limitations on the achievable control performance, even if the system to be controlled is linear [7].

  In MPC, time-domain constraints can be placed directly in the optimization problem in their original form, without doing any transformation. Such a direct and explicit handling of time-domain constraints leads to non-conservative or at least less conservative solutions. Moreover, because the future response of the system is predicted, early control action can be taken so as to avoid the violation of time-domain constraints (e.g., actuator saturation, safety constraints, emission regulation) while tracking, for example, a given reference trajectory with minimal tracking error. This performs somehow an active handling of time-domain constraints;

- **Online optimization.** An objective functional that specifies mathematically the desired control performance is minimized online at each sampling instance. A commonly used objective functional is an integrated weighted square error between predicted controlled variables and their desired references. There may, however, be different objectives to describe economic requirements. In consideration of time-domain constraints, a constrained dynamic optimization problem will be repeatedly solved online. The main reasons for an online repeated solution are listed as follows:
  - In general, we cannot find an analytic solution to the involved optimization problem. Numerical methods are used. A time-continuous input parameterization and/or the use of an infinite horizon may lead to an infinite-dimensional optimization problem that is numerically extremely demanding and often intractable. In order to get around that, the optimization problem is formulated with finite horizons. Through the moving horizon implementation, we can obtain the control action as the time goes;
  - Due to the existence of model uncertainties, the real dynamics is different from the predicted dynamics. The measurement available at each sampling time contains the information reflecting various uncertainties. Through repeating the whole procedure, prediction and optimization, the information is used to improve control performance;
  - Real systems suffer in general disturbances. If we want to achieve high performance for disturbance attenuation, strong control action is required, which may lead to the violation of time-domain constraints. Hence, we need a trade-off between satisfying constraints and achieving high performance. Through the online solution of the optimization problem, a performance adaptation is possible [8];
  - A detailed derivation shows that MPC admits a feed-forward and feedback structure [9]. The feed-forward information includes the measurable disturbance and the given reference over the prediction horizon, while the feedback information is the measured state/output.

D.1 Linear MPC

The terminology of linear MPC refers to MPC based on linear models, even if the existence of time-domain constraints renders the dynamics nonlinear. Since we can reformulate the step response model and impulse response model in the state space form [9], in the following we take the general form of state space model as an example. Given a basic form of linear discrete state space equations

$$x(k+1) = Ax(k) + B_u u(k) + B_d d(k), \tag{D.1a}$$

$$y_c(k) = C_c x(k), \tag{D.1b}$$

$$y_b(k) = C_b x(k), \tag{D.1c}$$

where $x \in \mathbb{R}^{n_x}$ is the system state, $u \in \mathbb{R}^{n_u}$ is the control input, $d \in \mathbb{R}^{n_d}$ is the measurable disturbance, $y_c \in \mathbb{R}^{n_c}$ is the controlled output, and $y_b \in \mathbb{R}^{n_b}$ is the constrained output.

It is well-known that the difference equation (D.1a) can be exactly obtained from the differential equation

$$\dot{x}(t) = A_c x(t) + B_{cu} u(t) + B_{cd} d(t) \tag{D.2}$$

by computing

$$A = e^{A_c \delta}, \tag{D.3a}$$

$$B_u = \int_0^{\delta} e^{A_c \tau}\, d\tau \cdot B_{cu}, \tag{D.3b}$$

$$B_d = \int_0^{\delta} e^{A_c \tau}\, d\tau \cdot B_{cd}, \tag{D.3c}$$

with δ being the sampling time.

In order to introduce the integral action to reduce offset, we rewrite (D.1a)–(D.1c) in the incremental form

$$\Delta x(k+1) = A \Delta x(k) + B_u \Delta u(k) + B_d \Delta d(k), \tag{D.4a}$$

$$y_c(k) = C_c \Delta x(k) + y_c(k-1), \tag{D.4b}$$

$$y_b(k) = C_b \Delta x(k) + y_b(k-1), \tag{D.4c}$$

where

$$\Delta x(k) = x(k) - x(k-1),$$

$$\Delta u(k) = u(k) - u(k-1),$$

$$\Delta d(k) = d(k) - d(k-1).$$

Assume that the state is measurable. If it is not the case, we can use an observer to estimate the state. Then, at time k, with the measured/estimated state $x(k)$, the optimization problem of linear MPC is formulated as

Problem D.1

$$\min_{\Delta U(k)} J\big(x(k), \Delta U(k), N_c, N_p\big) \tag{D.5}$$

subject to (D.4a)–(D.4c) and time-domain constraints

$$u_{\min} \le u(k+i|k) \le u_{\max}, \quad i = 0, 1, \ldots, N_c - 1, \tag{D.6a}$$

$$\Delta u_{\min} \le \Delta u(k+i|k) \le \Delta u_{\max}, \quad i = 0, 1, \ldots, N_c - 1, \tag{D.6b}$$

$$y_{\min}(k+i) \le y_b(k+i|k) \le y_{\max}(k+i), \quad i = 1, \ldots, N_p, \tag{D.6c}$$

$$\Delta u(k+i|k) = 0, \quad N_c \le i \le N_p \tag{D.6d}$$

where the objective functional is defined as

$$\begin{aligned} J\big(x(k), \Delta U(k), N_c, N_p\big) &= \sum_{i=1}^{N_p} \big\| \Gamma_{y,i}\big(y_c(k+i|k) - r(k+i)\big) \big\|^2 \\ &\quad + \sum_{i=0}^{N_c-1} \big\| \Gamma_{u,i} \Delta u(k+i|k) \big\|^2, \end{aligned} \tag{D.7}$$

or in the vector form,

$$J\big(x(k), \Delta U(k), N_c, N_p\big) = \big\| \Gamma_y\big(Y_c(k+1|k) - R(k+1)\big) \big\|^2 + \big\| \Gamma_u \Delta U(k) \big\|^2. \tag{D.8}$$

In the above, N_p and N_c are prediction and control horizons, respectively, satisfying $N_c \le N_p$, Γ_y and Γ_u are weights given as

$$\Gamma_y = \operatorname{diag}\{\Gamma_{y,1}, \Gamma_{y,2}, \ldots, \Gamma_{y,p}\}, \qquad \Gamma_{y,i} \in \mathbb{R}^{n_c \times n_c}, i = 1, 2, \ldots, N_p,$$

$$\Gamma_u = \operatorname{diag}\{\Gamma_{u,1}, \Gamma_{u,2}, \ldots, \Gamma_{u,m}\}, \qquad \Gamma_{u,j} \in \mathbb{R}^{n_u \times n_u}, j = 1, 2, \ldots, N_c,$$

$R(k+1)$ is the vector of the reference

$$R(k+1) = \begin{bmatrix} r(k+1) \\ r(k+2) \\ \vdots \\ r(k+N_p) \end{bmatrix}_{N_p \times 1}$$

and $\Delta U(k)$ is the vector form of the incremental control sequences defined as

$$\Delta U(k) = \begin{bmatrix} \Delta u(k|k) \\ \Delta u(k+1|k) \\ \vdots \\ \Delta u(k+N_c-1|k) \end{bmatrix}_{N_c \times 1}, \tag{D.9}$$

which is the independent variable of the optimization problem. The constraints on u and Δu come from actuator saturations. Note that although they are regarded as constant here, time-varying constraints can also be dealt with if only minor revision are added. Moreover, $y_c(k+i|k)$ and $y_b(k+i|k)$ are the controlled and constrained outputs predicted at time k, on the basis of the prediction model (D.4a)–(D.4c). They can be represented in the vector form of

$$Y_c(k+1|k) = \begin{bmatrix} y_c(k+1|k) \\ y_c(k+2|k) \\ \vdots \\ y_c(k+N_p|k) \end{bmatrix}_{N_p\times 1}, \qquad Y_b(k+1|k) = \begin{bmatrix} y_b(k+1|k) \\ y_b(k+2|k) \\ \vdots \\ y_b(k+N_p|k) \end{bmatrix}_{N_p\times 1}$$

for a clear formulation. By iterating the difference equation in (D.4a)–(D.4c), we get the prediction equations as follows:

$$Y_c(k+1|k) = \mathcal{S}_{x,c}\Delta x(k) + \mathcal{I}_c y_c(k) + \mathcal{S}_{d,c}\Delta d(k) + \mathcal{S}_{u,c}\Delta U(k), \tag{D.10a}$$

$$Y_b(k+1|k) = \mathcal{S}_{x,b}\Delta x(k) + \mathcal{I}_b y_b(k) + \mathcal{S}_{d,b}\Delta d(k) + \mathcal{S}_{u,b}\Delta U(k), \tag{D.10b}$$

where

$$\mathcal{S}_{x,c} = \begin{bmatrix} C_c A \\ C_c A^2 + C_c A \\ \vdots \\ \sum_{i=1}^{N_p} C_c A^i \end{bmatrix}_{N_p\times 1}, \qquad \mathcal{S}_{x,b} = \begin{bmatrix} C_b A \\ C_b A^2 + C_b A \\ \vdots \\ \sum_{i=1}^{N_p} C_b A^i \end{bmatrix}_{N_p\times 1},$$

$$\mathcal{I}_c = \begin{bmatrix} I_{n_c\times n_c} \\ I_{n_c\times n_c} \\ \vdots \\ I_{n_c\times n_c} \end{bmatrix}_{N_p\times 1}, \qquad \mathcal{I}_b = \begin{bmatrix} I_{n_b\times n_b} \\ I_{n_b\times n_b} \\ \vdots \\ I_{n_b\times n_b} \end{bmatrix}_{N_p\times 1},$$

$$\mathcal{S}_{d,c} = \begin{bmatrix} C_c B_d \\ C_c A B_d + C_c B_d \\ \vdots \\ \sum_{i=1}^{N_p} C_c A^{i-1} B_d \end{bmatrix}_{N_p\times 1}, \qquad \mathcal{S}_{d,b} = \begin{bmatrix} C_b B_d \\ C_b A B_d + C_b B_d \\ \vdots \\ \sum_{i=1}^{N_p} C_b A^{i-1} B_d \end{bmatrix}_{N_p\times 1},$$

$$\mathcal{S}_{u,c} = \begin{bmatrix} C_c B_u & \mathbf{0} & \mathbf{0}\ \dots & \mathbf{0} \\ \sum_{i=1}^{2} C_c A^{i-1} B_u & C_c B_u & \mathbf{0}\ \dots & \mathbf{0} \\ \vdots & \vdots & \vdots\ \vdots & \vdots \\ \sum_{i=1}^{N_c} C_c A^{i-1} B_u & \sum_{i=1}^{N_c-1} C_c A^{i-1} B_u & \dots\dots & C_c B_u \\ \vdots & \vdots & \vdots\ \vdots & \vdots \\ \sum_{i=1}^{N_p} C_c A^{i-1} B_u & \sum_{i=1}^{N_p-1} C_c A^{i-1} B_u & \dots\dots & \sum_{i=1}^{N_p-N_c+1} C_c A^{i-1} B_u \end{bmatrix}_{N_p\times N_c},$$

$$\mathcal{S}_{u,b} = \begin{bmatrix} C_b B_u & \mathbf{0} & \mathbf{0} \ldots & \mathbf{0} \\ \sum_{i=1}^{2} C_b A^{i-1} B_u & C_b B_u & \mathbf{0} \ldots & \mathbf{0} \\ \vdots & \vdots & \vdots \ldots & \vdots \\ \sum_{i=1}^{N_c} C_b A^{i-1} B_u & \sum_{i=1}^{N_c-1} C_b A^{i-1} B_u & \ldots\ldots & C_b B_u \\ \vdots & \vdots & \vdots \ldots & \vdots \\ \sum_{i=1}^{N_p} C_b A^{i-1} B_u & \sum_{i=1}^{N_p-1} C_b A^{i-1} B_u & \ldots\ldots & \sum_{i=1}^{N_p-N_c+1} C_b A^{i-1} B_u \end{bmatrix}_{N_p \times N_c} .$$

According to the basic of MPC, the optimization problem (Problem D.1) will be solved at each sampling time, updated with the new measurement. If we can find a solution of Problem D.1, denoted as $\Delta U^*(k)$,

the closed-loop control at time k is then defined as

$$u(k) := \Delta u^*(k|k) + u(k-1). \tag{D.12}$$

If we do not consider the time-domain constraints, then the optimization problem (Problem D.1) becomes

$$\min_{\Delta U(k)} \left\| \Gamma_y \big(Y_c(k+1|k) - R(k+1)\big) \right\|^2 + \left\| \Gamma_u \Delta U(k) \right\|^2 \tag{D.13}$$

with $Y_c(k+1|k)$ given by (D.10a). We can then obtain the solution by calculating the gradient of the objective function over the independent variable $\Delta U(k)$ and setting it to zero. The result reads

$$\Delta U^*(k) = \big(\mathcal{S}_{u,c}^T \Gamma_y^T \Gamma_y \mathcal{S}_{u,c} + \Gamma_u^T \Gamma_u\big)^{-1} \mathcal{S}_{u,c}^T \Gamma_y^T \Gamma_y E_p(k+1|k), \tag{D.14}$$

with $E_p(k+1|k)$ being calculated by

$$E_p(k+1|k) = R(k+1) - \mathcal{S}_{x,c} \Delta x(k) - \mathcal{I}_c y_c(k) - \mathcal{S}_{d,c} \Delta d(k). \tag{D.15}$$

According to the basic of MPC, we pick up the first element of $\Delta U^*(k)$ to build the closed-loop control as follows

$$\Delta u(k) = K_{\text{mpc}} E_p(k+1|k), \tag{D.16}$$

where K_{mpc} is calculated by

$$K_{\text{mpc}} = \begin{bmatrix} I_{n_u \times n_u} & \mathbf{0} & \ldots & \mathbf{0} \end{bmatrix}_{1 \times N_c} \big(\mathcal{S}_{u,c}^T \Gamma_y^T \Gamma_y \mathcal{S}_{u,c} + \Gamma_u^T \Gamma_u\big)^{-1} \mathcal{S}_{u,c}^T \Gamma_y^T \Gamma_y . \tag{D.17}$$

Substituting (D.15) into (D.16) leads to

$$\begin{aligned} \Delta u(k) = {} & K_{\text{mpc}} R(k+1) - K_{\text{mpc}} (\mathcal{S}_{x,c} + \mathcal{I}_c C_c) x(k) \\ & - K_{\text{mpc}} \mathcal{S}_{d,c} \Delta d(k) + K_{\text{mpc}} \mathcal{S}_{x,c} x(k-1). \end{aligned}$$

It is clear that

- $K_{\text{mpc}}R(k+1)$ represents a feed-forward depending on the future reference over the prediction horizon;
- $-K_{\text{mpc}}\mathcal{S}_{d,c}\Delta d(k)$ represents a feed-forward depending on the measurable disturbance;
- $-K_{\text{mpc}}(\mathcal{S}_{x,c}+\mathcal{I}_c C_c)x(k)+K_{\text{mpc}}\mathcal{S}_{x,c}x(k-1)$ represents a state feedback depending on the measurement.

Hence, MPC admits a feed-forward and feedback structure.

In the case of considering the time-domain constraints, the optimization problem can be formulated as a standard quadratic programming (QP) problem as follows:

$$\min_{\Delta U(k)} \Delta U(k)^T H \Delta U(k) - G(k+1|k)^T \Delta U(k) \tag{D.18a}$$

$$\text{subject to} \quad C_u \Delta U(k) \geq b(k+1|k), \tag{D.18b}$$

where

$$H = \mathcal{S}_{u,c}^T \Gamma_y^T \Gamma_y \mathcal{S}_{u,c} + \Gamma_u^T \Gamma_u,$$

$$G(k+1|k) = 2\mathcal{S}_{u,c}^T \Gamma_y^T \Gamma_y E_p(k+1|k),$$

$$C_u = \begin{bmatrix} -T \\ T \\ -L \\ L \\ -\mathcal{S}_{u,b} \\ \mathcal{S}_{u,b} \end{bmatrix},$$

$$b(k+1|k) = \begin{bmatrix} -\Delta u_{\max} \\ \vdots \\ -\Delta u_{\max} \\ \Delta u_{\min} \\ \vdots \\ \Delta u_{\min} \\ -u_{\max} + u(k-1) \\ \vdots \\ -u_{\max} + u(k-1) \\ u_{\min} - u(k-1) \\ \vdots \\ u_{\min} - u(k-1) \\ -Y_{\max}(k+1) + \mathcal{S}_{x,b}\Delta x(k) + \mathcal{I}_b y_b(k) + \mathcal{S}_{d,b}\Delta d(k) \\ Y_{\min}(k+1) - \mathcal{S}_{x,b}\Delta x(k) - \mathcal{I}_b y_b(k) - \mathcal{S}_{d,b}\Delta d(k) \end{bmatrix}$$

with

$$T = \begin{bmatrix} I_{n_u \times n_u} & \mathbf{0} & \cdots & \mathbf{0} & \mathbf{0} \\ \mathbf{0} & I_{n_u \times n_u} & \cdots & \mathbf{0} & \mathbf{0} \\ \vdots & \vdots & \ddots & \vdots & \vdots \\ \mathbf{0} & \mathbf{0} & \cdots & I_{n_u \times n_u} & \mathbf{0} \\ \mathbf{0} & \mathbf{0} & \cdots & \mathbf{0} & I_{n_u \times n_u} \end{bmatrix}_{N_c \times N_c},$$

$$L = \begin{bmatrix} I_{n_u \times n_u} & \mathbf{0} & \cdots & \mathbf{0} & \mathbf{0} \\ I_{n_u \times n_u} & I_{n_u \times n_u} & \cdots & \mathbf{0} & \mathbf{0} \\ \vdots & \vdots & \ddots & \vdots & \vdots \\ I_{n_u \times n_u} & I_{n_u \times n_u} & \cdots & I_{n_u \times n_u} & \mathbf{0} \\ I_{n_u \times n_u} & I_{n_u \times n_u} & \cdots & I_{n_u \times n_u} & I_{n_u \times n_u} \end{bmatrix}_{N_c \times N_c},$$

$$Y_{\min}(k+1) = \begin{bmatrix} y_{\min}(k+1) \\ y_{\min}(k+2) \\ \vdots \\ y_{\min}(k+N_p) \end{bmatrix}_{N_p \times 1}, \quad Y_{\max}(k+1) = \begin{bmatrix} y_{\max}(k+1) \\ y_{\max}(k+2) \\ \vdots \\ y_{\max}(k+N_p) \end{bmatrix}_{N_p \times 1}.$$

D.2 Nonlinear MPC (NMPC)

D.2.1 NMPC Based on Discrete-Time Model

Consider the following discrete nonlinear state-space equations:

$$x(k+1) = f\big(x(k), u(k)\big), \quad k \geq 0, \tag{D.20a}$$

$$y_c(k) = g_c\big(x(k), u(k)\big), \tag{D.20b}$$

$$y_b(k) = g_b\big(x(k), u(k)\big), \tag{D.20c}$$

where $x(k) \in \mathbb{R}^{n_x}$ is the system state, $u(k) \in \mathbb{R}^{n_u}$ is the control input, $y_c(k) \in \mathbb{R}^{n_c}$ is the controlled output, $y_b(k) \in \mathbb{R}^{n_b}$ is the constrained output. The constraints on input and output are represented as

$$u_{\min} \leq u(k) \leq u_{\max}, \quad \forall k \geq 0, \tag{D.21a}$$

$$\Delta u_{\min} \leq \Delta u(k) \leq \Delta u_{\max}, \quad \forall k \geq 0, \tag{D.21b}$$

$$y_{\min}(k) \leq y_b(k) \leq y_{\max}(k), \quad \forall k \geq 0. \tag{D.21c}$$

It is assumed that all states are measurable. If not all states are measurable, an observer has to be designed to estimate the state. Then at time k, based on the measured/estimated state $x(k)$, the optimization problem of discrete nonlinear MPC is formulated as

Problem D.2

$$\min_{U_k} J\big(x(k), U_k\big) \tag{D.22}$$

subject to (D.20a)–(D.20c) and time-domain constraints

$$u_{\min} \le \bar{u}(k+i) \le u_{\max}, \quad 0 \le i < N_c, \tag{D.23a}$$

$$\Delta u_{\min} \le \Delta\bar{u}(k+i) \le \Delta u_{\max}, \tag{D.23b}$$

$$\Delta\bar{u}(k+i) = \bar{u}(k+i) - \bar{u}(k+i-1), \tag{D.23c}$$

$$y_{\min}(k+i) \le \bar{y}_b(k+i) \le y_{\max}(k+i), \quad 0 < i \le N_p, \tag{D.23d}$$

$$\Delta\bar{u}(k+i) = 0, \quad N_c \le i \le N_p, \tag{D.23e}$$

where the objective functional is defined as

$$\begin{aligned} J\big(x(k), U_k\big) &= \sum_{i=1}^{N_p} \big\|\bar{y}_c(k+i) - r(k+i)\big\|_Q^2 \\ &\quad + \sum_{i=0}^{N_c-1} \Big(\big\|\bar{u}(k+i) - u_r(k+i)\big\|_R^2 + \big\|\Delta\bar{u}(k+i)\big\|_S^2\Big), \end{aligned} \tag{D.24}$$

and $\bar{y}_c(\cdot)$ and $\bar{y}_b(\cdot)$ as predicted controlled and constrained outputs, respectively, can be calculated through the following dynamic equations:

$$\bar{x}(i+1) = f\big(\bar{x}(i), \bar{u}(i)\big), \quad k \le i \le k+N_p, \ \bar{x}(k) = x(k), \tag{D.25a}$$

$$\bar{y}_c(i) = g_c\big(\bar{x}(i), \bar{u}(i)\big), \tag{D.25b}$$

$$\bar{y}_b(i) = g_b\big(\bar{x}(i), \bar{u}(i)\big). \tag{D.25c}$$

In the above description, N_p and N_c are the prediction and control horizons, respectively, satisfying $N_c \le N_p$; $(r(\cdot), u_r(\cdot))$ are the references of the controlled output and corresponding control input; (Q, R, S) are weighting matrices, allowed to be time varying; $\bar{u}(\cdot)$ is the predicted control input, defined as

$$\bar{u}(k+i) = \bar{u}_i, \quad i = 0, 1, \ldots, N_c - 1, \tag{D.26}$$

where $\bar{u}_0, \ldots, \bar{u}_{N_c-1}$ constitute the independent variables of the optimization problem, denoted as U_k

$$U_k \triangleq \begin{bmatrix} \bar{u}_0 \\ \bar{u}_1 \\ \vdots \\ \bar{u}_{N_c-1} \end{bmatrix}. \tag{D.27}$$

Note that $x(k)$, the system state, is also the initial condition of the prediction model (D.25a)–(D.25c), which is the key of MPC being a feedback strategy.

Assume that optimization problem D.2 is feasible at each sampling time and the solution is

$$U_k^* \triangleq \begin{bmatrix} \bar{u}_0^* \\ \bar{u}_1^* \\ \vdots \\ \bar{u}_{N_c-1}^* \end{bmatrix}, \tag{D.28}$$

then, according to the basics of MPC, the control input is chosen as

$$u(k) = \bar{u}_0^*. \tag{D.29}$$

Because U_k^* depends on the values of $x(k)$, (Q, S, R) and (N_c, N_p), $u(k)$ is an implicit function of these variables. Ignoring the dependence of (Q, S, R) and (N_c, N_p), denote $u(k)$ as

$$u(k) = \kappa\big(x(k)\big), \quad k \geq 0. \tag{D.30}$$

Substituting it into the controlled system (D.20a)–(D.20c), we have the closed system

$$x(k+1) = f\big(x(k), \kappa\big(x(k)\big)\big), \quad k \geq 0. \tag{D.31}$$

If not all states are measurable, only need to replace $x(k)$ with $\hat{x}(k)$.

D.2.2 NMPC Based on Continuous-Time Model

In general, it is difficult to achieve precise enough discretization of a nonlinear system. Hence nonlinear MPC on time continuous model is also investigated.

Consider the following continuous nonlinear system:

$$\dot{x}(t) = f\big(x(t), u(t)\big), \quad t \geq 0, \tag{D.32a}$$

$$y_c(t) = g_c\big(x(t), u(t)\big), \tag{D.32b}$$

$$y_b(t) = g_b\big(x(t), u(t)\big), \tag{D.32c}$$

where $x(t) \in \mathbb{R}^{n_x}$ is the state, $u(t) \in \mathbb{R}^{n_u}$ is the control input, $y_c(t) \in \mathbb{R}^{n_c}$ is the controlled output, $y_b(t) \in \mathbb{R}^{n_b}$ is the constrained output. The constraints on input and output are

$$u_{\min} \leq u(t) \leq u_{\max}, \quad \forall t \geq 0, \tag{D.33a}$$

$$du_{\min} \leq \dot{u}(t) \leq du_{\max}, \quad \forall t \geq 0, \tag{D.33b}$$

$$y_{\min}(t) \leq y_b(t) \leq y_{\max}(t), \quad \forall t \geq 0. \tag{D.33c}$$

At the present time t, based on the measured/estimated state $x(t)$ and ignoring the constraint on the change rate of the control input, the optimization problem is formulated as

Problem D.3

$$\min_{U_t} J\big(x(t), U_t\big) \tag{D.34}$$

subject to

$$u_{\min} \le \bar{u}(\tau) \le u_{\max}, \quad t \le \tau < t + T_c, \tag{D.35a}$$

$$y_{\min}(\tau) \le \bar{y}_b(\tau) \le y_{\max}(\tau), \quad t < \tau \le t + T_p, \tag{D.35b}$$

$$\bar{u}(\tau) = \bar{u}(t + T_c), t + T_c \le \tau \le t + T_p, \tag{D.35c}$$

where the objective functional is defined as

$$J\big(x(t), U_t\big) = \int_t^{t+T_p} \big(\big\|\bar{y}_c(\tau) - r(\tau)\big\|_Q^2 + \big\|\bar{u}(\tau) - u_r(\tau)\big\|_R^2\big)\, d\tau, \tag{D.36}$$

and $\bar{y}_c(\cdot)$ and $\bar{y}_b(\cdot)$ are the predicted controlled and constrained output, respectively, calculated through the following dynamic equation:

$$\dot{\bar{x}}(\tau) = f\big(\bar{x}(\tau), \bar{u}(\tau)\big), \quad t \le \tau \le t + T_p,\ \bar{x}(t) = x(t), \tag{D.37a}$$

$$\bar{y}_c(\tau) = g_c\big(\bar{x}(\tau), \bar{u}(\tau)\big), \tag{D.37b}$$

$$\bar{y}_b(\tau) = g_b\big(\bar{x}(\tau), \bar{u}(\tau)\big). \tag{D.37c}$$

In the above, T_c and T_p are the prediction and control horizons, satisfying $T_c \le T_p$; $(r(\cdot), u_r(\cdot))$ are the references of the controlled output and the corresponding control input; (Q, R) are weighting matrices, allowed to be time varying; $\bar{u}(\cdot)$ is the predicted control input, and for $\tau \in [t, t + T_c]$, it is defined as

$$\bar{u}(\tau) = \bar{u}_i,\ i = \mathrm{int}\left(\frac{\tau - t}{\delta}\right), \tag{D.38}$$

where δ is sampling time and $T_c = N_c\delta$. Then $\bar{u}_0, \ldots, \bar{u}_{N_c-1}$ constitute independent variables of the optimization problem, denoted as U_t

$$U_t \triangleq \begin{bmatrix} \bar{u}_0 \\ \bar{u}_1 \\ \vdots \\ \bar{u}_{N_c-1} \end{bmatrix}. \tag{D.39}$$

Note that $x(t)$, the system state, is also the initial condition of the prediction model (D.37a).

Remark D.1 By Eqs. (D.38) and (D.39), the control input is treated as constant during the sampling period, and thus the optimization problem is transferred into a problem with limited independent variables.

Remark D.2 The constraint on the change rate of control input can be taken into consideration through the following two methods. One is to add a penalty item into the objective function:

$$J\big(\hat{x}(t), U_t\big) = \int_t^{t+T_p} \big(\big\|\bar{y}_c(\tau) - r(\tau)\big\|_Q^2 + \big\|\bar{u}(\tau) - u_r(\tau)\big\|_R^2\big)\, d\tau + \sum_{i=0}^{N_c-1} \|\bar{u}_i - \bar{u}_{i-1}\|_S^2. \tag{D.40}$$

This is a somehow "soft" treatment. Another one is to use $\frac{\bar{u}_i - \bar{u}_{i-1}}{\delta}$ to approximate the change rate of the control input, and add the following constraint into (D.35a)–(D.35c):

$$du_{\min} \le \frac{\bar{u}_i - \bar{u}_{i-1}}{\delta} \le du_{\max}. \tag{D.41}$$

Assume that the optimization problem D.3 is feasible at each sampling time, and the solution is denoted as

$$U_t^* \triangleq \begin{bmatrix} \bar{u}_0^* \\ \bar{u}_1^* \\ \vdots \\ \bar{u}_{N_c-1}^* \end{bmatrix}, \tag{D.42}$$

then, according to the basics of MPC, the closed-loop control is defined as

$$u(\tau) = \bar{u}_0^*, \quad t \le \tau \le t + \delta. \tag{D.43}$$

Because U_t^* depends on the values of $x(t)$, (Q, R) and (T_c, T_p), $u(t)$ is an implicit function of these variables, denoted as

$$u(\tau) = \kappa\big(x(t)\big), \quad t \le \tau \le t + \delta, \tag{D.44}$$

where the dependence of (Q, R) and (T_c, T_p) is ignored for simplicity. By substituting it into (D.32a)–(D.32c), we have the closed-loop system

$$\dot{x}(\tau) = f\big(x(\tau), \kappa\big(x(t)\big)\big), \quad t \le \tau \le t + \delta,\ t \ge 0. \tag{D.45}$$

It is clear that predictive control has the characteristics of sampled-date systems.

The solution of NMPC is always summarized as solving a nonlinear programming (NLP) problem, and the procedure is shown in Fig. D.2. For a more detailed discussion on various MPC formulations, and theoretic issues as stability and robustness, we refer to, for example, [8, 10–19].

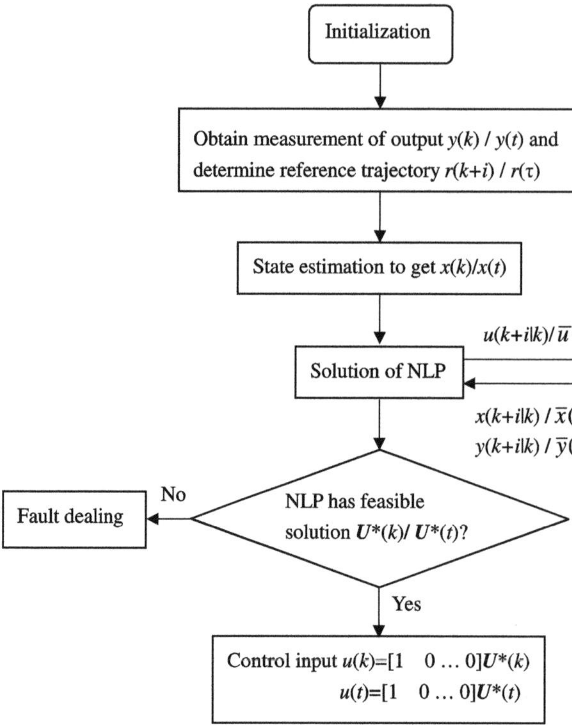

Fig. D.2 Schematic program of NMPC

References

1. Allgöwer F, Badgwell TA, Qin JS, Rawlings JB, Wright SJ (1999) Nonlinear predictive control and moving horizon estimation—an introductory overview. In: Frank PM (ed) Advances in control, highlights of ECC'99. Springer, Berlin, pp 391–449
2. Camacho EF, Bordons C (2004) Model predictive control. Springer, London
3. Maciejowski JM (2002) Predictive control: with constraints. Prentice Hall, New York
4. Mayne DQ, Rawlings JB, Rao CV, Scokaert POM (2000) Constrained model predictive control: stability and optimality. Automatica 36(6):789–814
5. Chen H (1997) Stability and robustness considerations in nonlinear model predictive control. Fortschr.-Ber. VDI Reihe 8, vol 674. VDI Verlag, Düsseldorf
6. Kröner A, Holl P, Marquardt W, Gilles ED (1989) DIVA—an open architecture for dynamic simulation. In: Eckermann R (ed) Computer application in the chemical industry. VCH, Weinheim, pp 485–492
7. Mayne DQ (1995) Optimization in model based control. In: Proc IFAC symposium dynamics and control of chemical reactors, distillation columns and batch processes, Helsingor, pp 229–242
8. Chen H, Scherer CW (2006) Moving horizon H_∞ control with performance adaptation for constrained linear systems. Automatica 42(6):1033–1040
9. Chen H (2013) Model predictive control. Science Press, Beijing. In Chinese
10. Bemporad A, Morari M, Dua V, Pistikopoulos EN (2002) The explicit linear quadratic regulator for constrained systems. Automatica 38(1):3–20
11. Chen H, Allgöwer F (1998) A quasi-infinite horizon nonlinear model predictive control scheme with guaranteed stability. Automatica 34(10):1205–1217

12. Chen H, Gao X-Q, Wang H (2006) An improved moving horizon H_∞ control scheme through Lagrange duality. Int J Control 79(3):239–248
13. Chisci L, Rossiter JA, Zappa G (2001) Systems with persistent disturbances: predictive control with restricted constraints. Automatica 37(7):1019–1028
14. Grimm G, Messina MJ, Tuna SE, Teel AR (2007) Nominally robust model predictive control with state constraints. IEEE Trans Autom Control 52(5):1856–1870
15. Grimm G, Messina MJ, Tuna SE, Teel AR (2004) Examples when nonlinear model predictive control is nonrobust. Automatica 40:1729–1738
16. Lazar M, Muñoz de la Peña D, Heemels W, Alamo T (2008) On input-to-state stabilizing of min–max nonlinear model predictive control. Syst Control Lett 57(1):39–48
17. Limón D, Álamo T, Salas F, Camacho EF (2006) Input to state stability of min–max MPC controllers for nonlinear systems with bounded uncertainties. Automatica 42(5):797–803
18. Mayne DQ, Kerrigan EC, van Wyk EJ, Falugi P (2011) Tube-based robust nonlinear model predictive control. Int J Robust Nonlinear Control 21(11):1341–1353
19. Mayne DQ, Seron MM, Rakovic SV (2005) Robust model predictive control of constrained linear systems with bounded disturbances. Automatica 41(2):219–224

Appendix E
Linear Matrix Inequality (LMI)

Many problems arising from control, identification and signal processing can be transformed into a few standard convex or quasi-convex (optimization or feasibility) problems involving linear matrix inequalities (LMIs) [1, 2] which can be solved efficiently in a numerical sense by the use of interior-point methods [3]. In this book, for example, we formulate the calculation of the observer gain in Chap. 2 and the solution of the feedback gain in Chap. 4 as convex optimization problems involving LMIs.

E.1 Convexity

Definition E.1 2 A set D in a linear vector space is said to be convex if

$$\{x_1, x_2 \in D\} \Rightarrow \big\{x : \alpha x_1 + (1-\alpha)x_2 \in D \text{ for all } \alpha \in (0,1)\big\}. \tag{E.1}$$

Geometrically, a set D is convex if the line segment between any two points in D lies in D.

Definition E.2 (Convex hull) The convex hull of a set D, denoted as $\text{Co}\{D\}$, is the intersection of all convex sets containing D. If D consists of a finite number of elements, then these elements are referred to as the vertices of $\text{Co}\{D\}$.

The convex hull of a finite point set forms a polytope and any polytope is the convex hull of a finite point set.

Definition E.3 A function $f : D \to \mathbb{R}$ is called convex if

- D is convex and
- For all $x_1, x_2 \in D$ and $\alpha \in (0, 1)$,

$$f\big(\alpha x_1 + (1-\alpha)x_2\big) \le \alpha f(x_1) + (1-\alpha) f(x_2). \tag{E.2}$$

H. Chen, B. Gao, *Nonlinear Estimation and Control of Automotive Drivetrains*,
DOI 10.1007/978-3-642-41572-2,

Moreover, f is called strictly convex if the inequality (E.2) is strict for $x_1, x_2 \in D$, $x_1 \neq x_2$ and $\alpha \in (0, 1)$.

Geometrically, (E.2) implies that the line segment between $(x_1, f(x_1))$ and $(x_2, f(x_2))$, i.e., the chord from x_1 to x_2, lies above the curve of f.

Moreover, a function $f : D \to \mathbb{R}$ is called affine if (E.2) holds with equality.

Definition E.4 (Local and global optimality) Let D be a subset of a normed space $\mathcal{X}$. An element $x_0 \in D$ is said to be a local optimal solution of $f : D \to \mathbb{R}$ if there exists $\epsilon > 0$ such that

$$f(x_0) \leq f(x) \tag{E.3}$$

for all $x \in D$ with $\|x - x_0\| < \epsilon$. It is called a global optimal solution if (E.3) holds for all $x \in D$.

Proposition E.1 *Suppose that $f : D \to \mathbb{R}$ is convex. If f has a local minimum at $x_0 \in D$, then $f(x_0)$ is also the global minimum of f. If f is strictly convex, then x_0 is, moreover, unique.*

Proposition E.2 (*Jensen's inequality*) *If f defined on D is convex, then for all $x_1, x_2, \ldots, x_r \in D$ and $\lambda_1, \lambda_2, \ldots, \lambda_r \geq 0$ with $\sum_{i=1}^{r} \lambda_i = 1$ one has*

$$f(\lambda_1 x_1 + \cdots + \lambda_r x_r) \leq \lambda_1 f(x_1) + \cdots + \lambda_r f(x_r). \tag{E.4}$$

E.2 Linear Matrix Inequalities

A linear matrix inequality (LMI) is an expression of the form

$$F(x) := F_0 + x_1 F_1 + \cdots + x_m F_m > 0, \tag{E.5}$$

where

- $x = (x_1, \ldots, x_m)$ is the decision variable;
- $F_0, \ldots, F_m$ are given real symmetric matrices, and
- The inequality $F(x) > 0$ means that $u^T F(x) u > 0$ for all $u \in \mathbb{R}^n$, $u \neq 0$.

While (E.5) is a strict LMI, we may also encounter non-strict LMIs which have the form of

$$F(x) \geq 0.$$

The linear matrix inequality (E.5) defines a convex constraint on x. That is, the set $\mathbb{F} := \{x | F(x) > 0\}$ is convex. Hence, optimization problems involving the minimization (or maximization) of a performance function $f : \mathbb{F} \to \mathbb{R}$ belong to the class of convex optimization problems, if the performance function f renders (E.2) satisfied for all $x_1, x_2 \in \mathbb{F}$ and $\alpha \in (0, 1)$. The full power of convex optimization theory can then be employed [4].

E.3 Casting Problems in an LMIs Setting

There are three generic problems related to LMIs [1, 2]:

(a) **Feasibility.** The test whether or not there exist solutions x of $F(x) > 0$ is called a feasibility problem. The LMI $F(x) > 0$ is said to be feasible if a solution exists, otherwise it is said to be infeasible.
(b) **Optimization.** Let $f : D \to \mathbb{R}$ be a convex objective function. The problem

$$\begin{aligned} &\min_{x \in D} f(x) \\ &\text{s.t.} \quad F(x) > 0 \end{aligned}$$

is called an optimization problem with an LMI constraint.
(c) **Generalized eigenvalue problem.** This problem amounts to minimizing the maximum generalized eigenvalue of a pair of matrices that depend affinely on a variable, subject to an LMI constraint. It admits the general form of

$$\begin{aligned} &\min \lambda \\ &\text{s.t.} \quad \lambda F(x) - G(x) > 0, \\ &\qquad F(x) > 0, \\ &\qquad H(x) > 0. \end{aligned}$$

Some control problems that can be easily casted in an LMI setting are given as follows:

Stability An example of the feasibility problem is to test if the linear system

$$\dot{x} = Ax \tag{E.6}$$

is asymptotically stable. This can be formulated as the following LMI feasibility problem:

$$P > 0, \quad A^T P + PA < 0 \tag{E.7}$$

with P as a variable. Indeed, with (E.7) feasible, we can easily show that the quadratic function $V(x) = x^T Px$ decreases along every nonzero trajectory of (E.6), and hence the stability property.

Moreover, if A is uncertain and varies in a polytope, i.e.,

$$A \in \text{Co}\{A_1, A_2, \ldots, A_r\},$$

then the stability test (E.7) becomes

$$P > 0, \quad A_i^T P + PA_i < 0, \quad i = 1, 2, \ldots, r. \tag{E.8}$$

If the uncertain system is described as diagonal norm-bounded Linear Differential Inclusions (LDIs) as follows

$$\dot{x} = A_0 x + B_p p, \tag{E.9a}$$

$$q = C_q x + D_{qp} p, \tag{E.9b}$$

$$p_i = \delta_i(t) q_i, \quad |\delta_i(t)| \le 1, \ i = 1, 2, \dots, n_q, \tag{E.9c}$$

then the stability test (E.7) becomes

$$P > 0, \quad \text{diagonal } \Lambda > 0, \quad \begin{pmatrix} A_0^T P + P A_0 + C_q^T \Lambda C_q & * \\ B_p^T P + D_{qp}^T \Lambda C_q & D_{qp}^T \Lambda D_{qp} - \Lambda \end{pmatrix} < 0, \tag{E.10}$$

where P and Λ are variable. The $\mathcal{S}$-procedure is used to get (E.10). A less conservative test can be obtained by the use of the full-block $\mathcal{S}$-procedure [2].

Decay Rate The decay rate of a system is defined to be the largest α such that

$$\lim_{t \to \infty} e^{\alpha t} \|x(t)\| = 0 \tag{E.11}$$

holds for all trajectories. In order to estimate the decay rate of system (E.6), we define $V(x) = x^T P x$ and require

$$\frac{dV(x)}{dt} \le -2\alpha V(x) \tag{E.12}$$

for all trajectories, which then leads to $\|x(t)\| \le e^{-\alpha t} |\frac{\lambda_{\max}(P)}{\lambda_{\min}(P)}|^{\frac{1}{2}} \|x(0)\|$. By exploring (E.12) for system (E.6), the decay rate problem can be casted in the following LMI optimization problem

$$\min_{\alpha, P} \alpha \tag{E.13a}$$

$$\text{s.t} \quad P > 0, \alpha > 0, A^T P + P A + 2\alpha P \le 0. \tag{E.13b}$$

Similarly, we can formulate the problem in an LMI setting for uncertain systems with polytopic description as follows:

$$\min_{\alpha, P} \alpha \tag{E.14a}$$

$$\text{s.t} \quad P > 0, \alpha > 0, A_i^T P + P A_i + 2\alpha P \le 0, i = 1, 2, \dots, r, \tag{E.14b}$$

and for uncertain systems with diagonal norm-bounded description as follows:

$$\min_{\alpha, P, \Lambda} \alpha \tag{E.15a}$$

$$\text{s.t} \quad P > 0, \alpha > 0, \text{diagonal } \Lambda > 0, \tag{E.15b}$$

$$\begin{pmatrix} A_0^T P + PA_0 + C_q^T \Lambda C_q + 2\alpha P & * \\ B_p^T P + D_{qp}^T \Lambda C_q & D_{qp}^T \Lambda D_{qp} - \Lambda \end{pmatrix} < 0. \qquad \text{(E.15c)}$$

Designing a Feedback Controller The problem of designing a feedback controller can be solved by the general procedure from analysis to synthesis [2]. For example, design a stabilizing state feedback gain, one can easily perform the following steps:

- Replace A in (E.7) by $A_K = A + B_u K$ to get

$$(A + B_u K)^T P + P(A + B_u K) < 0; \qquad \text{(E.16)}$$

- Define $Q = P^{-1}$ and right- and left-multiply with Q and Q^T, to obtain an equivalent inequality

$$QA^T + QK^T B_u + AQ + B_u KQ < 0; \qquad \text{(E.17)}$$

- And define $Y = KQ$ to cast the problem in the LMI of

$$QA^T + AQ + Y^T B_u^T + B_u Y < 0 \qquad \text{(E.18)}$$

  with $Q > 0$ and Y as variables.

A stabilizing feedback gain is then defined as $K = YQ^{-1}$.

Similarly, by the use of the procedure from analysis to synthesis, one can design a robust stabilizing feedback gain for the polytopic uncertainty, where (Q, Y) is a feasible solution of

$$Q > 0, \quad QA_i^T + A_i Q + Y^T B_{u,i}^T + B_{u,i} Y < 0, \quad i = 1, 2, \ldots, r, \qquad \text{(E.19)}$$

or for diagonal norm-bounded uncertainty, where (Q, Y, M) is a feasible solution of

$$Q > 0, \text{diagonal } M > 0, \qquad \text{(E.20a)}$$

$$\begin{pmatrix} QA_0^T + A_0 Q + B_p M B_p^T + Y^T B_u^T + B_u Y & * \\ D_{qp} M B_p^T + C_q Q + D_{qu} Y & D_{qp} M D_{qp}^T - M \end{pmatrix} < 0. \qquad \text{(E.20b)}$$

The following lemmas are useful for casting control, identification and signal processing problems in LMIs.

Lemma E.1 (*Schur Complement*) *For $Q(x), R(x), S(x)$ depending affinely on x and $Q(x), R(x)$ being symmetric, the LMI*

$$\begin{pmatrix} Q(v) & S(x) \\ S(v)^T & R(x) \end{pmatrix} > 0$$

is equivalent to

$$Q(v) > 0, \quad R(x) - S(v)^T Q(v)^{-1} S(v) > 0,$$

or to

$$R(v) > 0, \quad Q(x) - S(v)R(v)^{-1}S(v)^T > 0.$$

Lemma E.2 *($\mathcal{S}$-Procedure) Let $F_0, F_1, \ldots, F_p \in \mathbb{R}^{n \times n}$ be symmetric matrices. The requirement of*

$$\xi^T F_0 \xi > 0 \quad \text{for all } \xi \neq 0 \text{ such that } \xi^T F_i \xi \geq 0, \quad i = 1, 2, \ldots, p,$$

is satisfied if there exist $\lambda_1, \lambda_2, \ldots, \lambda_p \geq 0$ such that

$$F_0 - \sum_{i=1}^{p} \lambda_i F_i > 0.$$

For the case $p = 1$, the converse holds, provided that there is some ξ_0 such that $\xi_0^T F_1 \xi_0 > 0$.

References

1. Boyd S, El Ghaoui L, Feron E, Balakishnan V (1994) Linear matrix inequalities in system and control theory. SIAM, Philadelphia
2. Scherer CW, Weiland S (2000) Linear matrix inequalities in control. In: Delft center for systems and control. DISC lecture note, Dutch institute of systems and control
3. Nesterov Y, Nemirovsky A (1994) Interior point polynomial methods in convex programming. SIAM, Philadelphia
4. Boyd SP, Vandenberghe L (2004) Convex optimization. Cambridge University Press, Cambridge

Appendix F
Subspace Linear Predictor

For a linear time invariant (LTI) system in question, assume that it can be described in a state-space form as defined by the equations below:

$$x_{k+1} = Ax_k + Bu_k + Ke_k, \tag{F.1}$$

$$y_k = Cx_k + Du_k + e_k, \tag{F.2}$$

where $u_k \in \mathbb{R}^m$ is the input variable, $y_k \in \mathbb{R}^l$ is the output variable, $x_k \in \mathbb{R}^n$ is the state variable of the system and $e_k \in \mathbb{R}^l$ is the white noise. The matrices $A \in \mathbb{R}^{n\times n}$, $B \in \mathbb{R}^{n\times m}$, $C \in \mathbb{R}^{l\times n}$, $D \in \mathbb{R}^{l\times m}$, and $K \in \mathbb{R}^{l\times l}$ are the state, input, output, feed-through, and Kalman gain matrices of the system, respectively.

From the state equation in (F.1), for $t = k+1, k+2, \ldots, k+M$, we have

$$\begin{aligned}
x_{k+2} &= Ax_{k+1} + Bu_{k+1} + Ke_{k+1} \\
&= A(Ax_k + Bu_k + Ke_k) + Bu_{k+1} + Ke_{k+1} \\
&= A^2x_k + \begin{bmatrix} AB & B \end{bmatrix} \begin{bmatrix} u_k \\ u_{k+1} \end{bmatrix} + \begin{bmatrix} AK & K \end{bmatrix} \begin{bmatrix} e_k \\ e_{k+1} \end{bmatrix},
\end{aligned} \tag{F.3a}$$

$$\begin{aligned}
x_{k+3} &= Ax_{k+2} + Bu_{k+2} + Ke_{k+2} \\
&= A\big(A^2x_k + ABu_k + Bu_{k+1} + AKe_k + Ke_{k+1}\big) + Bu_{k+2} + Ke_{k+2} \\
&= A^3x_k + \begin{bmatrix} A^2B & AB & B \end{bmatrix} \begin{bmatrix} u_k \\ u_{k+1} \\ u_{k+2} \end{bmatrix} + \begin{bmatrix} A^2K & AK & K \end{bmatrix} \begin{bmatrix} e_k \\ e_{k+1} \\ e_{k+2} \end{bmatrix},
\end{aligned} \tag{F.3b}$$

$$\vdots$$

H. Chen, B. Gao, *Nonlinear Estimation and Control of Automotive Drivetrains*,
DOI 10.1007/978-3-642-41572-2,

$$x_{k+M} = A^M x_k + \begin{bmatrix} A^{M-1}B & A^{M-2}B & \ldots & B \end{bmatrix} \begin{bmatrix} u_k \\ u_{k+1} \\ \vdots \\ u_{k+M-1} \end{bmatrix} + \begin{bmatrix} A^{M-1}K & A^{M-2}K & \ldots & K \end{bmatrix} \begin{bmatrix} e_k \\ e_{k+1} \\ \vdots \\ e_{k+M-1} \end{bmatrix}, \tag{F.3c}$$

$$x_{k+M+\delta} = A^M x_{k+\delta} + \begin{bmatrix} A^{M-1}B & A^{M-2}B & \ldots & B \end{bmatrix} \begin{bmatrix} u_{k+\delta} \\ u_{k+\delta+1} \\ \vdots \\ u_{k+M+\delta-1} \end{bmatrix} + \begin{bmatrix} A^{M-1}K & A^{M-2}K & \ldots & K \end{bmatrix} \begin{bmatrix} e_{k+\delta} \\ e_{k+\delta+1} \\ \vdots \\ e_{k+M+\delta-1} \end{bmatrix}. \tag{F.4}$$

Thus, for $\delta = 0, 1, \ldots, N - M + 1$, we can then collect the state variables in a single block matrix equation as

$$\begin{aligned} &\begin{bmatrix} x_{k+M} & x_{k+M+1} & \ldots & x_{k+N+1} \end{bmatrix} \\ &= A^M \begin{bmatrix} x_k & x_{k+1} & \ldots & x_{k+N-M+1} \end{bmatrix} \\ &\quad + \begin{bmatrix} A^{M-1}B & A^{M-2}B & \cdots & B \end{bmatrix} \begin{bmatrix} u_k & u_{k+1} & \cdots & u_{k+N-M+1} \\ u_{k+1} & u_{k+2} & \cdots & u_{k+N-M+2} \\ \vdots & \vdots & \ddots & \vdots \\ u_{k+M-1} & u_{k+M} & \cdots & u_{k+N} \end{bmatrix} \\ &\quad + \begin{bmatrix} A^{M-1}K & A^{M-2}K & \ldots & K \end{bmatrix} \begin{bmatrix} e_k & e_{k+1} & \cdots & e_{k+N-M+1} \\ e_{k+1} & e_{k+2} & \cdots & e_{k+N-M+2} \\ \vdots & \vdots & \ddots & \vdots \\ e_{k+M-1} & e_{k+M} & \cdots & e_{k+N} \end{bmatrix}. \end{aligned} \tag{F.5}$$

Next we will look at the output equation (F.2) and develop recursively an output matrix equation. From (F.2), for $t = k + 1, k + 2, \ldots, k + M - 1$, we have

$$\begin{aligned} y_{k+1} &= Cx_{k+1} + Du_{k+1} + e_{k+1} \\ &= C(Ax_k + Bu_k + Ke_k) + Du_{k+1} + e_{k+1} \end{aligned}$$

$$= CA^1 x_k + \begin{bmatrix} CB & D \end{bmatrix} \begin{bmatrix} u_k \\ u_{k+1} \end{bmatrix} + \begin{bmatrix} CK & I \end{bmatrix} \begin{bmatrix} e_k \\ e_{k+1} \end{bmatrix}, \tag{F.6a}$$

$$y_{k+2} = Cx_{k+2} + Du_{k+2} + e_{k+2}$$

$$= C\left(A^2 x_k + ABu_k + Bu_{k+1} + AKe_k + Ke_{k+1}\right) + Du_{k+2} + e_{k+2}$$

$$= CA^2 x_k + \begin{bmatrix} CAB & CB & D \end{bmatrix} \begin{bmatrix} u_k \\ u_{k+1} \\ u_{k+2} \end{bmatrix} + \begin{bmatrix} CAK & CK & I \end{bmatrix} \begin{bmatrix} e_k \\ e_{k+1} \\ e_{k+2} \end{bmatrix}, \tag{F.6b}$$

$$\vdots$$

$$y_{k+M-1} = CA^{M-1} x_k + \begin{bmatrix} CA^{M-2}B & CA^{M-3}B & \dots & D \end{bmatrix} \begin{bmatrix} u_k \\ u_{k+1} \\ \vdots \\ u_{k+M-1} \end{bmatrix}$$

$$+ \begin{bmatrix} CA^{M-2}K & CA^{M-3}K & \dots & I \end{bmatrix} \begin{bmatrix} e_k \\ e_{k+1} \\ \vdots \\ e_{k+M-1} \end{bmatrix}. \tag{F.6c}$$

Compiling the above result for output equations into a single matrix equation gives us

$$\begin{bmatrix} y_k \\ y_{k+1} \\ y_{k+2} \\ \vdots \\ y_{k+M-1} \end{bmatrix} = \begin{bmatrix} C \\ CA \\ CA^2 \\ \vdots \\ CA^{M-1} \end{bmatrix} x_k$$

$$+ \begin{bmatrix} D & 0 & 0 & \dots & 0 \\ CB & D & 0 & \dots & 0 \\ CAB & CB & D & \dots & 0 \\ \vdots & \vdots & \vdots & \ddots & \vdots \\ CA^{M-2}B & CA^{M-3}B & CA^{M-4}B & \dots & D \end{bmatrix} \begin{bmatrix} u_k \\ u_{k+1} \\ u_{k+2} \\ \vdots \\ u_{k+M-1} \end{bmatrix}$$

$$+ \begin{bmatrix} I & 0 & 0 & \dots & 0 \\ CK & I & 0 & \dots & 0 \\ CAK & CK & I & \dots & 0 \\ \vdots & \vdots & \vdots & \ddots & \vdots \\ CA^{M-2}K & CA^{M-3}K & CA^{M-4}K & \dots & I \end{bmatrix} \begin{bmatrix} e_k \\ e_{k+1} \\ e_{k+2} \\ \vdots \\ e_{k+M-1} \end{bmatrix}. \tag{F.7a}$$

Adjusting the time for the variables in (F.7a) by an arbitrary discrete time δ will result in a similar matrix equation

$$\begin{bmatrix} y_{k+\delta} \\ y_{k+1+\delta} \\ y_{k+2+\delta} \\ \vdots \\ y_{k+M+\delta-1} \end{bmatrix} \tag{F.8}$$

$$= \begin{bmatrix} C \\ CA \\ CA^2 \\ \vdots \\ CA^{M-1} \end{bmatrix} x_{k+\delta}$$

$$+ \begin{bmatrix} D & 0 & 0 & \dots & 0 \\ CB & D & 0 & \dots & 0 \\ CAB & CB & D & \dots & 0 \\ \vdots & \vdots & \vdots & \ddots & \vdots \\ CA^{M-2}B & CA^{M-3}B & CA^{M-4}B & \dots & D \end{bmatrix} \begin{bmatrix} u_{k+\delta} \\ u_{k+1+\delta} \\ u_{k+2+\delta} \\ \vdots \\ u_{k+M+\delta-1} \end{bmatrix}$$

$$+ \begin{bmatrix} I & 0 & 0 & \dots & 0 \\ CK & I & 0 & \dots & 0 \\ CAK & CK & I & \dots & 0 \\ \vdots & \vdots & \vdots & \ddots & \vdots \\ CA^{M-2}K & CA^{M-3}K & CA^{M-4}K & \dots & I \end{bmatrix} \begin{bmatrix} e_{k+\delta} \\ e_{k+\delta+1} \\ e_{k+\delta+2} \\ \vdots \\ e_{k+M+\delta-1} \end{bmatrix}. \tag{F.9}$$

Therefore, by collecting the variables for $\delta = 0, 1, \dots, N - M + 1$, we can compose the output equations in a single block matrix equations as shown below:

$$\begin{bmatrix} y_k & y_{k+1} & \cdots & y_{k+N-M+1} \\ y_{k+1} & y_{k+2} & \cdots & y_{k+N-M+2} \\ y_{k+2} & y_{k+3} & \cdots & y_{k+N-M+3} \\ \vdots & \vdots & \ddots & \vdots \\ y_{k+N-M+1} & y_{k+M} & \cdots & y_{k+N} \end{bmatrix}$$

$$= \begin{bmatrix} C \\ CA \\ CA^2 \\ \vdots \\ CA^{M-1} \end{bmatrix} \begin{bmatrix} x_k & x_{k+1} & \dots & x_{k+N-M+1} \end{bmatrix}$$

$$+\begin{bmatrix} D & 0 & 0 & \dots & 0 \\ CB & D & 0 & \dots & 0 \\ CAB & CB & D & \dots & 0 \\ \vdots & \vdots & \vdots & \ddots & \vdots \\ CA^{M-2}B & CA^{M-3}B & CA^{M-4}B & \dots & D \end{bmatrix}$$

$$\times \begin{bmatrix} u_k & u_{k+1} & \cdots & u_{k+N-M+1} \\ u_{k+1} & u_{k+2} & \cdots & u_{k+N-M+2} \\ u_{k+2} & u_{k+3} & \cdots & u_{k+N-M+3} \\ \vdots & \vdots & \ddots & \vdots \\ u_{k+N-M+1} & u_{k+M} & \cdots & u_{k+N} \end{bmatrix}$$

$$+\begin{bmatrix} I & 0 & 0 & \dots & 0 \\ CK & I & 0 & \dots & 0 \\ CAK & CK & I & \dots & 0 \\ \vdots & \vdots & \vdots & \ddots & \vdots \\ CA^{M-2}K & CA^{M-3}K & CA^{M-4}K & \dots & I \end{bmatrix}$$

$$\times \begin{bmatrix} e_k & e_{k+1} & \cdots & e_{k+N-M+1} \\ e_{k+1} & e_{k+2} & \cdots & e_{k+N-M+2} \\ e_{k+2} & e_{k+3} & \cdots & e_{k+N-M+3} \\ \vdots & \vdots & \ddots & \vdots \\ e_{k+N-M+1} & e_{k+M} & \cdots & e_{k+N} \end{bmatrix}. \tag{F.10}$$

From the derivation of Eqs. (F.5) and (F.10), we can write the subspace I/O matrix equations in the field of subspace system identification [1] as follows:

$$Y_p = \Gamma_M X_p + H_M^d U_p + H_N^s E_p, \tag{F.11}$$

$$Y_f = \Gamma_M X_f + H_M^d U_f + H_N^s E_f, \tag{F.12}$$

$$X_f = A^M X_p + \Delta_M^d U_p + \Delta_M^s E_p, \tag{F.13}$$

where the subscripts p and f denote the 'past' and 'future' matrices of the respective variables, the superscripts d and s stand for the deterministic and stochastic part of the system, respectively. Open-loop data u_k and y_k, $k \in \{0, 1, \dots, N\}$ are available for identification. Therefore, for the definition in (F.11)–(F.13), the past and future data matrices are constructed as follows:

$$Y_p = \begin{bmatrix} y_1 & y_2 & \cdots & y_{N-2M+1} \\ y_2 & y_3 & \cdots & y_{N-2M+2} \\ \vdots & \vdots & \ddots & \vdots \\ y_M & y_{M+1} & \cdots & y_{N-M} \end{bmatrix}, \quad Y_f = \begin{bmatrix} y_{M+1} & y_{M+2} & \cdots & y_{N-M+1} \\ y_{M+2} & y_{M+3} & \cdots & y_{N-M+2} \\ \vdots & \vdots & \ddots & \vdots \\ y_{2M} & y_{2M+1} & \cdots & y_N \end{bmatrix}, \tag{F.14}$$

$$U_p = \begin{bmatrix} u_1 & u_2 & \dots & u_{N-2M+1} \\ u_2 & u_3 & \dots & u_{N-2M+2} \\ \vdots & \vdots & \ddots & \vdots \\ u_M & u_{M+1} & \dots & u_{N-M} \end{bmatrix}, \quad U_f = \begin{bmatrix} u_{M+1} & u_{M+2} & \dots & u_{N-M+1} \\ u_{M+2} & u_{M+3} & \dots & u_{N-M+2} \\ \vdots & \vdots & \ddots & \vdots \\ u_{2M} & u_{2M+1} & \dots & u_N \end{bmatrix}, \tag{F.15}$$

$$E_p = \begin{bmatrix} e_1 & e_2 & \dots & u_{N-2M+1} \\ e_2 & e_3 & \dots & e_{N-2M+2} \\ \vdots & \vdots & \ddots & \vdots \\ e_M & e_{M+1} & \dots & e_{N-M} \end{bmatrix}, \quad E_f = \begin{bmatrix} e_{M+1} & e_{M+2} & \dots & e_{N-M+1} \\ e_{M+2} & e_{M+3} & \dots & e_{N-M+2} \\ \vdots & \vdots & \ddots & \vdots \\ e_{2M} & e_{2M+1} & \dots & e_N \end{bmatrix}. \tag{F.16}$$

The matrices Γ_M, Δ_M^d and Δ_M^s are defined as follows:

- Extended observability matrix

$$\Gamma_M = \begin{bmatrix} C \\ CA \\ CA^2 \\ \dots \\ CA^{M-1} \end{bmatrix}; \tag{F.17}$$

- Reversed extended controllability matrix (deterministic)

$$\Delta_M^d = \begin{bmatrix} A^{M-1}B & A^{M-2}B & \dots & B \end{bmatrix}; \tag{F.18}$$

- Reversed extended controllability matrix (stochastic)

$$\Delta_M^s = \begin{bmatrix} A^{M-1}K & A^{M-2}K & \dots & K \end{bmatrix}. \tag{F.19}$$

The lower-triangular Toeplitz matrices H_M^d and H_M^s are given below:

$$H_M^d = \begin{bmatrix} D & 0 & 0 & \dots & 0 \\ CB & D & 0 & \dots & 0 \\ CAB & CB & D & \dots & 0 \\ \vdots & \vdots & \vdots & \ddots & \vdots \\ CA^{M-2}B & CA^{M-3}B & CA^{M-4}B & \dots & D \end{bmatrix}, \tag{F.20}$$

$$H_M^s = \begin{bmatrix} I & 0 & 0 & \dots & 0 \\ CK & I & 0 & \dots & 0 \\ CAK & CK & I & \dots & 0 \\ \vdots & \vdots & \vdots & \ddots & \vdots \\ CA^{M-2}K & CA^{M-3}K & CA^{M-4}K & \dots & I \end{bmatrix}. \tag{F.21}$$

Furthermore, the past and future state matrices are also defined by

$$X_p = \begin{bmatrix} x_1 & x_2 & \dots & x_{N-2M+1} \end{bmatrix}, \tag{F.22}$$

$$X_f = \begin{bmatrix} x_{M+1} & x_{M+2} & \dots & x_{N-M+1} \end{bmatrix}. \tag{F.23}$$

Taking Eq. (F.11) and solving for X_p will render us

$$X_p = \Gamma_M^{\dagger}(Y_p - H_M^d U_p - H_M^s E_p), \tag{F.24}$$

where the subscript "†" denotes Moore–Penrose pseudoinverse of a matrix [2]. Substituting Eq. (F.24) into (F.13) will then give

$$\begin{aligned} X_f &= A^M(\Gamma_M^{\dagger}(Y_p - H_M^d U_p - H_M^s E_p)) + \Delta_M^d U_p + \Delta_M^s E_p \\ &= A^M \Gamma_M^{\dagger} Y_p + (\Delta_M^d - A^M \Gamma_M^{\dagger} H_M^d) U_p + (\Delta_M^s - A^M \Gamma_M^{\dagger} H_M^s) E_p. \end{aligned} \tag{F.25}$$

Therefore, substituting Eq. (F.25) into (F.12) will result in an equation for future output as given below:

$$\begin{aligned} Y_f &= \Gamma_M(A^M \Gamma_M^{\dagger} Y_p + (\Delta_M^d - A^M \Gamma_M^{\dagger} H_M^d) U_p \\ &\quad + (\Delta_M^s - A^M \Gamma_M^{\dagger} H_M^s) E_p) + H_M^d U_f + H_N^s E_f \\ &= \Gamma_M A^M \Gamma_M^{\dagger} Y_p + \Gamma_M(\Delta_M^d - A^M \Gamma_M^{\dagger} H_M^d) U_p \\ &\quad + H_M^d U_f + \Gamma_M(\Delta_M^s - A^M \Gamma_M^{\dagger} H_M^s) E_p + H_N^s E_f. \end{aligned} \tag{F.26}$$

Due to the effect of E_f which is stationary white noise, and by the virtue of the stability of a Kalman filter, for a set of measurements that is sufficiently large, Eq. (F.26) above can then be written to give an optimal prediction of Y_f as follows:

$$\hat{Y}_f = \begin{bmatrix} L_w & L_u \end{bmatrix} \begin{bmatrix} W_p \\ U_f \end{bmatrix} = L_w W_p + L_u U_f \tag{F.27}$$

where "$\widehat{\ }$" denotes the estimate and

$$W_p = \begin{bmatrix} U_p \\ Y_p \end{bmatrix}. \tag{F.28}$$

Equation (F.27) is thus known as the subspace linear predictor equation, with L_w being the subspace matrix that corresponds to the past input and output data matrix W_p, and L_u is the subspace matrix that corresponds to the future input data matrix U_f.

In order to calculate the subspace linear predictor coefficients L_w and L_u from the Hankel data matrices U_p, Y_p and U_f, we will solve the following least squares problem, thus giving us the prediction equation for Y_f:

$$\min_{L_w, L_u} \left\| Y_f - \begin{bmatrix} L_w & L_u \end{bmatrix} \begin{bmatrix} W_p \\ U_f \end{bmatrix} \right\|^2, \tag{F.29}$$

where

$$\begin{aligned}\begin{bmatrix} L_w & L_u \end{bmatrix} &= Y_f \begin{bmatrix} W_p \\ U_f \end{bmatrix}^{\dagger} \\ &= Y_f \begin{bmatrix} W_p^T & U_f^T \end{bmatrix} \left(\begin{bmatrix} W_p \\ U_f \end{bmatrix} \begin{bmatrix} W_p^T & U_f^T \end{bmatrix} \right)^{-1}. \end{aligned} \tag{F.30}$$

In the control implementation, only the leftmost column of the matrix will be used for the prediction of future output values. Therefore, after the subspace linear predictor coefficients L_w and L_u are found from the identification data, we can then streamline equation (F.27) by taking only the leftmost column of matrices $\hat{Y}_f$, Y_p, U_p and U_f by defining

$$\begin{aligned} \hat{y}_f &= \begin{bmatrix} y_{t+1} \\ y_{t+2} \\ \vdots \\ y_{t+M} \end{bmatrix}, \quad y_p = \begin{bmatrix} y_{t-M+1} \\ y_{t-M+2} \\ \vdots \\ y_t \end{bmatrix}, \\ u_p &= \begin{bmatrix} u_{t-M+1} \\ u_{t-M+2} \\ \vdots \\ u_t \end{bmatrix}, \quad u_f = \begin{bmatrix} u_{t+1} \\ u_{t+2} \\ \vdots \\ u_{t+M} \end{bmatrix}, \end{aligned} \tag{F.31}$$

and

$$w_p = \begin{bmatrix} u_p \\ y_p \end{bmatrix}, \tag{F.32}$$

then we will arrive at a streamlined subspace-base linear predictor equation, namely

$$\hat{y}_f = L_w w_p + L_u u_f. \tag{F.33}$$

According to Eq. (F.33), we can predict the output of the system by the past input and output data as well as the future input data that is applied. This result will be utilized in the implementation of model predictive control algorithm, that is, data-driven predictive control algorithm.

References

1. Overschee PV, Moor BD (1996) Subspace identification for linear systems: theory, implementation, applications. Kluwer Academic, Norwell
2. Van Overschee P, De Moor B (1995) A unifying theorem for three subspace system identification algorithms. Automatica 31(12):1853–1864

MIX
Papier aus verantwortungsvollen Quellen
Paper from responsible sources
FSC® C105338
FSC
www.fsc.org

If you have any concerns about our products,
you can contact us on
ProductSafety@springernature.com

In case Publisher is established outside the EU,
the EU authorized representative is:
Springer Nature Customer Service Center GmbH
Europaplatz 3, 69115 Heidelberg, Germany

Printed by Libri Plureos GmbH
in Hamburg, Germany